PHYSICS RESEARCH AND TECHNOLOGY

COMPUTER PHYSICS

PHYSICS RESEARCH AND TECHNOLOGY

Additional books in this series can be found on Nova's website
under the Series tab.

Additional E-books in this series can be found on Nova's website
under the E-books tab.

COMPUTER SCIENCE, TECHNOLOGY AND APPLICATIONS

Additional books in this series can be found on Nova's website
under the Series tab.

Additional E-books in this series can be found on Nova's website
under the E-books tab.

PHYSICS RESEARCH AND TECHNOLOGY

COMPUTER PHYSICS

BRIAN S. DOHERTY
AND
AMY N. MOLLOY
EDITORS

Nova Science Publishers, Inc.

New York

LIBRARY OF CONGRESS CATALOGING-IN-PUBLICATION DATA

Computer physics / editors, Brian S. Doherty and Amy N. Molloy.
   p. cm.
  Includes index.
  ISBN 978-1-61324-790-7 (hardcover)
  1. Nuclear physics--Data processing. I. Doherty, Brian S. II. Molloy, Amy N.
  QC783.3.C655 2011
  539.70285--dc23
                  2011017295

Published by Nova Science Publishers, Inc. † New York

CONTENTS

PREFACE

In this book, the authors present topical research from across the globe in the study of computer physics, including multidimensional experimental data processing in nuclear physics; solving the vibrational Schrodinger equation using a combined neural network collocation approach; the numerical solution of nonlinear partial differential equations using symplectic geometric integrators and grid computing multiple shooting algorithms for extended phase space sampling and long time propagation in molecular dynamics.

Chapter 1 is devoted to the problems connected with the processing of multidimensional experimental data in nuclear physics. It presents the extensions and generalizations of the conventional processing methods for multidimensional data and the definitions of new original approaches, methods, and algorithms.

The first step of event processing is their selection or separation - sorting. The sorting operation is based on "gates" or "conditions". Properly proposed gating methods lead to an improvement in spectral quality, in particular, the peak-to-background ratio.

One of the basic problems in the analysis of the spectra (both one- and multidimensional) is the separation of useful information contained in peaks from the useless information (background, noise). We present an extension of the algorithms of background estimation up to n-dimensional spectra. We generalized the existing basic algorithms for additional parameters and possibilities that make it possible to improve substantially the quality of the background estimation.

One of the most delicate problems of any spectrometric method is that related to the extraction of the correct information out of the spectra sections, where due to the limited resolution of the equipment, signals coming from various sources are overlapping. The deconvolution methods represent an efficient tool to improve the resolution in the nuclear spectra. We developed modifications and extensions of iterative deconvolution algorithms as well as new regularization technique.

The quality of the analysis of nuclear spectra consists in the correct identification of the existence of peaks and subsequently in the good estimation of their positions. We extended the conventional algorithms up to n-dimensional spectra. We also proposed the algorithm of ridges identification used in nuclear multifragmentation and a new pattern recognition algorithm applied for the determination of rings in two-dimensional spectra from RICH detectors.

The positions of found peaks can be fed as initial estimate into a fitting procedure. The basic problem of the fitting of multidimensional γ-ray spectra resides in the existence of a large number of peaks and thus large number of fitted parameters. We proposed several modifications of the algorithm without matrix inversion that allow increasing its efficiency.

With increasing dimensionality of histograms (nuclear spectra), the requirements in developing of multidimensional scalar visualization techniques become striking. Conventional isosurfacing, volume rendering and glyph-based techniques are not extendible to higher dimensions. Therefore, we have suggested a new algorithm of hypervolume visualization that is based on particle scattering display mode and is extendible to any dimension.

In Chapter 2 we review approaches for solving the vibrational Schrödinger equation focusing on collocation and neural network (NN) methods that do not require computing integrals. A Radial Basis Neural Network (RBNN) - based method we developed is given particular attention. It allows one to compute several vibrational levels with an accuracy of the order of 1 cm^{-1} by expanding the wavefunction in an extremely small and flexible basis. This approach combines non-linear optimization of basis (neuron) parameters with a linear rectangular matrix method. It improves dimensionality scaling and permits computing several levels from a single RBNN. The algorithm avoids the calculation of integrals and of a potential energy function – the vibrational spectrum may be obtained directly from ab initio data. Owing to the fact that the construction of a PES is bypassed, it is possible to directly use this approach with an ab initio code to compute vibrational spectra. When normal model coordinates are used, the method is molecule-independent. It is particularly useful for computing vibrational spectra of molecules adsorbed at surfaces and finally provides surface scientists with a black-box and easily usable tool to compute spectra of adsorbate complexes without neglecting anharmonicity and coupling effects. Applications to synthetic problems of different dimensionalities, the water molecule, and to H_2O on Pt(111) are reviewed.

The numerical solution of nonlinear partial differential equations (PDEs) using symplectic geometric integrators has been the subject of many studies in recent years. Many nonlinear partial differential equations can be formulated as an infinite dimensional Hamiltonian system. After semi-discretization in the space variable, a system of Hamiltonian ordinary differential equations (ODEs) is obtained, for which various symplectic integrators can be applied.

Numerical results show that symplectic schemes have superior performance, especially in long time simulations.

The concept of multisymplectic PDEs and multisymplectic schemes can be viewed as the generalization of symplectic schemes. In the last decade, many multisymplectic methods have been proposed and applied to nonlinear PDEs, like to nonlinear wave equation, nonlinear Schrodinger equation, Korteweg de Vries equation, Dirac equation, Maxwell equation and sine-Gordon equation. In Chapter 3, recent results of multisymplectic integration on the coupled nonlinear PDEs, the coupled nonlinear Schrodinger equation, the modified complex Korteweg de Vries equation and the Zakharov system will be given. The numerical results are discussed with respect to the stability of the schemes, accuracy of the solutions, conservation of the energy and momentum, preservation of dispersion relations.

Grid computing refers to a well established computational platform of geographically distributed computers that offer a seamless, integrated, computational and collaborative environment. It provides the means for solving highly demanded, in computer time and storage media problems of molecular dynamics. However, because of the rather high latency network, to exploit the unprecedented amount of computational resources of the Grid, it is necessary to develop new or to adapt old algorithms for investigating dynamical and statistical molecular behaviors at the desired temporal and spatial resolution. In Chapter 4 we review methods that assist one to harness the current computational Grid infrastructure for carrying out extended samplings of phase space and integrating the classical mechanical equations of motion for long times. Packages that allow to automatically submit and propagate trajectories in the Grid and to check and store large amounts of intermediate data are described. We report our experience in employing the European production Grid infrastructure for investigating the dynamics and free energy hypersurfaces of enzymes such as Cytochrome c Oxidases. Time autocorrelation functions of dynamical variables yield vibrational spectra of the molecule and reveal the localization of energy in specific bonds in the active site of the enzyme. Dynamical calculations and free energy landscapes of the Cytochrome c Oxidase protein interacting with gases like O_2, CO and NO reveal the pathways for the molecules to penetrate in the cavities of the enzyme and how they reach the active site where the reactions take place. The discussed methods can be adopted in any intensive computational campaign, which involves the scheduling of a large number of long term running jobs.

In Chapter 5, numerical schemes for solving the Vlasov-Maxwell system of equations are presented. Our Universe is filled with collisionless plasma, which is a dielectric medium with nonlinear interactions between charged particles and electromagnetic fields. Thus computer simulations play an essential role in studies of such highly nonlinear systems. The full kinetics of collisionless plasma is described by the Vlasov-Maxwell equations. Since the Vlasov equation treats charged particles as position-velocity phase-space distribution functions in hyper dimensions, huge supercomputers and highly-scalable parallel codes are essential. Recently, a new parallel Vlasov-Maxwell solver is developed by adopting a stable but less-dissipative scheme for time integration of conservation laws, which has successfully achieved a high scalability on massively parallel supercomputers with multi-core scalar processors. The new code has applied to 2P3V (two dimensions for position and three dimensions for velocity) problems of cross-scale plasma processes such as magnetic reconnection, Kelvin-Helmholtz instability and the interaction between the solar wind and an asteroid.

In: Computer Physics   ISBN: 978-1-61324-790-7
Editors: B.S. Doherty and A.N. Molloy, pp. 1-236   © 2012 Nova Science Publishers, Inc.

Chapter 1

MULTIDIMENSIONAL EXPERIMENTAL DATA PROCESSING IN NUCLEAR PHYSICS

Miroslav Morháč[1] and Vladislav Matoušek

Institute of Physics, Slovak Academy of Sciences,
Dúbravská cesta 9, 845 11 Bratislava, Slovakia

Abstract

This chapter is devoted to the problems connected with the processing of multidimensional experimental data in nuclear physics. It presents the extensions and generalizations of the conventional processing methods for multidimensional data and the definitions of new original approaches, methods, and algorithms.

The first step of event processing is their selection or separation - sorting. The sorting operation is based on "gates" or "conditions". Properly proposed gating methods lead to an improvement in spectral quality, in particular, the peak-to-background ratio.

One of the basic problems in the analysis of the spectra (both one- and multidimensional) is the separation of useful information contained in peaks from the useless information (background, noise). We present an extension of the algorithms of background estimation up to n-dimensional spectra. We generalized the existing basic algorithms for additional parameters and possibilities that make it possible to improve substantially the quality of the background estimation.

One of the most delicate problems of any spectrometric method is that related to the extraction of the correct information out of the spectra sections, where due to the limited resolution of the equipment, signals coming from various sources are overlapping. The deconvolution methods represent an efficient tool to improve the resolution in the nuclear spectra. We developed modifications and extensions of iterative deconvolution algorithms as well as new regularization technique.

The quality of the analysis of nuclear spectra consists in the correct identification of the existence of peaks and subsequently in the good estimation of their positions. We extended the conventional algorithms up to n-dimensional spectra. We also proposed the algorithm of ridges identification used in nuclear multifragmentation and a new pattern recognition algorithm applied for the determination of rings in two-dimensional spectra from RICH detectors.

---

[1] E-mail address: Miroslav.Morhac@savba.sk

The positions of found peaks can be fed as initial estimate into a fitting procedure. The basic problem of the fitting of multidimensional γ-ray spectra resides in the existence of a large number of peaks and thus large number of fitted parameters. We proposed several modifications of the algorithm without matrix inversion that allow increasing its efficiency.

With increasing dimensionality of histograms (nuclear spectra), the requirements in developing of multidimensional scalar visualization techniques become striking. Conventional isosurfacing, volume rendering and glyph-based techniques are not extendible to higher dimensions. Therefore, we have suggested a new algorithm of hypervolume visualization that is based on particle scattering display mode and is extendible to any dimension.

1. Introduction

Spectrometers, such as GAMMASPHERE [1], EUROGAM [2], and GASP [3], permit one to obtain very interesting physical information where high-fold γ-rays coincidence events are registered. Progress in the understanding of nuclear structure depends on the ability to analyze these multidimensional spectra correctly. Obviously, the lack of methods for analyzing such sets of data can create a major obstacle to extracting physical information. The information contained in the high-fold coincidence spectra is overwhelming and so extracting all the detailed information contained in multidimensional peaks proves difficult and could pose serious technical problems. The classical techniques developed to analyze one-parameter spectra cannot be directly extended to higher-fold data. Moreover, some operations are extremely time-consuming. The problems become relevant for big sizes and for multidimensional data, where the number of operations grows exponentially with the sizes. Analysis of information-rich high-fold coincidence data requires sophisticated methods and algorithms to extract the physically interesting information from the raw data.

Basic tasks connected with the processing of multiparameter data in nuclear physics can be divided into the following subjects

- sorting of events (gating), histogramming
- elimination of background and noise in the spectra
- resolution improvement in the spectra using deconvolution methods
- identification of peaks in the spectra
- fitting and analysis of the spectra
- visualization of both raw data and calculated results.

The measurements of data in nuclear physics experiments are oriented towards gathering a large amount of data. Nuclear physics is one of the scientific fields where the requirements to higher dimensions and better resolutions are continuously increasing. From the viewpoint of the investigated physical phenomena, the multidimensional space can be divided to relevant and irrelevant events. Consequently, the space of events can be divided to the subspace of events that are accepted for further processing and the subspace that are ignored and can be excluded from subsequent processing. The operation of the sorting is based on "gates" or "conditions". When storing the events in "list mode" the sorting operation can substantially reduce storage volume necessary. Very frequently, the gates are used to create slices of the integral spectra of lower dimensionality. If high-fold coincidence events are unfolded into a lower histogram dimension, the statistics can become artificially high so that peak positions and statistics may be wrong. Properly proposed gating methods lead to an improvement in spectral quality in particular the peak-to-background ratio, and to a decrease of uncorrelated events in the resulting spectrum [4].

One of the basic problems in the analysis of the spectra (both one- and multidimensional) is the separation of useful information contained in peaks from the useless information (background, noise). The accuracy and reliability of the background analysis depend critically on the treatment in order to resolve strong peak overlaps, to account for continuum background contributions, and to distinguish artifacts to the responses of some detector types. A background approximation must pay particular attention to the reliable estimation of the background continuum under low-statistics peaks, both isolated and in close proximity to strong peaks. In order to process data from numerous analyses efficiently and reproducibly, the background approximation must be free of user-adjustable parameters [5]. In multidimensional spectra, the algorithm must be able to recognize not only continuous background but also to include all the combinations of coincidences of the background in some dimensions and the peaks in the other ones.

However, after the elimination of the useless information contained in the background and lower order coincidences the peaks as the main carrier of spectrometric information are very frequently positioned close to each other. One of the most delicate problems of any spectrometric method is that related to the extraction of the correct information out of the spectra sections, where due to the limited resolution of the equipment, signals coming from various sources are overlapping [6], [7], [8] and [9]. Deconvolution and restoration are the names given to the endeavor to improve the resolution of an experimental measurement by mathematically removing the smearing effects of an imperfect instrument, using its known resolution function. To devise reliable methods for doing this has been a long-term endeavor by many scientists and mathematicians [10]. The deconvolution methods are widely applied in various fields of data processing. Recently many applications have been found in various domains of experimental science, e.g. image and signal restoration, the determination of thickness of multilayer structures, tomography, magnetic resonance imaging, crystallography, geophysics, etc. [11]. The deconvolution methods can be successfully applied for the determination of positions and intensities of peaks and for the decomposition of multiplets in γ-ray spectroscopy [12]. Error sources that negatively influence the result can be divided into two groups

- inaccurate input values (e.g. measured noise);
- errors generated during the calculation resulting from the algorithm employed and from the finite computer word length, respectively.

Various deconvolution algorithms very frequently utilize the factorization property of the Fourier transform. According to this property, the operation of convolution is factorized to the product of the corresponding elements of the transformed arrays of the input signal and the response function. The errors introduced by dividing the transformed coefficients and by the direct and the inverse Fourier transforms become evident by oscillations of the solution. Therefore, the development of error-free algorithms to solve the deconvolution problem is emerging.

From a numerical point of view, the deconvolution is so called ill-posed problem, which means that many different functions solve a convolution equation within error bounds of experimental data. The estimates of the solution are sensitive to rounding-off errors as well as to errors in the input data [13]. When employing the standard algorithms to solve a convolution system, small errors or noise can cause enormous oscillations in the result. This implies that the regularization must be employed. Tikhonov first treated this problem on a strict mathematical basis by introducing the regularization theory and methods [14], [15]. The regularization encompasses a class of solution techniques that modifies an ill-posed problem into a well-posed one, by approximation, so that a physically acceptable approximate solution can be obtained and the solution is sufficiently stable from the computational viewpoint [16]. Therefore, the methods of regularization can be classified from several aspects:

- quality of the deconvolved signal (smoothness, positive solution, oscillations, etc.);
- computational complexity;
- convergence speed, etc.

In the presented work, we analyze successively one-, two-, and multidimensional convolution systems up to the time (energy) dependent linear systems.

After possible background removal and eventual resolution improvement of close lying and/or even overlapping peaks using the above-mentioned deconvolution methods, the identification of peaks is the first step of the analysis of the nuclear spectrum. The identification of relatively narrow peaks either on smoothly varying background or after its removal is an ostensible simple task. Although there are many different algorithms and approaches to detect peaks in data, there is no one perfect method. In fact, when it comes down to it the best peak finder in the world is actually the human eye. The main problem lies in distinguishing true peaks from statistical fluctuations, Compton edges and other undesired spectrum features [17]. The use of the convolution method to locate the peaks is a well-established approach and is utilized in many of the available algorithms. It removes statistical fluctuations and is insensitive to linear background under Gaussians [6]. However, application of these methods cannot be directly extended to searching for peaks in multidimensional spectra. In a simplified way, one can think of a one-dimensional spectrum as a composition of background, peaks and noise. For instance, two-dimensional spectrum, besides of these components, contains also one-fold coincidences (ridges as a result of peak-background coincidences) in both directions. The algorithm to search for peaks in n-dimensional spectra must be able to distinguish between intersections of lower-fold coincidences and n-fold coincidence.

Hence, the analysis of peaks in spectra consists in the determination of peaks positions and subsequent fitting, which results in the estimate of peak shape parameters. The positions of found peaks are then fed as initial estimate into a fitting procedure.

Fitting procedures are currently utilized in a large set of computational problems. A lot of algorithms have been developed (Gauss-Newton, Levenberg-Marquart, conjugate gradients, etc.) and more or less implemented into programs for analysis of complex spectra, e.g. [17], [18], and [19]. Good survey and comparison of different χ^2 statistics and maximum likelihood estimation methods applied to Gauss as well as to Poison distributed data is given in [20]. Both χ^2 and the maximum likelihood turn into well-known linearization and successive iterative solution of the system of linear equations. Matrix inversion can impose appreciable convergence difficulties mainly for a large number of fitted parameters.

Peaks can be fitted either separately, each peak (or multiplets) in a region or together all peaks in a spectrum. To fit separately each peak one needs to determine the fitted region. However, it can happen that the regions of neighboring peaks are overlapping (mainly in multidimensional γ-ray spectra). Then the results of the fitting are very poor. On the other hand, when fitting all peaks found in a γ-ray spectrum one needs to have a method that is stable (converges) and fast enough to carry out fitting in reasonable time. The gradient methods based on the inversion of large matrices are not applicable because of two reasons:

- calculation of an inverse matrix is extremely time consuming; and
- due to accumulation of truncation and rounding-off errors, the result of the gradient methods based on the inversion of very large matrices can become worthless.

According to numerical analysis, it should be emphasized that the direct inversion of large matrices should be avoided wherever possible [21]. The methods based on Slavic's algorithm without matrix inversion [8] allow fitting large blocks of data and large number of parameters (several hundreds or thousands) in reasonable time.

In today's nuclear and high energy physics experiments the number of detectors being included in the measurements is going up to one hundred or more. The results of such measurements, however, generate such large data sets as to be nearly incomprehensible. Scanning these large sets of numbers to determine trends and relationships is a tedious and ineffective process. To address this problem the physicists have turned to visualization of experimental data. If the data are converted to a visual form, the trends are often immediately apparent. The goal of visualization of experimental data is an improved understanding of the result of the information gathered during experiment. It is one of the most powerful and direct ways how the huge amount of information can be conveyed in a form comprehensible to a human eye. The visualization techniques presented in this work make it possible to display either raw experimental spectra, processed data or to make slices of the same or lower dimensionality in an interactive way. They allow obtaining an imagination about event distribution and correlations in coincidence spectra up to five-dimensional space.

2. Event Data Sorting

2.1. Introduction

Today's generation of γ-ray spectrometers such as Gammasphere [1] are capable of collecting large quantities of high-fold data in which many γ-rays are detected in coincidence in the same event [22], [23]. To utilize the optimum resolving power of the instruments, it is necessary to analyze the data with the optimum gating fold. The optimum gating fold depends on experimental factors such as the average multiplicity of γ-ray cascades and the amount of data collected. Typical high-fold analysis requires gating folds between 2–5, but high-fold events of over 10 energies per event occur in the experiments. The question is how to sort such high-fold coincidence data in order to produce spectra of various fixed dimensions under various selection criteria.

Therefore, after taking events from detectors the first step of event processing is their selection or separation. The sorting operation is based on "gates" or "conditions". When storing the events in "list mode", the sorting operation can substantially reduce the storage volume necessary. Very frequently, the gates are used to create slices of the integral spectra of lower dimensionality also. If high-fold coincidence events are unfolded into lower dimensional histograms, the statistics can become artificially high so that peak positions and statistics may be wrong [23]. Properly proposed gating methods lead to an improvement in spectral quality, in particular, the peak-to-background ratio, and to a decrease of uncorrelated events in the resulting spectrum.

The events space can be divided to the:

- subspace of events that are accepted for further processing, and
- subspace that are ignored and can be excluded from subsequent processing.

The sorting operation can substantially reduce the storage volume necessary (e.g., when storing the events in "list mode"). The gates can also be used to create slices of the integral spectra of lower dimensionality.

2.2. Sorting of High-Fold Data from Γ-ray Detector Arrays

Sorting methods involve combinatorial decomposing each n-fold event into a number of lower m-fold sub-events ($m \leq n$) [23].

Each n-fold event generates

$$\binom{n}{m} = \frac{n!}{m!(n-m)!} = \frac{n(n-1)\cdots(n-m+1)}{1\cdot 2\cdot 3\cdots m}$$

m - fold sub-events where each sub-event contains m energies from the original event. Common difficulties using the sorting methods of analysis are that:

- if unfolded m-fold sub-events are assumed to be independent and can be further decomposed into lower fold events – this is not correct, it can introduce spikes appearing in the spectra, [23]
- if high-fold coincidence events are unfolded into a lower histogram dimension, the statistics can become artificially high so that peak positions and statistics may be wrong.

To avoid these spikes appearing in the spectra each m-fold correlation can only be used once in a spectrum incrementation. Properly proposed gating methods lead to an improvement in spectral quality in particular the peak-to-background ratio, and to a decrease of uncorrelated events in the resulting spectrum.

2.3. Proposed Sorting Methods

The basic element for data sorting is a gate (condition). To satisfy typical experimental needs we have proposed the following types of gates / conditions:

- rectangular window,
- polygon,
- arithmetic function,
- spherical gates,
- ellipsoids,
- composed gate.

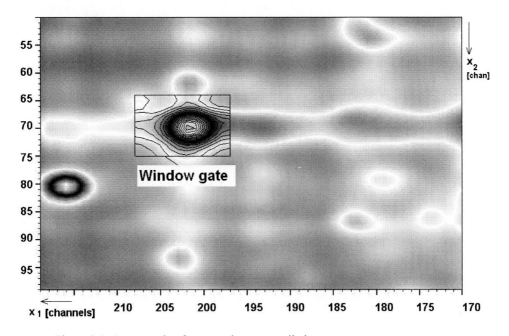

Figure 2.1. An example of rectangular gate applied to a two-parameter spectrum.

2.3.1. Rectangular Window

It specifies a set of event variables with lower and upper channels determining the region of event acceptance. This is a classical commonly used gating method. Let us take the simplest two-fold events and let us assume we desire to analyze a variable, but only those events that satisfy the condition

$$g_{l1} \leq x_1 \leq g_{u1},$$

where g_{l1}, g_{u1} are lower and upper limits of the gate, respectively. Analogously for three and more fold events, a two-dimensional region can be set e.g.

$$g_{l1} \leq x_1 \leq g_{u1} \quad AND \quad g_{l2} \leq x_2 \leq g_{u2}$$

In general, for n-fold rectangular window gate one can write

$$
\begin{aligned}
& g_{l1} \leq x_1 \leq g_{u1} \quad \text{AND} \quad g_{l2} \leq x_2 \leq g_{u2} \quad \text{AND} \quad \cdots \\
& \text{AND} \quad g_{ln} \leq x_n \leq g_{un}
\end{aligned}
\tag{2.1}
$$

where $x_1, x_2, \ldots, x_n$ are event variables. An example of rectangular gate applied to a two-parameter spectrum is shown in Fig. 2.1. The gate is located around the peak area.

2.3.2. Polygon

An efficient and simple way to choose the region of event acceptance in a two-dimensional space of event variables is by setting interactively the appropriate closed polygon. The polygon gate consists of sequence of coordinates of two event variables. It allows one to separate various parts of the spectrum. The advantage of this kind of gate is that one can design easily any irregular shape. Its disadvantage is that it cannot be extended to higher dimensions and that it must be set manually. An example of polygon gate applied to two-parameter spectrum is shown in Fig. 2.2. The gate in the figure is defined by a set of segments, which form the closed polygon around the irregular shape.

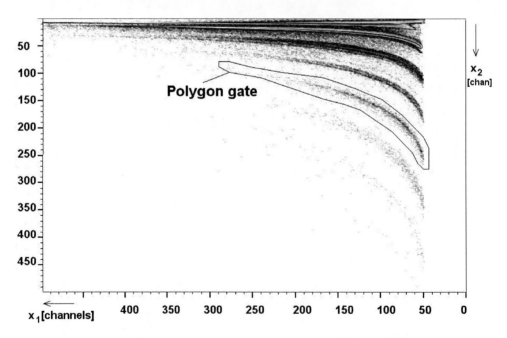

Figure 2.2. An example of polygon gate applied to two-parameter spectrum.

2.3.3. Arithmetic Function

This type of gate is represented by a mathematical function of event variables (detection lines) $x_1, x_2, \ldots, x_n$

$$
f(x_1, x_2, \ldots, x_n) \leq 0
$$

The allowed operators are: +, -, *, /, ^, sqr, log, exp, cos, sin. The built-in syntax analyzer is able to recognize the expressions written in Fortran-like style using names of event variables for operands, above-given operators and parentheses. During the sorting, for each event, the value of the function is calculated. If the value is less or equal to zero, the event is accepted—i.e., the logical value of the condition is "true".

By employing a suitable analytical function, one can specify more exactly the region of interesting parts in the spectrum.

2.3.4. Spherical Gates

Photopeak events are distributed approximately with a Gaussian distribution. It means that in the hyperspace they have a spherical symmetry. When choosing the function with elliptic base for two variables one can reduce the uncorrelated background

$$\left(\frac{x_1-c_1}{r_1}\right)^2 + \left(\frac{x_2-c_2}{r_2}\right)^2 - 1 \le 0,$$

where c_1, c_2 determine the position of the photopeak, r_1, r_2 are proportional to σ_1, σ_2 (see Fig. 2.3).

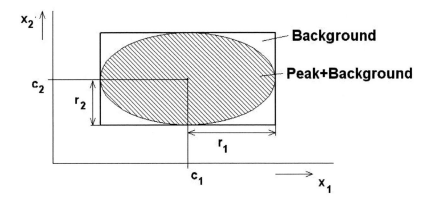

Figure 2.3. Elliptical gate.

The extension of this function to the n-dimensional case is straightforward. In general, one can define any free arithmetic function using the above-given set of operators.

When sorting events with a Gaussian or quasi Gaussian distribution the gates with elliptic base are of special interest. The radii of the ellipses are proportional to the standard deviations σ_i or to the $\text{FWHM}_i = \sqrt{2\log 2}\sigma_i$ (full width at half maximum) of the photopeak distribution. Then for symmetrical n-dimensional spherical gates one can write

$$\sum_{i=1}^{n}\left(\frac{x_i-c_i}{0.5FWHM_i}\right)^2 - R^2 \le 0$$

However, due to various effects in detectors the peaks exhibit left-hand tailing. One can introduce the special gates reflecting the tails in spectrum peaks. In [4] special gates reflecting the tails in spectrum peaks were proposed. The problem is to extend the gate shape in the tail direction. To explain it, we resort to two-dimensional asymmetrical spherical gates by introducing the tailing coefficient T in the variable x_2

$$\left(\frac{x_1-c_1}{0.5\cdot FWHM_1}\right)^2 + \left(\frac{x_2-c_2}{0.5\cdot FWHM_2\cdot T}\right)^2 - R^2 \leq 0$$

where $T \geq 1$. To introduce tailings to both directions it turns out that we have to use a logical OR function so that the gate is an intersection between two-tailed spherical gates

$$\left(\frac{x_1-c_1}{0.5\cdot FWHM_1}\right)^2 + \left(\frac{x_2-c_2}{0.5\cdot FWHM_2\cdot T}\right)^2 - R^2 \leq 0 \quad OR$$

$$\left(\frac{x_1-c_1}{0.5\cdot FWHM_1\cdot T}\right)^2 + \left(\frac{x_2-c_2}{0.5\cdot FWHM_2}\right)^2 - R^2 \leq 0.$$

It is straightforward to extend this idea to the n-dimensional spherical gates [4]

$$\left(\frac{x_1-c_1}{0.5\cdot FWHM_1\cdot T}\right)^2 + \left(\frac{x_2-c_2}{0.5\cdot FWHM_2}\right)^2 + \ldots + \left(\frac{x_n-c_n}{0.5\cdot FWHM_n}\right)^2 - R^2 \leq 0 \quad OR$$

$$\left(\frac{x_1-c_1}{0.5\cdot FWHM_1}\right)^2 + \left(\frac{x_2-c_2}{0.5\cdot FWHM_2\cdot T}\right)^2 + \ldots + \left(\frac{x_n-c_n}{0.5\cdot FWHM_n}\right)^2 - R^2 \leq 0 \quad OR \quad (2.2)$$

$$\vdots$$

$$\left(\frac{x_1-c_1}{0.5\cdot FWHM_1}\right)^2 + \left(\frac{x_2-c_2}{0.5\cdot FWHM_2}\right)^2 + \ldots + \left(\frac{x_n-c_n}{0.5\cdot FWHM_n\cdot T}\right)^2 - R^2 \leq 0$$

By using the spherical gates, the significant volume reductions over the cuboidal gates can be achieved as shown in Tab. 2.1.

Table. 2.1. The ratios of volume reduction achieved when using the spherical gates over the cuboidal gates

Dimension	Volume of hypersphere	Volume of hypercube	Ratio
1	$2\,r$	$2\,r$	1.00
2	$\pi\,r^2$	$4\,r^2$	1.27
3	$^4/_3\,\pi r^3$	$8\,r^3$	1.91
4	$^1/_2\,\pi^2 r^4$	$16\,r^4$	3.04
5	$^8/_{15}\,\pi^2 r^5$	$32\,r^5$	8.11
10	$^1/_{120}\,\pi^5 r^{10}$	$1024\,r^{10}$	401.54

From the table we can see that the volume of processed (and consequently stored) events can be significantly decreased if an ellipsoidal gate is used instead of a cuboidal gate.

By application of a suitable spherical gate to raw data, the experimenter can separate the events contributing to the background from the spectrum [24] and [25]. This can be achieved by designing the shape of the gate in coincidence with the shape created by the distribution of events of our interest in the spectrum.

In the next few examples, we present examples of gates for two- and three-dimensional data. An example of two-dimensional spherical gate is shown in Fig. 2.4. An example of another two-dimensional γ-ray spectrum with spherical balls gate is shown in Fig. 2.5. To illustrate the use of three-dimensional spherical gate, let us consider the three-dimensional γ-ray spectrum as shown in Fig. 2.6. We have chosen the spherical gate similar to that shown in Fig. 2.7. When applying this gate to the spectrum from Fig. 2.6, we get the three-dimensional spectrum shown in Fig. 2.8. The gate is shown as shaded balls inside of the spectrum, i.e., the shaded channels satisfy the condition connected with the gate. An example of three-dimensional γ-γ-γ-ray spectrum with a spherical gate with different widths of Gaussians in different directions is shown in Fig. 2.9.

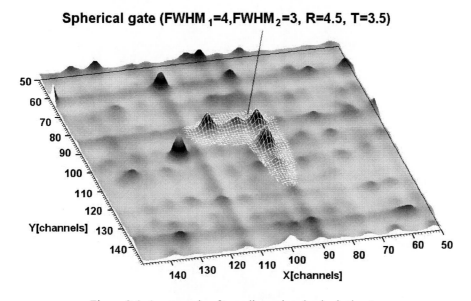

Figure 2.4. An example of two-dimensional spherical gate.

The proposed model of this type of gates allows us to change the width of spherical Gaussians. By changing the parameters, we can select any part of the spectrum. However, according to this model, one cannot change independently tailings for individual dimensions and angles between the Gaussians.

2.3.5. Ellipsoids

To change independently the tailings for individual dimensions and angles between the Gaussians and to make the gate shape design more flexible we propose ellipsoid gates. In parametric form, the ellipse can be expressed as

$$x_1 = r_1 \cos(t) + c_1,$$
$$x_2 = r_2 \sin(t) + c_2,$$

where c_1, c_2 determine the position of the ellipse, r_1, r_2 are radii and are independent of t.

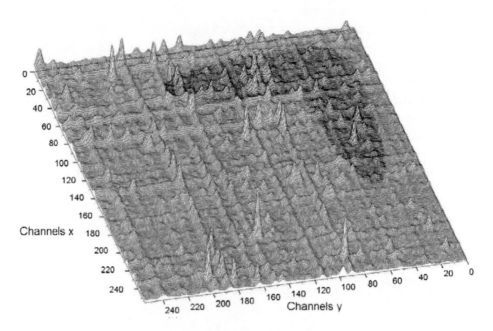

Figure 2.5. Two-dimensional γ-ray spectrum with spherical balls gate.

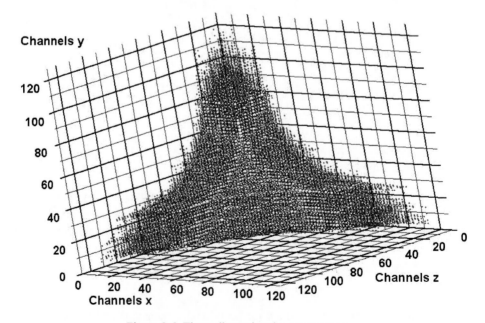

Figure 2.6. Three-dimensional γ-ray spectrum.

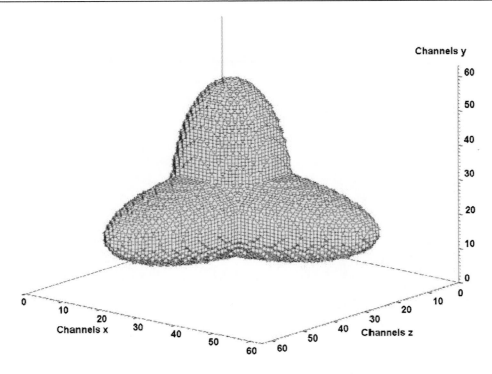

Figure 2.7. Spherical gate shown as surface.

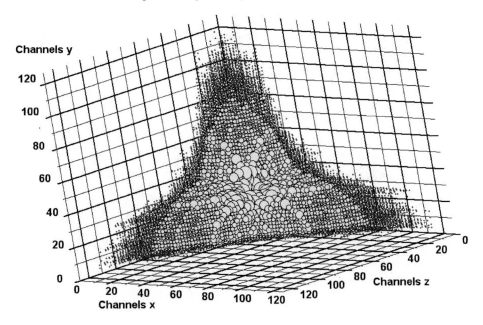

Figure 2.8. An example of the application of three-dimensional spherical gate to three-dimensional spectrum. The gate is shown as shaded balls inside of the spectrum, i.e., the shaded channels satisfy the condition connected with the gate.

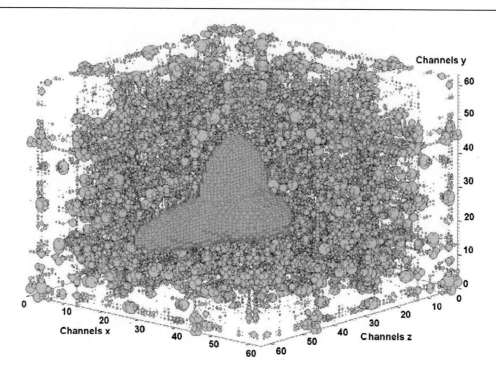

Figure 2.9. An example of three-dimensional γ-γ-γ-ray spectrum with a spherical gate with different widths of Gaussians in different directions.

One can change freely the lengths of half-axes, the widths of ellipsoids as well as the angle between them. Let us express n-dimensional k-th ellipsoid E_k with center positioned at $c_1, c_2, ..., c_n$ and lengths of half-axes $a_1, b_1, a_2, b_2, ..., a_n, b_n$,

$$x_1 = b_1 \cos(t_{n-1})\cos(t_{n-2})....\cos(t_1) + c_1$$
$$\vdots$$
$$x_{k-1} = b_k \cos(t_{n-1})\cos(t_{n-2})....\cos(t_{k-3})\sin(t_{k-2}) + c_{k-1}$$
$$x_k = A_k(t_{k-1})\cos(t_{n-1})\cos(t_{n-2})....\cos(t_{k-2})\sin(t_{k-1}) + c_k \qquad (2.3)$$
$$x_{k+1} = b_{k+1}\cos(t_{n-1})\cos(t_{n-2})....\cos(t_{k-1})\sin(t_k) + c_{k+1}$$
$$\vdots$$
$$x_n = b_n \sin(t_{n-1}) + c_n$$

In order to describe tailings in the gate let us change the radii as a function of the parameters t_i

for $i = 1$

$$A_1(t_1) = \frac{a_1 + b_1}{2} + \frac{a_1 - b_1}{2}\cos(2t_1) \quad \text{for} \quad t_1 \in \langle 0, \frac{\pi}{2}\rangle \text{ AND } \langle\frac{3\pi}{2}, 2\pi\rangle$$
$$A_1(t_1) = b_1 \qquad\qquad\qquad\qquad\qquad \text{otherwise}$$

for $i > 1$

$$A_i(t_{i-1}) = \frac{a_i + b_i}{2} + \frac{a_i - b_i}{2}\sin(2t_{i-1}) \quad \text{for} \quad t_{i-1} \in \langle 0, \pi \rangle$$

$$A_i(t_{i-1}) = b_i \qquad\qquad\qquad\qquad\qquad \text{otherwise}$$

where $t_i = u_i + s_i$, $u_i \in \langle 0, 2\pi \rangle$ and s_i is angle shift of the axis. Finally, the n-dimensional gate can be expressed as logical OR of the individual ellipsoids

$$E_1 \quad \text{OR} \quad E_2 \quad \text{OR} \ldots \text{OR} \quad E_n.$$

Using this gate, one can define in a more flexible way the region of interest.

An example of a two-dimensional ellipsoid gate applied to a nuclear multifragmentation spectrum is shown in Fig. 2.10, [26]. The ellipsoid gates have been positioned to the areas of interest, i.e. during the sorting process, the events which form the ridges, are excluded. An example of three-dimensional ellipsoid gate is shown in Fig. 2.11. The figure demonstrates that one can change freely the lengths of half-axes, the widths of ellipsoids as well as the angle between them.

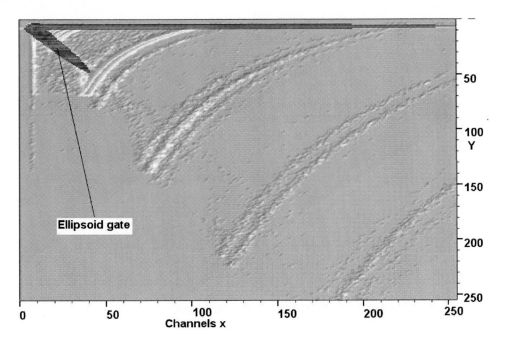

Figure 2.10. Two-dimensional ellipsoid gate applied to the spectrum of nuclear multifragmentation.

2.3.6. Composed Gates

The result of the application of any of the above-defined gates (conditions) is either "true", i.e., the event is accepted for further processing or "false" (event is ignored). By applying the logical operators (AND, OR, and NOT) to operands (previously defined gates) and using parentheses one can write very complex logical expressions defining the shape of the composed gate. The shape can be very complicated and can define even the composition of disjoint subsets.

Let us denote the gates shown in Figs. 2.12(a), (b) and (c) as *A*, *B* and *C*, respectively. Then, the resulting composed gate defined by logical function (*A* OR *B*) AND *C* is shown in Fig. 2.12(d).

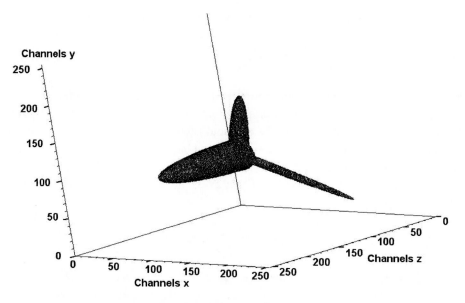

Figure 2.11. An example of three-dimensional ellipsoid gate. One can change freely the lengths of half-axes, the widths of ellipsoids as well as the angle between them.

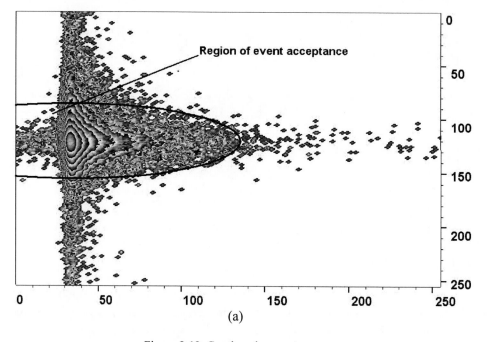

(a)

Figure 2.12. Continued on next page.

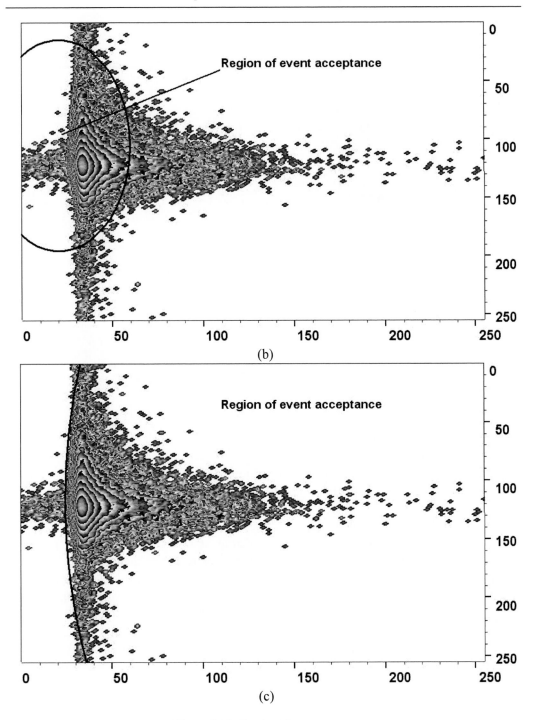

Figure 2.12. Continued on next page.

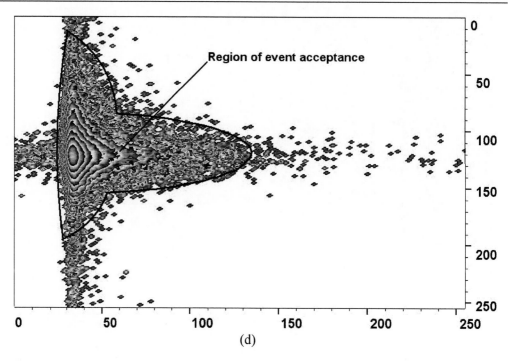

(d)

Figure 2.12. Two-dimensional gate A (a), gate B (b), gate C (c) and resulting composed gate (A OR B) AND C (d).

2.4. Examples and Discussion

Let us illustrate the influence of gating on spectral quality in γ-ray spectra from Gammasphere [1] [22]. In principle, the gating technique can be applied to any kind of coincidence nuclear data. In Fig. 2.13 (a), (b) we show rectangular and circle gate applied to two-dimensional peak, respectively.

In Fig. 2.14 we present unfolded one-dimensional spectrum without gating (upper spectrum) from 3-fold events and spectrum with application of rectangular gate from Fig. 2.13(a) (lower spectrum). One can observe considerable decrease of the background in the spectrum with gating (lower spectrum). In Fig. 2.15, we present a detail of both spectra. Besides of the decrease of the background one can see also an improvement in resolution.

Let us now compare the influence of gate shape on the resulting spectra. In Fig. 2.16 we show a detail of the spectrum slice with application of rectangular gate from Fig. 2.13(a) (thick line) and with application of circle gate from Fig. 2.13(b) (thin line). The peak-to-background ratios are

peak #1	p/b=1.298 – for rectangular gate,
peak #1	p/b=1.511 – for circle gate,
peak #2	p/b=1.712 – for rectangular gate,
peak #2	p/b=1.843 – for circle gate.

As one can see for both peaks, the circle gate increases and, thus, improves peak-to-background ratios.

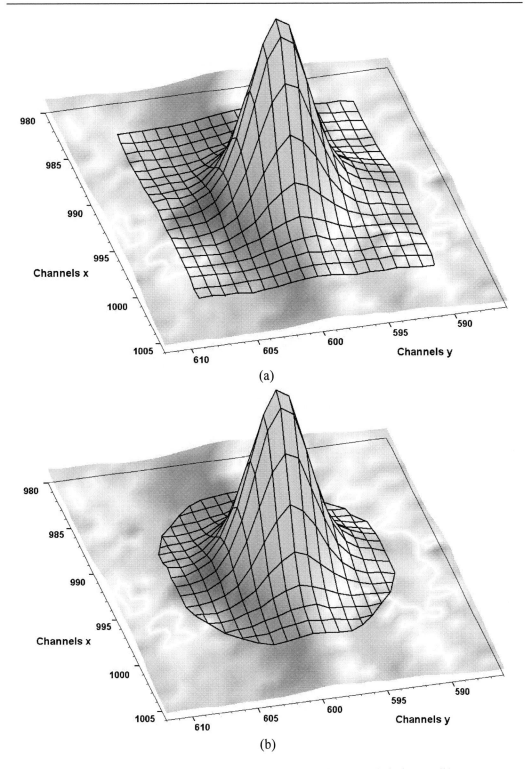

Figure 2.13. Two-dimensional peak with rectangular (a), and circle gate {b).

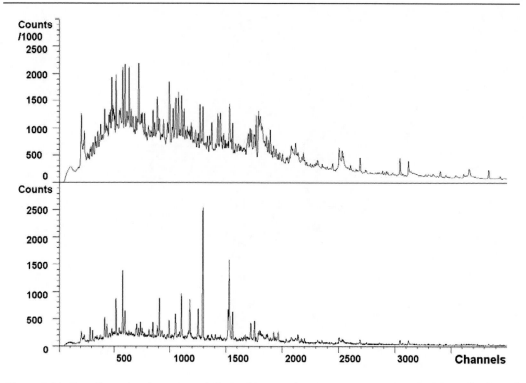

Figure 2.14. One-dimensional spectrum (slice) without gating (upper spectrum) and with application of rectangular gate from Fig. 2.13 (a), (lower spectrum).

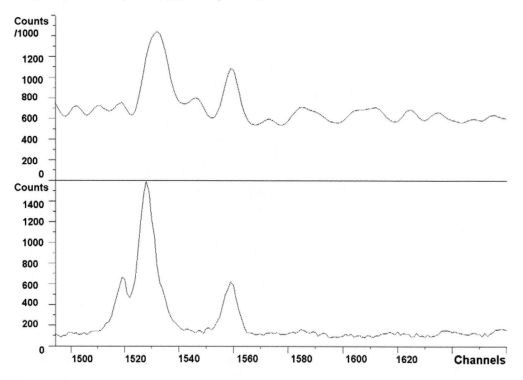

Figure 2.15. Detail of the slice from Fig. 2.14 without gating (upper spectrum) and with application of rectangular gate from Fig. 2.13 (a), (lower spectrum).

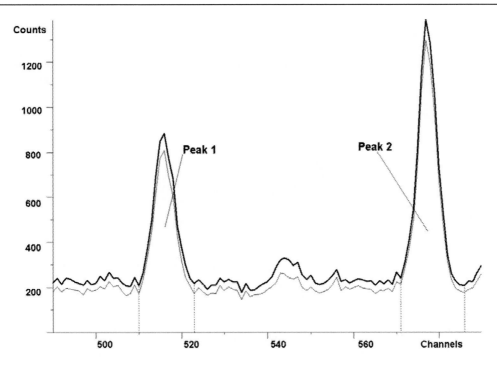

Figure 2.16. Detail of the spectrum slice with application of rectangular gate from Fig. 2.13 (a), (thick line) and with application of circle gate from Fig. 2.13 (b), (thin line).

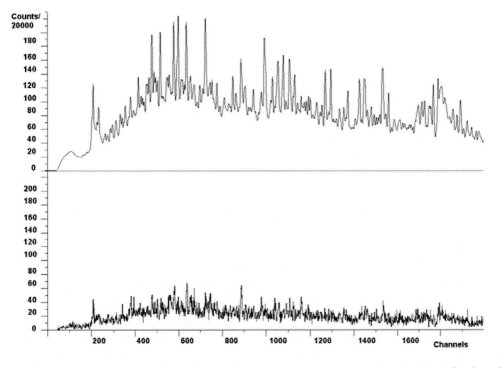

Figure 2.17. One-dimensional spectrum (slice) without gating (upper spectrum) and with application of rectangular gate $x, y, z \in \langle 444, 507 \rangle$ (lower spectrum).

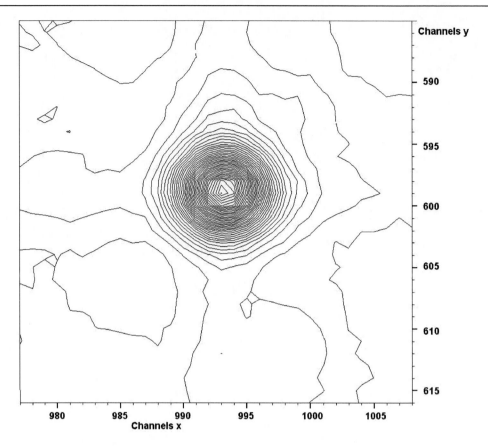

Figure 2.18. Projection of two-dimensional spectrum to the x, y plane shown in contours display mode.

Some gating techniques, e.g. rectangular gates (2.1) or spherical gates (2.2) can be extended to any dimension. There is no limitation in dimensionality. Let us proceed to three-dimensional gate applied to raw γ-ray events. In Fig. 2.17, we present unfolded one-dimensional spectrum without gating (upper spectrum) from 4-fold events and spectrum with application of rectangular gate $x, y, z \in \langle 444, 507 \rangle$ (lower spectrum).

The sorting methods decrease substantially the volume of the data. The number of events satisfying the condition given by rectangular gate from Fig. 2.13(a) is 0.0452% of all events. When applying circle gate from Fig. 2.13(b), the number of events satisfying this condition decreases even more to 0.039% of all events. Naturally, for more dimensional data this efficiency is even higher. In the example given in Fig. 2.17, the number of events satisfying the condition $x, y, z \in \langle 444, 507 \rangle$ is 0.0046% of all events.

For some kinds of data, the determination of shape and parameters of gates being employed is empirical (e.g. data in Fig. 2.2). On the other hand, in some studies, e.g. study of transitions of isotopes accompanying fission process the association between isotopes and their corresponding peaks in one-dimensional spectra is known. For more dimensional gates, their parameters can be well estimated for example from projections of data to two-dimensional planes. Such a projection for data from Fig. 2.13 is shown in Fig. 2.18.

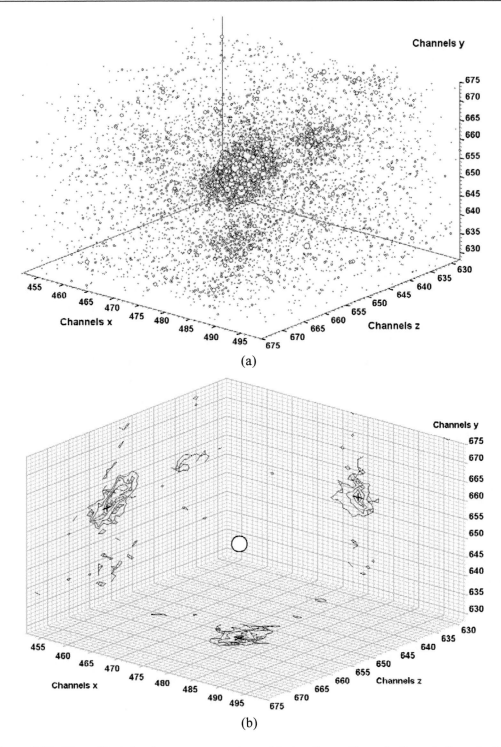

Figure 2.19. Three-dimensional γ-γ-γ-ray spectrum (a), together with projection of slices in its maximum to two-dimensional planes shown in contours display mode (b). Positions of the maximum are outlined by crosses.

Apparently, the most suitable is circle gate. From Fig. 2.18, one can easily determine its position and radius or eventually other parameters (see (2.2)).

An example of three-dimensional γ-γ-γ-ray spectrum together with projection of slices in its maximum to two-dimensional planes are shown in Figs. 2.19a and b, respectively. Again, the parameters of the three-dimensional spherical gate (2.2) can be determined from these projections.

3. Background Estimation in Spectroscopic Data

3.1. Introduction

The determination of the position and net areas of peaks, e.g. due to γ-ray emissions, requires the accurate estimation of the spectral background. In the majority of prior works [27], [28], [29] and [30], the peak-free regions had to be identified in order to construct the background beneath peaks. Once background regions have been located an approximation to the shape of the spectral background must be found, both in background and in peak containing regions of the spectrum. For instance, the methods based on polynomial fits [31], can be applied. In [32] a background estimation method, which employs a zero-area digital filter to identify peak-free regions of γ-ray spectra, is employed. The filter used in this work consists of three segments of equal widths placed symmetrically about the channel of interest. The high and low energy segments are given negative sign and unit weight. The central segment is given positive sign and a weight of two. Where the spectrum is flat (peak-free regions) or varying slowly compared to the size of the filter, therefore, the total area of the filter is zero. The filter can be expressed as

$$y'(i) = \sum_{j=i-n}^{i+n} \left\{ 2y(j) - y\left[j - (2n+1)\right] - y\left[j + (2n+1)\right] \right\} \qquad (3.1)$$

with $y(i)$ = data in channel i, $y'(i)$ = filtered data for channel i and n is chosen so that $(2n+1) \geq fwhm$ (full width at half-maximum) for peaks in the spectrum. From the filtered data, one can determine lower and upper boundaries of the peak regions. Once upper and lower boundaries have been obtained, the background value in channel i can be calculated

$$y = A\left\{ \sum_{j=b_1}^{i} \left[y(j) - y(b_1) \right] \right\} + y(b_1) \qquad (3.2)$$

where

$$A = \left[y(b_2) - y(b_1) \right] \Big/ \left\{ \sum_{k=b_1}^{b_2} \left[y(k) - y(b_1) \right] \right\}$$

b_1 = boundary channel number with lower count

b_2 = boundary channel number with higher count

Various approaches to the background treatment cover Fourier transform techniques [33], digital filters, families of parabolas [34], iterative line stripping [35], peak clipping [36], and the treatment of the first and the second derivative spectra [37]. Very efficient method of background estimation has been developed in [5]. The method is based on Statistics-sensitive Non-linear Iterative Peak-clipping algorithm. In order to compress the dynamic range of the channel counts the LLS operator [38] is applied to the spectrum data

$$v(i) = \log\left[\log\left(\sqrt{y(i)+1}+1\right)+1\right] \tag{3.3}$$

The log operator allows one to work over a few orders of magnitude and the square-root operator selectively enhances small peaks. From $v(i)$ we calculate step by step the vectors $v_1(i)$, $v_2(i)$ up to $v_m(i)$ where m is given parameter. It should be chosen so that $2m+1 = w$, where w is the width of a preserved object. The new value in the channel i in the p-th iteration is obtained by

$$v_p(i) = \min\left\{v_{p-1}(i), \frac{1}{2}\left[v_{p-1}(i+p)+v_{p-1}(i-p)\right]\right\} \tag{3.4}$$

After the vector $v_m(i)$ is calculated and the inverse LLS operator is applied, we obtain the resulting baseline spectrum. On the contrary, to the previously mentioned methods, SNIP method implicitly determines peak regions and peak-free regions.

However, no one of these methods can be straightforwardly extended to two-fold and in general to higher-fold γ-ray coincidence spectra. In this case, the goal is to estimate not only slowly changing continuous two-dimensional background but also coincidences peak in one-dimension and the background in the other one in both directions. These one-fold coincidences are in the two-dimensional spectrum represented by ridges. The situation is complicated even more by the fact that two-fold coincidence peaks are located at the crossing points of one-fold coincidence ridges. Therefore, the algorithm must be able to recognize what is crossing point of the ridges and what is real two-fold coincidence peak.

Several attempts have been carried out to cope with the problem of the background estimate in γ-ray coincidence spectra [39], [40] and [41]. The backgrounds are parameterized as the cross products of lower-dimensional projections of the data and a one-dimensional background spectrum. A somewhat novel method of correcting for a mixture of different reaction channels in the complete data sets, by making use of one or more gates on background channels, is presented in [42]. The principle of the method is as follows. Let the counts in such a data set be represented by $y(i,j)$ where i and j indices are channels corresponding to the γ-ray energies. The one-dimensional projections of the matrix is then

$$P(i) = \sum_j y(i,j) \tag{3.5}$$

Let us divide the counts in this projection into a „background" spectrum $b(i)$ and a "peak" spectrum $p(i)$ such that

$$P(i) = b(i) + p(i) \tag{3.6}$$

This can be done using one of the above-mentioned one-dimensional methods. Then the estimated value of the two-dimensional background spectrum is

$$B(i,j) = \frac{1}{T}b(i)P(j) = \frac{1}{T}\big(b(i)b(j) + b(i)p(j)\big), \tag{3.7}$$

where

$$T = \sum_{i,j} y(i,j)$$

is the total number of counts in the matrix. However, this does not remove the background counts in coincidence with the peak of the gate. In order to do this we extend the procedure and get the following two-dimensional background

$$B(i,j) = \frac{1}{T}\big[b(i)b(j) + b(i)p(j) + b(j)p(i)\big] = \frac{1}{T}\big[P(i)P(j) - p(i)p(j)\big]. \tag{3.8}$$

In the contribution [42], the method has been extended to higher-fold coincidence spectra. The method is very simple but on the other hand, it has its limitations, which result in generation of the false data. The aim of background estimation is to separate useless information (background and combinations of coincidences of the background with peaks - ridges) from useful information contained in n-fold coincidence peaks of n-parameter spectrum.

In further investigations, the sensitive nonlinear iterative peak clipping (SNIP) algorithm has been taken as the basis of the methods for estimation of the background in multidimensional spectra. First in the paper [43], we extended the SNIP algorithm to two-dimensional spectra. For two-dimensional spectra, the algorithm should remove not only continual component of the background but also one-fold coincidences, i.e., the coincidences peak - background in both dimensions. The algorithm should recognize the crossing point of two one-fold coincidences - ridges from a two-fold coincidence, two-dimensional peak. Moreover, the situation is complicated by the fact that two-fold coincidence peak is located in the vicinity of the crossing point of the ridges. The algorithm is based on the method of successive comparisons of the average value in the vertices of investigated square with the point in the center of the square side.

Furthermore, in the paper we extended the proposed algorithm to three-dimensional spectra. Now in addition to the continual component, the algorithm should include all one-fold coincidences (peak – background - background), two-fold coincidences (peak - peak - background) and all their combinations into the estimated background. The interesting information is located in the three-fold coincidence peaks. The algorithm is again based on the method of successive comparisons of average values in vertices of the investigated cube with the values in centers of edges of the cube and subsequently with the values in the centers of sides of the cube. Finally, in the paper we generalized the proposed algorithm for n-dimensional spectra. Nevertheless, the algorithms derived in the paper [43] suffer from several imperfections. Improved versions of the algorithms have been presented in [44].

3.2. One-Dimensional Spectra

3.2.1. Order of the Clipping Filter

The basic SNIP algorithm was proposed so as to include automatically the linear component under the peak into the estimated background. Implicitly, the linear background was assumed under the peak. This is well eliminated by the clipping filter of 2-nd order. The full energy peak results from photoelectric effect and has a shape that is very close to Gaussian function. However, besides of that also the events with energy less than full energy occur rather frequently resulting in various interactions with different energy dependence. Therefore, different components, e.g. Compton edge, Compton background, backscatter peak, escape, and annihilation peaks appear in the measured spectrum. In these cases however, the background under the peak is much more complicated. It is desired to include these elements into estimated background. For example, the Compton edge can be, to some extent, approximated by cubic function, which can be removed by clipping filter of 4-th order. The peaks (Gaussians), which we want to preserve in the spectrum, are quasi-symmetrical functions. Hence, we shall consider only even orders of the clipping filters. In general, the algorithm of background estimation for clipping filter of the 2-nd, 4-th, 6-th, 8-th order can be expressed in the following way. Let $v(i)$ be the input spectrum and

$$v_0(i) = v(i); \ i \in \langle 0, size - 1 \rangle$$

- 2-nd order filter

$$a = v_{p-1}(i), \tag{3.9}$$

$$b = \frac{v_{p-1}(i-p) + v_{p-1}(i+p)}{2}, \tag{3.10}$$

$$v_p(i) = \min(a,b), \tag{3.11}$$

- 4-th order filter

$$a = v_{p-1}(i),$$ (3.12)

$$b = \frac{v_{p-1}(i-p) + v_{p-1}(i+p)}{2},$$ (3.13)

$$c = \frac{-v_{p-1}(i-p) + 4v_{p-1}(i-p/2) + 4v_{p-1}(i+p/2) - v_{p-1}(i+p)}{6},$$ (3.14)

$$v(i) = \min(a, \max(b,c))$$ (3.15)

- 6-th order filter

a, b, c are calculated according to Eqs. 3.12, 3.13, 3.14 and 3.15

$$d = \frac{v_{p-1}(i-p) - 6v_{p-1}(i-2p/3) + 15v_{p-1}(i-p/3)}{20} +$$
$$+ \frac{15v_{p-1}(i+p/3) - 6v_{p-1}(i+2p/3) + v_{p-1}(i+p)}{20};$$ (3.16)

$$v(i) = \min(a, \max(b,c,d));$$ (3.17)

- 8-th order filter

a, b, c are calculated according to Eqs. 3.12, 3.13, 3.14, d according to Eq. 3.16 and

$$e = \frac{-v_{p-1}(i-p) + 8v_{p-1}(i-3p/4) - 28v_{p-1}(i-p/2) + 56v_{p-1}(i-p/4)}{70} +$$
$$+ \frac{56v_{p-1}(i+p/4) - 28v_{p-1}(i+p/2) + 8v_{p-1}(i+3p/4) - v_{p-1}(i+p)}{70};$$ (3.18)

$$v(i) = \min(a, \max(b,c,d,e)),$$ (3.19)

where for all filters $p \in \langle 1, m \rangle$, $i \in \langle p, size - p \rangle$, $m > 0$ is given clipping window and size is the length of processed spectrum. The coefficients in Eqs. 3.14, 3.16, and 3.18 originate from formulas for higher order differences. In fact, they can be represented by binomial coefficients of Pascal's triangle with changing signs. Generalization of the algorithm for higher order filters is straightforward.

Now let us illustrate the influence of the width of clipping window W on the estimated background. We take γ-ray spectrum with very complicated background and with many peaks (Fig. 3.1). In the example we employed the algorithm with decreasing window for W = 10, 20, 30, 40. Apparently, by changing the width one can substantially influence the estimated background.

The other way to improve the estimate of the background is to employ higher order clipping filters as proposed in this section. The example of a portion of γ-ray spectrum for clipping filters of 2-nd, 4-th, 6-th, 8-th orders for W =40 are shown in Fig. 3.2.

Judging from Figs. 3.1 and 3.2 it could seem that the higher-order filters give similar results to a shorter clipping length. As an example, let us take γ-ray spectrum of ^{56}Co. Besides of narrow photopeaks, the spectrum contains wide Compton edges and backscattering peaks. They should be included into the estimated background. In Fig. 3.3, one can see the influence of the clipping window lengths on the estimated background. Choosing the length $w = 20$ one can observe that the rests of background close to the photopeaks (mainly their left sides) as well as wide Compton edges and backscattering peaks remain included in the resulting spectrum. When decreasing the length up to $w = 5$ the estimation of undesired wide components (Compton edges and backscattering peaks) is improved. On the other hand, the estimated background goes under the photopeaks and interferes with their areas. It can strongly deform the shapes of the photopeaks.

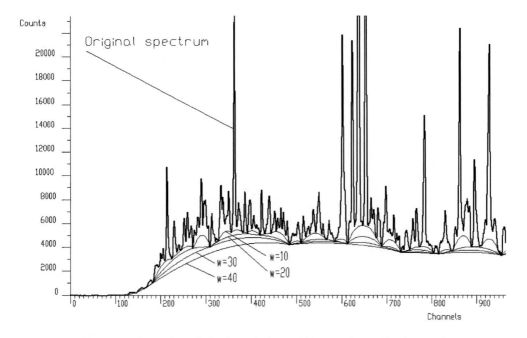

Figure 3.1. Illustration of clipping window width on estimated background.

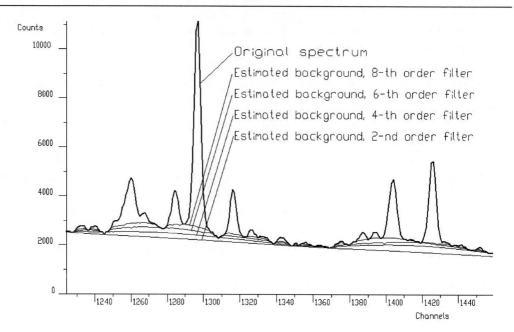

Figure 3.2. Illustration of the influence of clipping filter order on estimated background.

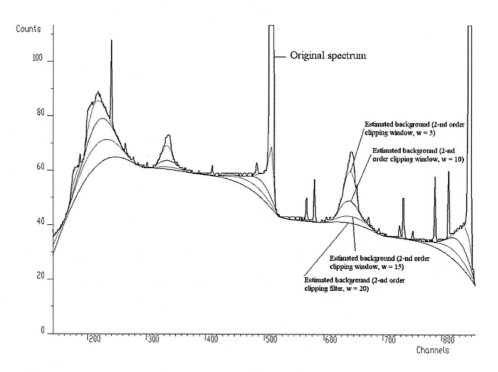

Figure 3.3. Detail of the γ-ray spectrum of ^{56}Co illustrating the influence of window widths on estimated background using second order clipping filter.

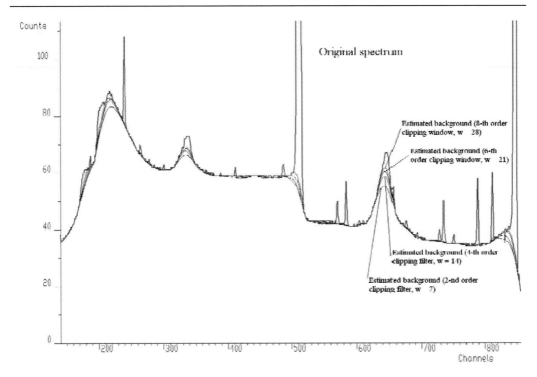

Figure 3.4. Detail of the γ-ray spectrum of ^{56}Co illustrating the influence of clipping filter order with appropriate window widths.

The remedy of the above mentioned undesired effects consists in using higher-order clipping filters. In Fig. 3.4 we illustrate the estimation of the background of the same γ-ray spectrum using 2-nd, 4-th, 6-th and 8-th clipping filter orders. According to Eqs. 3.14, 3.16, and 3.18 the lengths of clipping window should be increased by factor 2, 3 and 4. As one can see with increasing clipping order the background is estimated more precisely.

3.2.2. Increasing versus Decreasing Clipping Window

Initially, the SNIP algorithm was proposed for increasing clipping window, i.e. the window was changing starting from the value 1 up to the given value m. Due to the presence of the noise, the estimated background at the edges of peaks goes under peaks. Since the operation of comparison in the above-proposed algorithm is nonlinear, the result achieved by employing the algorithm with decreasing window is rather different. The algorithm with decreasing clipping window removes the above-mentioned defect. The implementation of this very simple idea gives substantially smoother estimated background. Hence, the background estimation can be done with either

- increasing window $p = 1, 2, 3, ..., m$ or
- decreasing window $p = m, m-1, m-2, ..., 1$.

The examples given below prove in favor of the algorithm with decreasing window. The example of application of the algorithm with increasing clipping window to γ-ray spectrum with rather complicated background is presented in Fig. 3.5. In the estimated background, one can notice lobes at the edges of big peaks due to fluctuations present in the spectrum. Naturally, after subtraction of the estimated background from the spectrum the shapes of peaks are deformed.

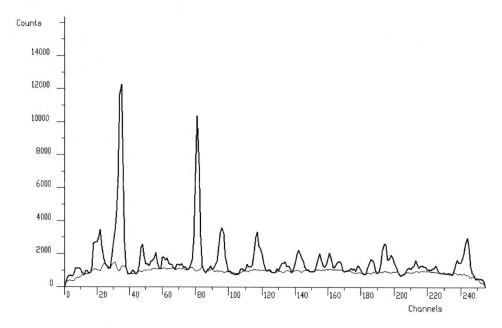

Figure 3.5. One-dimensional γ-ray spectrum and estimated background using increasing clipping window algorithm.

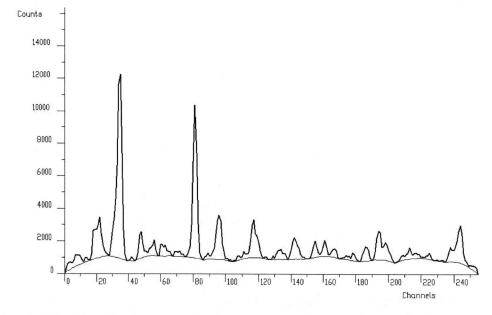

Figure 3.6. One-dimensional γ-ray spectrum and estimated background using decreasing clipping window algorithm.

When we switch the direction of the change of the clipping window to decreasing one we get the result presented in Fig. 3.6. The estimated background is smoother and consequently the results of the analysis are better. The width of the clipping window W in both examples was set to 6.

3.2.3. Background Estimation with Simultaneous Smoothing

For noisy data, the operation of comparison in Eqs. 3.11, 3.15, 3.17 and 3.19 is a source of systematic error. The estimated background copies the minimum values of noise spikes. Therefore, we have developed a more sophisticated background estimation algorithm with simultaneous smoothing. The new value in the channel i in the p-th iteration step is determined according to the following algorithm

1) Let us define

$$s_{p-1}(k) = \sum_{j=-w}^{j=w} f(j) v_{p-1}(k+j),$$ (3.20)

where $f(j)$ is a smoothing filter and w is a smoothing window. Various smoothing filters were presented [38], [45] and [46]. Let us consider the following methods of filtration:

- averaging in window w;
- LLS averaging;
- n-point smoothing;
- convolution with Gaussian.

2) Read out the value a using Eq. 3.9 or 3.12.

3) Due to the given order of the clipping filter in formulas 3.10, 3.13, 3.14, 3.16, and 3.18 replace $y_{p-1}(k)$ by approximate smoothed value $s_{p-1}(k)$ and calculate the values b, c, d, e.

4) For
2-nd order filter

$$\text{if } b < a \text{, set } v_{p-1}(i) = b$$
$$\text{else, calculate } s_{p-1}(i) \text{ and set } v_{p-1}(i) = s_{p-1}(i);$$

4-th, 6-th or 8-th order filter determine

$$b = \max(b,c), \ b = \max(b,c,d) \text{ or } b = \max(b,c,d,e), \text{ respectively.}$$
$$\text{If } b < a \text{, set } v_{p-1}(i) = b$$

else, calculate $s_{p-1}(i)$ and set $v_{p-1}(i) = s_{p-1}(i)$, where $i \in \langle p, size - p \rangle$.

The principle of the algorithm is outlined in Fig. 3.7. Let us denote the investigated point as O and the points entering the background estimation in the iteration step p as P_1, P_2.

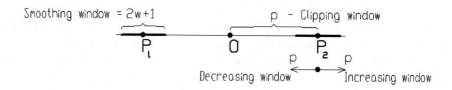

Figure 3.7. Principle of background estimation in one-dimensional spectra with simultaneous smoothing.

In other words, we compare the average value of the smoothed spectrum in the points P_1, P_2 with the value (non-smoothed) in the point O. If the value in the point O is bigger, it is replaced by the average of smoothed values of the points P_1, P_2, i.e., the point O is situated inside of a peak region. Otherwise, the point O is situated outside of the peak region and we replace it by its smoothed value.

The example of the background estimates with both non-smoothing and smoothing algorithm is given in Fig. 3.8. We employed standard 5-point smoothing. Apparently, the estimate of the background can be influenced by noise present in the spectrum. One can eliminate this influence through the use of the smoothing algorithm proposed in this section. In the original algorithm without smoothing, the estimated background snatches the lower spikes in the noise. Consequently, the areas of peaks are biased by this error.

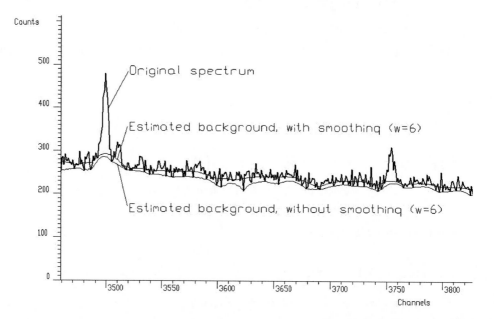

Figure 3.8. Illustration of non-smoothing and smoothing algorithm of background estimation.

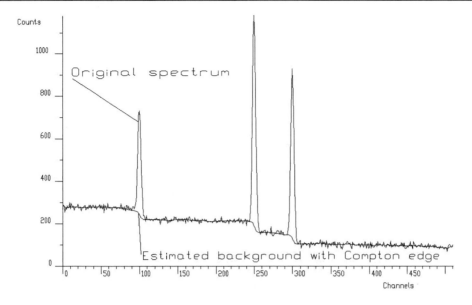

Figure 3.9. Estimation of background with Compton edges.

During the analysis of the experimental data, one can combine above-given parameters. The example of such a synthetic spectrum with estimated background including Compton edges (decreasing clipping window $W = 10$, 8-th order clipping filter, 5-point smoothing) is shown in Fig. 3.9.

3.2.4. Background Estimation in Spectra with Non-symmetrical Peaks

So far, we assumed that the peaks in processed spectra are symmetrical. However, this is not the case for some types of spectroscopic data, e.g. electron or positron annihilation spectra. Consequently, to estimate background in this kind of data we had to modify the above-proposed SNIP algorithms.

Without loss of generality, we shall assume we use increasing window strategy. The left and right clipping window widths are denoted as m_1, m_2. At the beginning, we increase both widths of the clipping window simultaneously in both directions until smaller of them reaches the minimum of both values m_1, m_2 according to Eqs. 3.9, 3.10, and 3.11. Then we stop increasing of window in the direction of smaller width and increase only the clipping window in the other direction. The algorithm can be expressed formally in the following way

$$m = \max\left(m_1, m_2\right)$$

$$p = 1, 2, 3, ..., m$$

$$p_1 = 1, 2, ..., \min\left(p, m_1\right), \ p_2 = 1, 2, ..., \min\left(p, m_2\right).$$

The situation for decreasing window is analogous. The illustrative example for electron spectrum is given in Fig. 3.10.

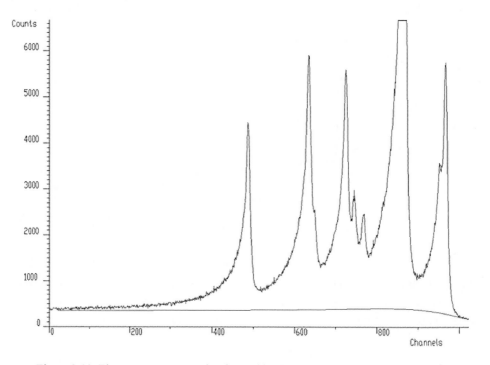

Figure 3.10. Electron spectrum and estimated background for $m_1 = 80$ and $m_2 = 20$.

3.3. Two-Dimensional Spectra

In the paper [43], we derived the original algorithm of background estimation in two-dimensional coincidence γ-ray spectra. The algorithm is based on the method of successive comparisons that consists of several steps. The operation of comparison is, mainly for noisy data, a source of systematic errors, which appear in the form of artifacts (ridges) around peaks. After analyzing the algorithm more deeply we have found that the algorithm can be merged into one step only. This algorithm removes the above-mentioned imperfection.

3.3.1. Order of the Clipping Filter

Let us extend the proposed 2, 4, 6, 8-th order filter algorithm to two-dimensional spectra. Let $v(i_1, i_2)$ be the input spectrum and $v_0(i_1, i_2) = v(i_1, i_2)$; $i_1 \in \langle 0, size_1 - 1 \rangle$, $i_2 \in \langle 0, size_2 - 1 \rangle$. Furthermore, let us define vectors

$$c_2 = [-2,1]^T$$
$$c_4 = [-6,4,-1]^T$$
$$c_6 = [-20,15,-6,1]^T$$
$$c_8 = [-70,56,-28,8,-1]^T , \qquad (3.21)$$

which in fact represent difference operators for the calculation of 2, 4, 6, 8-th order differences. Analogously to binomial coefficients for the elements of these vectors, one can write

$$c_{2n}(i) = \binom{2n}{n-i}(-1)^{i+1},$$

where $i \in \langle 0,n \rangle$ and $n = 1,2,3,4$.

- 2-nd order filter

$$a = v_{p-1}(i_1,i_2) ; \qquad (3.22)$$

$$b = \frac{v_{p-1}(i_1-p,i_2-p) - 2v_{p-1}(i_1-p,i_2) + v_{p-1}(i_1-p,i_2+p) - 2v_{p-1}(i_1,i_2-p)}{4} +$$
$$+ \frac{-2v_{p-1}(i_1,i_2+p) + v_{p-1}(i_1+p,i_2-p) - 2v_{p-1}(i_1+p,i_2) + v_{p-1}(i_1+p,i_2+p)}{4} \qquad (3.23)$$

$$v_p(i_1,i_2) = \min(a,b) \qquad (3.24)$$

Using vector c_2 the relation 3.23 can be written

$$b = \left[\sum_{\substack{j_1=-1 \\ j_1 \neq 0, j_2 \neq 0}}^{1} \sum_{j_2=-1}^{1} v_{p-1}(i_1+j_1 p, i_2+j_2 p) c_2(|j_1|) c_2(|j_2|) \right] \Big/ |c_2(0)|^2 ; \qquad (3.25)$$

- 4-th order filter

a, b are calculated according to 3.22, 3.25 and

$$c = \left[\sum_{\substack{j_1=-2 \\ j_1 \neq 0, j_2 \neq 0}}^{2} \sum_{j_2=-2}^{2} v_{p-1}(i_1+j_1 \, p/2, i_2+j_2 \, p/2) c_4(|j_1|) c_4(|j_2|) \right] \Big/ |c_4(0)|^2 ; \qquad (3.26)$$

$$v_p\left(i_1,i_2\right) = \min\left(a,\max\left(b,c\right)\right);\qquad(3.27)$$

- 6-th order filter

$a,\ b,\ c$ are calculated according to 3.22, 3.25, 3.26 and

$$d = \left[\sum_{\substack{j_1=-3 \\ j_1\neq 0, j_2\neq 0}}^{3}\sum_{j_2=-3}^{3} v_{p-1}\left(i_1 + j_1\, p/3, i_2 + j_2\, p/3\right)c_6\left(\left|j_1\right|\right)c_6\left(\left|j_2\right|\right)\right]\Bigg/\left|c_6\left(0\right)\right|^2;\quad(3.28)$$

$$v_p\left(i_1,i_2\right) = \min\left(a,\max\left(b,c,d\right)\right)\ ;\qquad(3.29)$$

- 8-th order filter

$a,\ b,\ c,\ d$ are calculated according to Eqs. 3.22, 3.25, 3.26, 3.28 and

$$e = \left[\sum_{\substack{j_1=-4 \\ j_1\neq 0, j_2\neq 0}}^{4}\sum_{j_2=-4}^{4} v_{p-1}\left(i_1 + j_1\, p/4, i_2 + j_2\, p/4\right)c_8\left(\left|j_1\right|\right)c_8\left(\left|j_2\right|\right)\right]\Bigg/\left|c_8\left(0\right)\right|^2,\quad(3.30)$$

$$v_p\left(i_1,i_2\right) = \min\left(a,\max\left(b,c,d,e\right)\right),\qquad(3.31)$$

where $i_1 \in \langle p, size_1 - p\rangle$, $i_2 \in \langle p, size_2 - p\rangle$, $p = 1,2,...,m$ for increasing window strategy or $p = m, m-1,...,1$ for decreasing strategy and m is given clipping window.

For data with higher level of noise when using the algorithm based on successive comparisons in the slices around the investigated point the artificial ridges emerge in the estimated background. Let us take synthetic two-dimensional noisy spectrum given in Fig. 3.11. After we determined the background employing the algorithm based on successive comparisons and subtracted it from original data, we got spectrum in Fig. 3.12. This spectrum contains non-natural ridges due to several comparison operations in the algorithm.

In this section, we proposed a new modified algorithm, which removes the above-given problem. The algorithm is based on two-dimensional filtering approach and comparison is carried out only once. When applying it to the spectrum from Fig. 3.11, we get the result without artificial ridges (Fig. 3.13).

Now analogously to one-dimensional data let us study the influence of additional parameters to the quality of background estimate. We take two-dimensional γ-ray spectrum with rather complicated background shown in Fig. 3.14. After elimination of the background estimated by successive comparisons based algorithm with the increasing clipping window of the 2-nd order we get the result with artificial ridges given in Fig. 3.15.

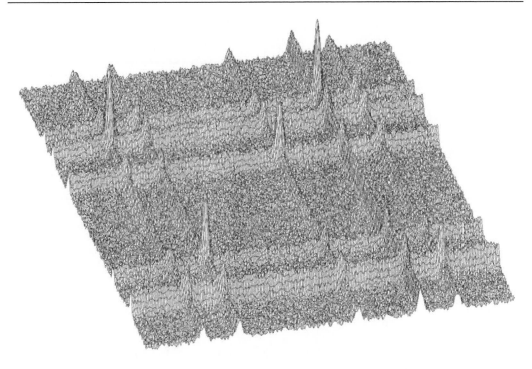

Figure 3.11. Two-dimensional synthetic noisy spectrum.

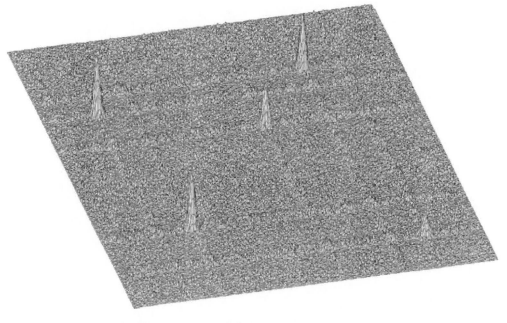

Figure 3.12. Spectrum from Fig. 3.11 after elimination of the background estimated using the algorithm based successive comparisons.

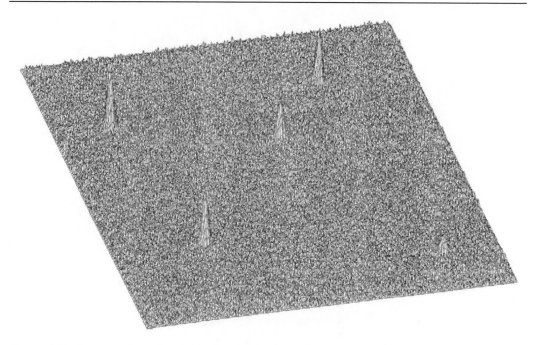

Figure 3.13. Spectrum from Fig. 3.11 after elimination of background estimated using the algorithm based on two-dimensional filtering.

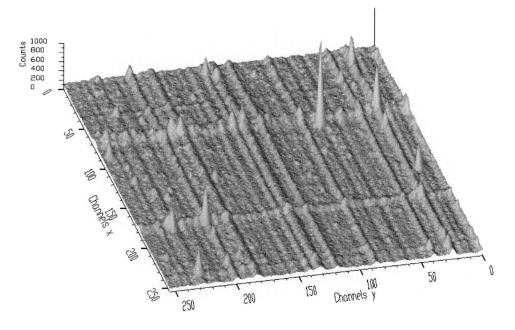

Figure 3.14. Two-dimensional γ-ray spectrum with complicated background.

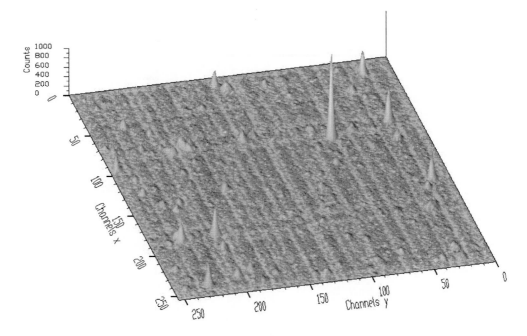

Figure 3.15. Spectrum from Fig. 3.14 after elimination of background estimated using the algorithm based on successive comparisons with increasing clipping window.

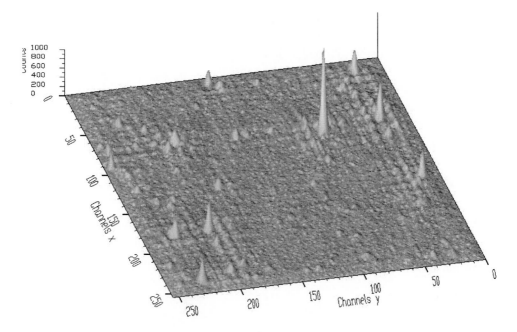

Figure 3.16. Spectrum from Fig. 3.14 after elimination of background estimated using the approach based on one-step filtering algorithm with decreasing window of the 2-nd order.

Next, if we change the clipping window of the 2-nd order to the decreasing one and employ the one-step filtering algorithm we get the result given in Fig. 3.16. Finally, we changed the order of the clipping window to 4 (keeping other parameters unchanged) and we obtained the result given in Fig. 3.17.

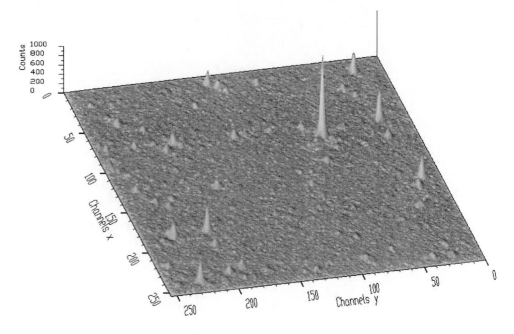

Figure 3.17. Spectrum from Fig. 3.14 after elimination of background estimated using the approach based on one-step filtering algorithm with decreasing window of the 4-th order.

3.3.2. Background Estimation with Simultaneous Smoothing

According to high number of combinations of coincidences the smoothing of the spectra during the process of background estimation is for two and multidimensional spectra even more urgent than for one-dimensional spectra. Let us illustrate the algorithm of background estimation with simultaneous smoothing (Eq. 3.18). We shall assume that we estimate the background in the point O. The points in the vertices of the investigated square entering the estimate are denoted as P_1, P_2, P_3, P_4 and the points in the centers of square sides are denoted as S_1, S_2, S_3, S_4.

While in the vicinity of points P_1, P_2, P_3, P_4 the smoothing is carried out in the outlined squares, in the vicinity of points S_1, S_2, S_3, S_4 the smoothing is carried out in the outlined line segments. This situation corresponds to the expected crossing of ridges together with two-fold peak located in the point O. In 3.23 the values $v_{p-1}(i_1, i_2)$ are replaced by smoothed values $s_{p-1}(i_1, i_2)$

$$b = \left[\sum_{\substack{j_1=-1 \\ j_1 \neq 0}}^{1} \sum_{\substack{j_2=-1 \\ j_2 \neq 0}}^{1} s_{p-1}(i_1 + j_1 p, i_2 + j_2 p) c_2(|j_1|) c_2(|j_2|) \right] \Big/ |c_2(0)|^2. \qquad (3.32)$$

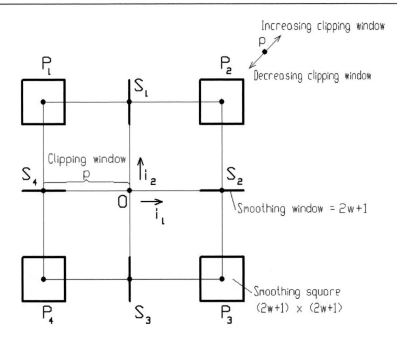

Figure 3.18. Principle of background estimation in two-dimensional spectra with simultaneous smoothing.

If $j_1 \neq 0$ and $j_2 \neq 0$ (points P_1, P_2, P_3, P_4)

$$s_{p-1}\left(i_1 + j_1 p, i_2 + j_2 p\right) = \sum_{l_1=-w}^{l_1=w} \sum_{l_2=-w}^{l_2=w} f\left(l_1, l_2\right) v_{p-1}\left(i_1 + j_1 p + l_1, i_2 + j_2 p + l_2\right) \quad (3.33)$$

else if $j_1 \neq 0$ and $j_2 = 0$ (points S_2, S_4)

$$s_{p-1}\left(i_1 + j_1 p, i_2\right) = \sum_{l_1=-w}^{l_1=w} f\left(l_1, 0\right) v_{p-1}\left(i_1 + j_1 p + l_1, i_2\right) \quad (3.34)$$

else if $j_1 = 0$ and $j_2 \neq 0$ (points S_1, S_3)

$$s_{p-1}\left(i_1, i + j_2 p\right) = \sum_{l2=-w}^{l_2=w} f\left(0, l_2\right) v_{p-1}\left(i_1, i_2 + j_2 p + l_2\right), \quad (3.35)$$

where $f\left(l_1, l_2\right)$ is a two-dimensional filter analogous to the one-dimensional filter and w is a width of smoothing window. The Eqs. 3.33, 3.34 and 3.35 can be merged into one formula

$$s_{p-1}\left(i_1 + j_1 p, i_2 + j_2 p\right) = \sum_{l_1=-z_1 w}^{z_1 w} \sum_{l_2=-z_2 w}^{z_2 w} f\left(l_1, l_2\right) v_{p-1}\left(i_1 + j_1 p + l_1, i_2 + j_2 p + l_2\right), \quad (3.36)$$

where

$$z_1 = \left\langle \begin{array}{l} 0 \text{, if } j_1 = 0 \\ 1, \text{if } j_1 \neq 0 \end{array} \right. \quad ; \quad z_2 = \left\langle \begin{array}{l} 0 \text{, if } j_2 = 0 \\ 1, \text{if } j_2 \neq 0 \end{array} \right.$$

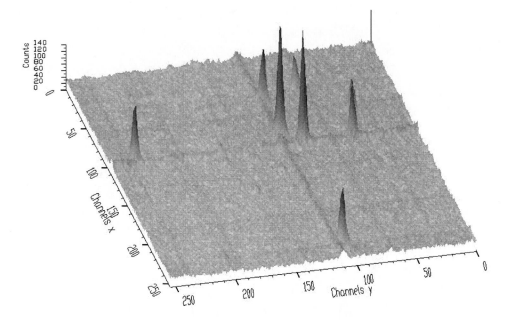

Figure 3.19. Original noisy two-dimensional γ-ray spectrum.

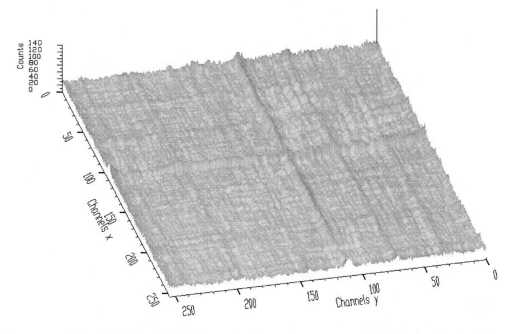

Figure 3.20. Estimated background of the spectrum from Fig. 3.19 using the algorithm without smoothing.

The generalization of Eqs. 3.26, 3.28, 3.30 for 4-th, 6-th, and 8-th order clipping window for the estimation of two-dimensional background with simultaneous smoothing is then apparent.

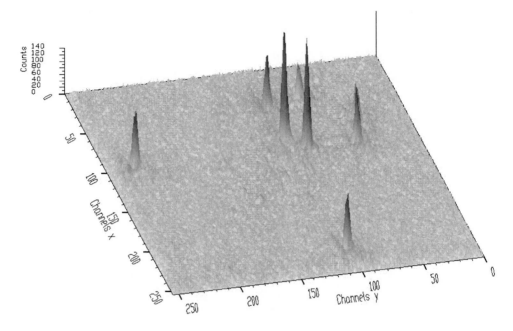

Figure 3.21. Spectrum from Fig. 3.19 after background subtraction (Fig. 3.20).

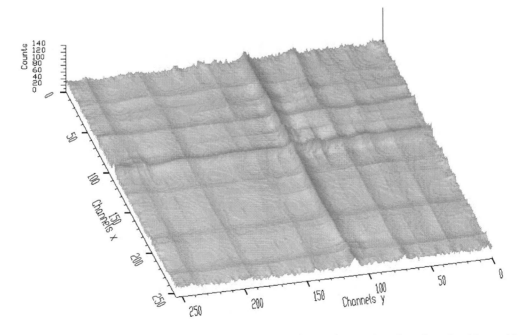

Figure 3.22. Estimated background of the spectrum from Fig. 3.19 using the algorithm with simultaneous smoothing.

The proposed algorithm with simultaneous smoothing allows improving the quality of the estimated spectrum. In Figs. 3.19, 3.20 and 3.21 we present original noisy two-dimensional γ-ray spectrum, estimated background and the spectrum after elimination of the background estimated without smoothing, respectively. The spectrum in Fig. 3.21 contains neither quasi-constant background nor rectangular ridges. However, it contains considerable level of noise.

When we employed the algorithm with simultaneous smoothing (LLS operator), we obtained the estimated background and after subtraction pure peaks shown in Fig. 3.22 and Fig. 3.23, respectively. Even small peaks are recognizable in the spectrum. The noise in the spectrum is suppressed and thus the analysis of the data is more convenient.

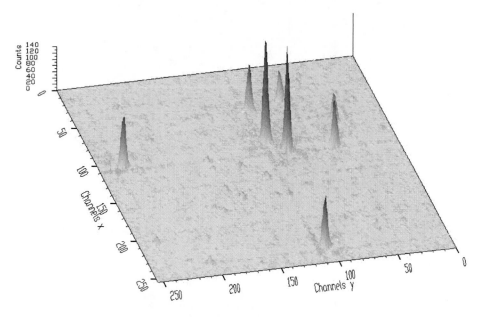

Figure 3.23. Spectrum from Fig. 3.19 after subtraction of background estimated using the algorithm with simultaneous smoothing (Fig. 3.22).

3.3.3. Different Widths of Clipping Window

So far, we assumed that the width of peaks is the same for all dimensions. Presumably, this assumption is satisfied in the case of $\gamma - \gamma$ coincidences. However for some kinds of data one has to consider different widths of separated objects, peaks (e.g. in the case of (γ, X) spectra). For two-dimensional spectrum, let us denote the widths of clipping window as m_1, m_2. For the sake of simplicity, let us assume for the moment that we use increasing window strategy. At the beginning, let us increase the widths of the clipping window simultaneously in both dimensions up to the moment when it reaches the minimum of both values m_1, m_2. Then we stop increasing of window in the direction of smaller width and increase the clipping window only in the other direction. The algorithm can be expressed formally in the following way

$$m = \max\left(m_1, m_2\right)$$

$$p = 1, 2, 3, ..., m$$
$$p_1 = 1, 2, ..., \min(p, m_1), \ p_2 = 1, 2, ..., \min(p, m_2)$$

Then, in relations of the section 3.3 instead of the variable p, we can use the appropriate values p_1, p_2. Subsequent generalization of the algorithm for decreasing window and for multidimensional systems is straightforward.

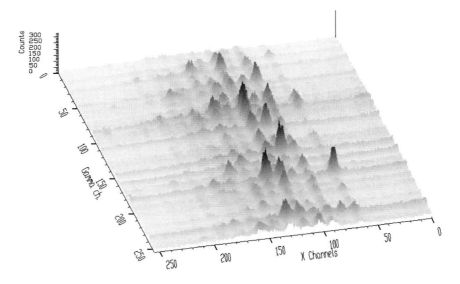

Figure 3.24. Original two-dimensional $\gamma - X$ spectrum.

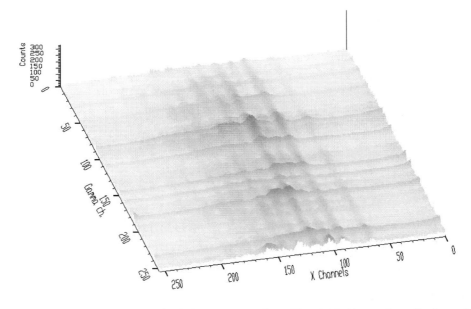

Figure 3.25. Estimated background of the spectrum from Fig. 3.24 (decreasing clipping window $w_\gamma = 6$, $w_X = 12$).

In Figs. 3.24, 3.25 and 3.26 we show an illustrative example of the original $\gamma - X$ -ray spectrum, estimated background and the peaks after background elimination, respectively (decreasing clipping window W_γ =6, W_X =12).

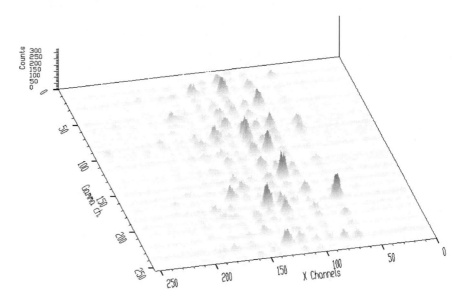

Figure 3.26. Spectrum from Fig. 3.24 after background subtraction (shown in Fig. 3.25).

3.3.4. Skew Ridges

In all algorithms presented so far, we assumed that undesired components of lower-fold coincidences (ridges) are perpendicular to coordinate system. In some experiments however, in two-dimensional spectra, the perpendicular ridges are accompanied also by skew ridges and consequently the requirement of their separation or inclusion into estimated background is emerging. The skew ridges are generated by Compton scattering of a γ-ray from one of the detectors into a coincident detector where it is fully absorbed, so that the sum of the energies deposited in both detectors is equal to the original energy of the γ-ray. This motivated us to development of a new algorithm, which would be able to separate ridges of this kind. Let us illustrate the example of such data in Fig. 3.27. In one point we assume the crossing of maximum two ridges.

Let us denote the points that contribute to the background estimation as I_1 up to I_8. Now the expected crossing of the ridges in the point O needs not to be necessarily perpendicular (in Fig. 3.25 we have illustrated crossing of perpendicular and skew ridges). In this example the points at the tops of ridges I_3, I_4, I_7, I_8 correspond to analogous points at the tops of perpendicular ridges (S_1, S_2, S_3, S_4) according to Eq. 3.18. Thus, in the proposed algorithm we divide the values in the points I_1 up to I_8 into two groups according to their magnitudes (using e.g. bubble-sorting algorithm). The points of the first group correspond to points S_1 up to S_4 and the points of the second one correspond to points P_1

up to P_4. To the points assigned in this way, we can apply the algorithm presented in the section 3.1 (see Eq. 3.25).

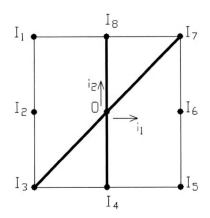

Figure 3.27. Principle of the algorithm of estimation skew ridges.

In Fig. 3.28, we present the results of the estimation of continuous two-dimensional background together with perpendicular and skew ridges in a synthetic spectrum.

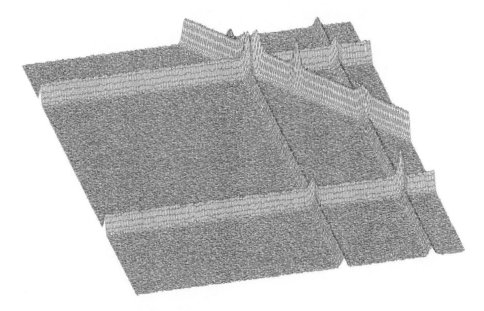

Figure 3.28. Two-dimensional synthetic spectrum containing skew ridges.

The goal is to remove perpendicular as well as skew ridges from the spectrum and to leave only two-dimensional coincidence peaks. After applying background elimination procedure and subtraction, we get the two-dimensional peaks presented in Fig. 3.29.

In Fig. 3.30 and Fig. 3.31 we present the experimental spectrum with skew ridges and estimated background, respectively.

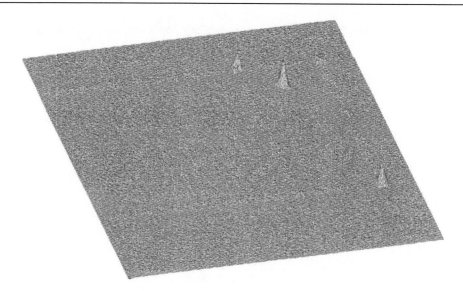

Figure 3.29. Spectrum from Fig. 3.28 after subtraction of background.

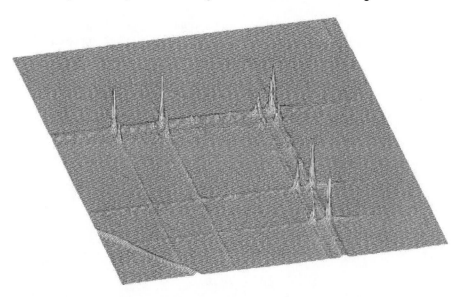

Figure 3.30. Experimental two-dimensional spectrum containing skew ridges.

3.3.5. Nonlinear Ridges

Let us proceed to the most general shape of ridges. In some experiments (e.g. nuclear multifragmentation), the perpendicular ridges are accompanied by nonlinear ones with the exponential shape. The origin of nonlinear ridges results from evolution of energy losses in two subsequent detectors (Si-CsI telescope) for various fragments at various energies [47]. To include the ridges of exponential shape requires development of a special algorithm. Here we propose the algorithm that is schematically outlined in Fig. 3.32 for $p = 4$. Again, in one point we assume crossing of maximum two ridges. In this algorithm, we investigate all points entirely along the perimeter of the square (rectangle).

We find maximum and minimum values on the upper side of the square (according to Eq. 3.18 they correspond to points S_1, P_1), followed by maximum and minimum values on the right side of the square (they correspond to points S_2, P_2), then by maximum and minimum values on the bottom side of the square (they correspond to points S_3, P_3) and finally by maximum and minimum values on the left side of the square (they correspond to points S_4, P_4).

Figure 3.31. Estimated background (with skew ridges) of the spectrum from Fig. 3.30.

To the points assigned in this way, we can apply the algorithm presented in the section 3.3.1 (see Eq. 3.24).

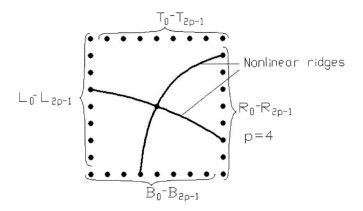

Figure 3.32. Principle of the algorithm of estimation nonlinear ridges.

To illustrate the data of such a kind we present synthetic spectrum shown in Fig. 3.33. The estimated background is given in Fig. 3.34. Pure Gaussian after subtracting the background from the original spectrum is shown in Fig. 3.35.

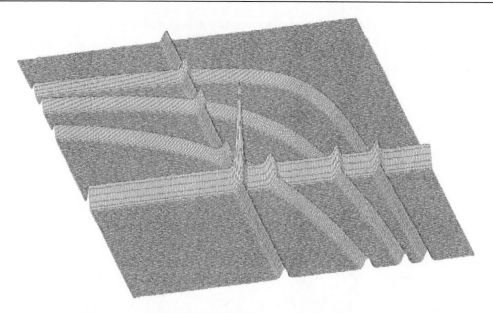

Figure 3.33. Synthetic two-dimensional spectrum containing rectangular as well as nonlinear ridges.

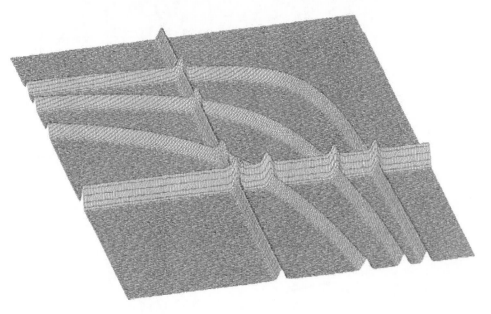

Figure 3.34. Estimated background (with nonlinear ridges) of the spectrum from Fig. 3.33.

3.4. Three-Dimensional Spectra

In the paper [43], we derived the original algorithm of background estimation in three-dimensional coincidence γ-ray spectra as well. Here again we have employed the method of successive comparisons. Due to the above-mentioned reasons, we extend the one-step filtration algorithm also to three-dimensional case.

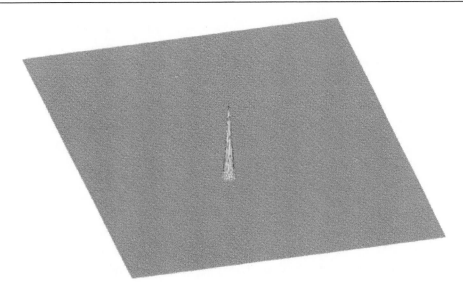

Figure 3.35. Two-dimensional peak after subtraction of background (Fig. 3.34) from the spectrum shown in Fig. 3.33.

3.4.1. Order of the Clipping Filter

Let $v\left(i_1,i_2,i_3\right)$ be the three-dimensional input spectrum and

$$v_0\left(i_1,i_2,i_3\right)=v\left(i_1,i_2,i_3\right);\ i_1\in\left\langle 0,size_1-1\right\rangle;\ i_2\in\left\langle 0,size_2-1\right\rangle;\ i_3\in\left\langle 0,size_3-1\right\rangle$$

- 2-nd order filter

$$a=v_{p-1}\left(i_1,i_2,i_3\right);\tag{3.37}$$

$$b=\left[\sum_{j_1=-1}^{1}\sum_{j_2=-1}^{1}\sum_{j_3=-1}^{1}v_{p-1}\left(i_1+j_1p,i_2+j_2p,i_3+j_3p\right)c_2\left(\left|j_1\right|\right)c_2\left(\left|j_2\right|\right)c_2\left(\left|j_3\right|\right)\right]\Big/\left|c_2\left(0\right)\right|^3;\tag{3.38}$$

$$v_p\left(i_1,i_2,i_3\right)=\min\left(a,b\right)\tag{3.39}$$

- 4-th order filter

$a,\ b$ are calculated according to Eqs. 3.37, 3.38

$$c=\left[\sum_{j_1=-2}^{2}\sum_{j_2=-2}^{2}\sum_{j_3=-2}^{2}v_{p-1}\left(i_1+j_1p/2,i_2+j_2p/2,i_3+j_3p/2\right)c_4\left(\left|j_1\right|\right)c_4\left(\left|j_2\right|\right)c_4\left(\left|j_3\right|\right)\right]\Big/\left|c_4\left(0\right)\right|^3;\tag{3.40}$$

$$v_p\left(i_1,i_2,i_3\right)=\min\left(a,\max\left(b,c\right)\right)\tag{3.41}$$

- 6-th order filter

a, b ,c are calculated according to Eqs.3.37, 3.38 and 3.40

$$d = \left[\sum_{j_1=-3}^{3} \sum_{j_2=-3}^{3} \sum_{j_3=-3}^{3} v_{p-1} \left(i_1 + j_1 p/3, i_2 + j_2 p/3, i_3 + j_3 p/3 \right) c_6 \left(|j_1| \right) c_6 \left(|j_2| \right) c_6 \left(|j_3| \right) \right] \Big/ \left| c_6 (0) \right|^3 ; \quad (3.42)$$

$$v_p \left(i_1, i_2, i_3 \right) = \min \left(a, \max \left(b, c, d \right) \right) \tag{3.43}$$

- 8-th order filter

a, b ,c ,d are calculated according to Eqs. 3.37, 3.38, 3.40 and 3.42

$$e = \left[\sum_{j_1=-4}^{4} \sum_{j_2=-4}^{4} \sum_{j_3=-4}^{4} v_{p-1} \left(i_1 + j_1 p/4, i_2 + j_2 p/4, i_3 + j_3 p/4 \right) c_8 \left(|j_1| \right) c_8 \left(|j_2| \right) c_8 \left(|j_3| \right) \right] \Big/ \left| c_8 (0) \right|^3 \quad (3.44)$$

$$v_p \left(i_1, i_2, i_3 \right) = \min \left(a, \max \left(b, c, d, e \right) \right), \tag{3.45}$$

where $i_1 \in \langle p, size_1 - p \rangle$, $i_2 \in \langle p, size_2 - p \rangle$, $i_3 \in \langle p, size_3 - p \rangle$, $p = 1, 2, ..., m$ for increasing strategy or $p = m, m-1, ..., 1$ for decreasing strategy, m is given clipping window.

To illustrate the generalization of the background estimation algorithms, we proceed to three-fold coincidence $\gamma - \gamma - \gamma$-ray spectra. In this case, we have to include to the estimated background its continuous component as well as one-, and two-fold coincidences in all dimensions. The example of such a spectrum is given in Fig. 3.36. We remind that the position of a sphere in the picture determines the position of a channel and the size of the sphere is proportional to the contents of the channel. Estimated background and resulting spectrum for decreasing clipping window (W =5) are given in Figs. 3.37 and 3.38, respectively.

3.4.2. Background Estimation with Simultaneous Smoothing

Let us illustrate the algorithm for the estimation of the background with simultaneous smoothing in three-dimensional spectra (Fig. 3.39). We shall assume that we estimate the background in the point O. Then, all the points in the vertices of the investigated cube P_1 up to P_8 (smoothing is carried out in small smoothing cubes), the points in the centers of cube edges S_1 up to S_{12} (smoothing is carried out in smoothing squares) and the points in the centers of cube sides R_1 up to R_6 (smoothing is carried out in smoothing line segments) contribute to the estimated background.

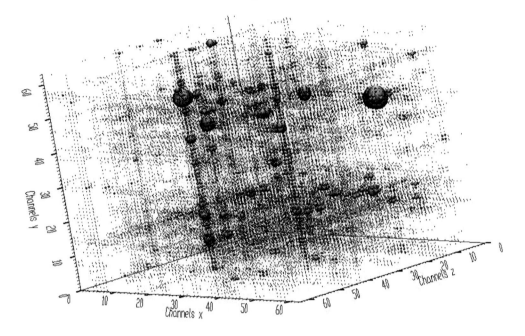

Figure 3.36. Original γ-γ-γ-ray spectrum.

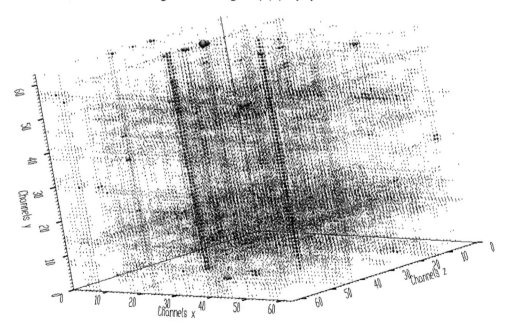

Figure 3.37. Estimated background (for decreasing clipping window, $w = 5$) of the spectrum from Fig. 3.36.

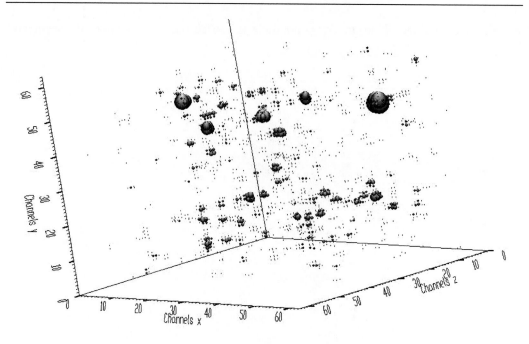

Figure 3.38. Spectrum from Fig. 3.36 after background subtraction.

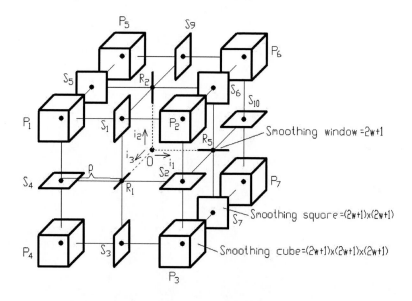

Figure 3.39. Principle of background estimation in three-dimensional spectra with simultaneous smoothing.

Analogously to two-dimensional spectrum in Eq. 3.38, the values $v_{p-1}(i_1, i_2, i_3)$ are replaced by smoothed values. Taking into account Eq. 3.36 for three-dimensional data one can write

$$s_{p-1}\left(i_1 + j_1 p, i_2 + j_2 p, i_3 + j_3 p\right) =$$

$$(3.46)$$

$$= \sum_{l_1=-z_1}^{z_1 w} \sum_{l_2=-z_2}^{z_2 w} \sum_{l_3=-z_3}^{z_3 w} f\left(l_1, l_2, l_3\right) v_{p-1}\left(i_1 + j_1 p + l_1, i_2 + j_2 p + l_2, i_3 + j_3 p + l_3\right)$$

where

$$z_1 = \left\langle \begin{matrix} 0 \ \text{if } j_1 = 0 \\ 1 \ \text{else} \end{matrix} \right. ; \ z_2 = \left\langle \begin{matrix} 0 \ \text{if } j_2 = 0 \\ 1 \ \text{else} \end{matrix} \right. ; \ z_1 = \left\langle \begin{matrix} 0 \ \text{if } j_3 = 0 \\ 1 \ \text{else} \end{matrix} \right. ,$$

and $f\left(l_1, l_2, l_3\right)$ is three-dimensional filter. Further one can continue in accordance to point 4) for one-dimensional spectra.

3.5. Four-Dimensional Spectra

We have implemented the algorithm of background determination even for four-dimensional spectra. The algorithm allows one to separate four-fold coincidences from continuous background as well as from lower order coincidences. In Fig. 3.40, we present synthetic four-dimensional spectrum. Three parameters determine the position of the center of the slice in the 4-th dimension. The channels of the slice are changing with the angle starting from 9 o'clock position. 2π angle of the slice is divided to appropriate number of channels in the 4-th dimension. The sizes of lines from the centre of the slice are proportional to the contents of channels. In Fig. 3.41 we present the estimated spectrum for decreasing clipping window $W = 4$ and in Fig. 3.42 the resulting spectrum of four-dimensional peaks (Gaussians).

3.6. Generalization for N-dimensional Γ-ray Coincidence Spectra

Now we generalize the relations derived in previous sections for n-dimensional case.

3.6.1. Order of the Clipping Filter

Let $v\left(i_1, i_2, ..., i_n\right)$ be the n-dimensional input spectrum and

$$v_0\left(i_1, i_2, ..., i_n\right) = v\left(i_1, i_2, ..., i_n\right); \ i_1 \in \left\langle 0, size_1 - 1\right\rangle, ..., i_n \in \left\langle 0, size_n - 1\right\rangle.$$

- 2-nd order filter

$$a = v_{p-1}\left(i_1, i_2, ..., i_n\right); \tag{3.47}$$

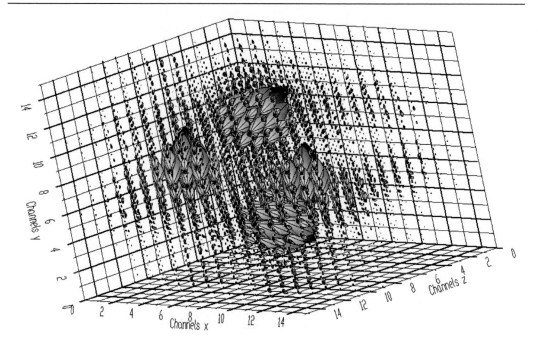

Figure 3.40. Synthetic four-dimensional spectrum.

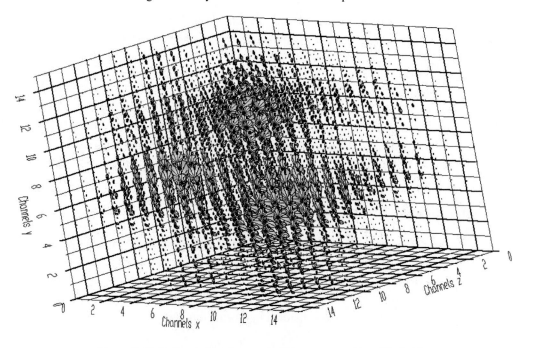

Figure 3.41. Estimated background of the spectrum from Fig.3.40.

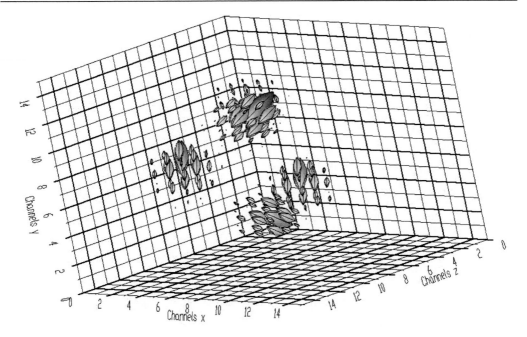

Figure 3.42. Spectrum from Fig. 3.40 after background elimination.

$$t_2 = \left[\sum_{j_1=-1}^{1} \sum_{j_2=-1}^{1} ... \sum_{\substack{j_n=-1 \\ j_n \neq 0 \& ... \& j_n \neq 0}}^{1} v_{p-1}\left(i_1 + j_1 p,...,i_n + j_n p\right) \prod_{k=1}^{n} c_2\left(|j_k|\right) \right] \bigg/ \left|c_2(0)\right|^n \quad (3.48)$$

$$v_p\left(i_1, i_2, ... i_n\right) = \min\left(a, t_2\right); \qquad (3.49)$$

- 4-th order filter

a, t_2 are calculated according to Eqs. 3.48 and 3.49

$$t_4 = \left[\sum_{j_1=-2}^{2} \sum_{j_2=-2}^{2} ... \sum_{\substack{j_n=-2 \\ j_n \neq 0 \& ... \& j_n \neq 0}}^{2} v_{p-1}\left(i_1 + j_1 p/2,...,i_n + j_n p/2\right) \prod_{k=1}^{n} c_4\left(|j_k|\right) \right] \bigg/ \left|c_4(0)\right|^n$$

$$v_p\left(i_1, i_2, ... i_n\right) = \min\left(a, \max\left(t_2, t_4\right)\right) \qquad (3.50)$$

- $2f$-th order filter

Generally, for the value t_{2f} of the filter order $2f\left(f = 1, 2, 3, 4, ...\right)$ for n-dimensional
data we get

$$t_{2f} = \left[\sum_{j_1=-f}^{f} \sum_{j_2=-f}^{f} \cdots \sum_{\substack{j_n=-f \\ j_n \neq 0 \& \cdots \& j_n \neq 0}}^{f} v_{p-1}\left(i_1 + j_1 p/f, \ldots, i_n + j_n p/f\right) \prod_{k=1}^{n} c_{2f}\left(|j_k|\right) \right] \Big/ \left| c_{2f}(0) \right|^n \quad (3.51)$$

$$v_p\left(i_1, i_2, \ldots i_n\right) = \min\left(a, \max\left(t_2, t_4, \ldots, t_{2f}\right)\right), \quad (3.52)$$

where $i_1 \in \langle p, size_1 - p \rangle, \ldots, i_n \in \langle p, size_n - p \rangle$, $p = 1, 2, \ldots, m$ for increasing window, $p = m, m-1, \ldots, 1$ for decreasing window, m is given clipping window. The filter coefficients can be obtained by generalization of the vectors in Eq. 3.21 and using appropriate relations for differences.

3.6.2. Background Estimation with Simultaneous Smoothing

Let us generalize the relation 3.47 for n-dimensional spectrum and replace the values $v_{p-1}\left(i_1, i_2, \ldots i_n\right)$ in Eq. 3.48 by smoothed values

$$s_{p-1}\left(i_1 + j_1 p, \ldots, i_n + j_n p\right) = \sum_{l_1=-z_1 w}^{z_1 w} \cdots \sum_{l_n=-z_n w}^{z_n w} f\left(l_1, \ldots, l_n\right) v_{p-1}\left(i_1 + j_1 p + l_1, \ldots, i_n + j_n p + l_n\right) \quad (3.53)$$

where

$$z_1 = \left\langle \begin{array}{l} 0 \ if \ j_1 = 0 \\ \\ 1 \ else \end{array} \right., \ldots, z_n = \left\langle \begin{array}{l} 0 \ if \ j_n = 0 \\ \\ 1 \ else \end{array} \right.$$

and $f\left(l_1, \ldots, l_n\right)$ is n-dimensional filter. Then we get

$$v_p\left(i_1, i_2, \ldots, i_n\right) = \left\langle \begin{array}{l} s_{p-1}\left(i_1, i_2, \ldots, i_n\right) \ if \ a < \max\left(t_2, t_4, \ldots, t_{2f}\right) \\ \\ \max\left(t_2, t_4, \ldots, t_{2f}\right) \ else. \end{array} \right.$$

4. Deconvolution

4.1. Introduction

The accuracy and reliability of the analysis depend critically on the treatment in order to resolve strong peak overlaps. The peaks as the main carrier of spectrometric information are very frequently positioned close to each other. The extraction of the correct information out of the spectra sections, where due to the limited resolution of the equipment, signals coming from various sources are overlapping, is a very complicated problem. Deconvolution and restoration are the names given to the endeavor to improve the resolution of an experimental measurement by mathematically removing the smearing effects of an imperfect instrument, using its resolution function. Devising reliable methods for doing this has been a long-term endeavor by many scientists and mathematicians.

The deconvolution methods are widely applied in various fields of data processing. Recently, many applications have been found in various domains of experimental science, e.g. image and signal restoration, astronomy, determination of thickness of multilayer structures, tomography, magnetic resonance imaging, crystallography, geophysics, etc.— [10], [11], [13], [48-51]. The deconvolution methods can be successfully applied for the determination of positions and intensities of peaks and for the decomposition of multiplets in γ-ray spectroscopy.

From a numerical point of view, the deconvolution is so called ill-posed problem, which means that many different functions solve a convolution equation within error bounds of experimental data. The estimates of the solution are sensitive to rounding-off errors and to errors in the input data [52]. While the first kind of errors can be removed by employing error-free algorithms [53-55], to suppress the influence of the second kind of errors we have to employ more sophisticated tools. When employing standard algorithms to solve a convolution system, small errors or noise can cause enormous oscillations in the result. This implies that the regularization must be employed. Tikhonov first treated this problem on a strict mathematical basis by introducing the regularization theory and methods [14, 15]. The regularization encompasses a class of solution techniques that modifies an ill-posed problem into a well-posed one, by approximation, so that a physically acceptable approximate solution can be obtained and the solution is sufficiently stable from the computational viewpoint [16]. Therefore, the methods of regularization can be classified from several aspects

- quality of the deconvolved signal (smoothness, positive solution, oscillations, etc.);
- computational complexity;
- convergence speed, etc.

The output spectrum registered by analyzer can be thought of being composed of shifted response (instrument) functions. So at the input of the detector and electronic system it can be imagined as a sum of δ- functions. At the output of the system, the spectrum represents a linear combination of the δ-functions with various amplitudes positioned in various channels, which are blurred by the system response function. Our endeavor in the deconvolution operation is to remove the influence of the response function to the utmost, i.e., to deblur the data and in the ideal case to obtain a spectrum consisting of δ -functions like "peaks".

4.2. Brief Overview of Deconvolution Methods

Stationary system that satisfies the superposition principle can be described by convolution integral

$$y(t) = \int_{-\infty}^{t} x(\tau)h(t-\tau)d\tau = x(t)*h(t)+n(t),$$

where $x(t)$ is the input into the system, $h(t)$ is its impulse function (response), $y(t)$ is the output from the system, $n(t)$ is additive noise. Analogously for discrete systems one can write

$$y(i) = \sum_{k=0}^{i} x(k)h(i-k) + n(i) = x(i)*h(i)+n(i), \quad i = 0,1,...,N-1 \quad (4.1)$$

where N is number of samples and the mark * denotes the operation of the convolution. This system can be in matrix form written

$$y = Hx + n \quad (4.2)$$

where the matrix H has dimension $N \times M$, the vectors y, n have length N and the vector x has length M, while $N \geq M$ (overdetermined system).

Singular Value Decomposition (SVD). Any $N \times M$ overdetermined system matrix H can be written as the product of an $N \times M$ column orthogonal matrix U, an $M \times M$ diagonal matrix W with positive or zero elements (the singular values) and the transpose of an $M \times M$ orthogonal matrix V [56], i.e.,

$$H = U \cdot \left[diag(w_j) \right] \cdot V^T, \quad (4.3)$$

where $j \in \langle 1, M \rangle$. Then from 4.2 one can write (without considering the noise vector n)

$$x = V \cdot [diag(1/w_j)] \cdot U^T y. \quad (4.4)$$

Very popular is truncated singular value decomposition [57], where in 4.4 we replace $1/w_j$ for

$$w_j \leq l \quad (4.5)$$

where l is given non-negative limit, with zeroes.

Least Square Solution. To find least square solution of above given system of linear equations the functional

$$\| H\hat{x} - y \|^2$$

should be minimized. Direct unconstrained least squares estimate of the vector $\boldsymbol{x}$ is

$$\hat{\boldsymbol{x}} = (H^T H)^{-1} H^T \boldsymbol{y} \qquad (4.6)$$

where $A = H^T H$ is the Toeplitz matrix. When employing this algorithm to solve a convolution system small errors or noise can cause enormous oscillations in the result. The problem of finding $\boldsymbol{x}$, where H, $\boldsymbol{y}$ are known, is a discrete ill-posed problem [58-60], and requires regularization techniques to get adequate solution.

Several methods to regularize the solution of Eq. 4.2 were developed. Most methods used in inverse problems adopt both an extreme criterion to unfold data (for instance, those of least squares or the maximum entropy) and a regularization method to reduce the very large fluctuations of the unfolded spectrum. Three types of regularization methods are very often used [61]:

- smoothing,
- constraints imposition (for example only non-negative data are accepted),
- choice of a prior information probability function - Bayesian statistical approach.

Tikhonov-Miller Regularization. At the beginning, let us follow the classical way of solving Eq. 4.2, i.e., the Tikhonov-Miller regularization procedure [60], [62]. To find a regularized approximate solution of Eq. 4.2, the functional

$$\| H\hat{\boldsymbol{x}} - \boldsymbol{y} \|^2 + \alpha \| Q\hat{\boldsymbol{x}} \|^2 \qquad (4.7)$$

is minimized, Q and α being the regularization matrix and parameter, respectively. This solution can be obtained by solving the equation

$$\hat{\boldsymbol{x}} = \left(H^T H + \alpha Q^T Q \right)^{-1} H^T \boldsymbol{y} \qquad (4.8)$$

where superscript T denotes transpose of the matrix. There exist various ways how to solve this system of linear equations, e.g. Refs. 11 and 63. The solution of Eq. 4.8 for $\alpha = 0$ is often called the principal solution. It is a limiting case of what is called zero-th order or Tikhonov regularization for $Q = E$ - unit matrix. Then we get

$$\hat{\boldsymbol{x}} = \left(H^T H + \alpha \right)^{-1} H^T \boldsymbol{y}. \qquad (4.9)$$

Together with χ^2 also the sum of squares of elements of the estimated vector $\hat{\boldsymbol{x}}$ is minimized. There exist also some recommendations how to choose α $(0 < \alpha < \infty)$. In [56] it is suggested to try

$$\alpha = \text{Trace}(H^T H) / \text{Trace}(Q^T Q).$$

Tikhonov-Miller Regularization of Squares of Negative Values (MSNV) [64]. Let us define iterative deconvolution algorithm with Tikhonov-Miller regularization

$$\widehat{\boldsymbol{x}}^{(k)} = \left(H^T H + \alpha Q^{(k)T} Q^{(k)}\right)^{-1} H^T \boldsymbol{y}^{(k)}$$

$$\boldsymbol{y}^{(k+1)} = H\widehat{\boldsymbol{x}}^{(k)}, \tag{4.10}$$

where k is iteration step and $\boldsymbol{y}^{(0)} = \boldsymbol{y}$. If we set the regularization matrix to be diagonal with elements

$$Q^{(k)}(i,j) = \begin{cases} 1 \ if \ x^{(k)}(i) < 0 \ and \ i = j \\ 0 \ else. \end{cases} \tag{4.11}$$

Besides of χ^2 we minimize also the sum of squares of negative elements of the estimated vector $\widehat{\boldsymbol{x}}$. We do not care about positive values of the resulting vector.

Riley Algorithm. This algorithm is commonly called iterated Tikhonov regularization. To obtain smoother solution one may use the Eq. 4.8 with iterative refinement

$$\boldsymbol{x}^{(n+1)} = \boldsymbol{x}^{(n)} + \left(H^T H + \alpha Q^T Q\right)^{-1}\left(H^T \boldsymbol{y} - H^T H\boldsymbol{x}^{(n)}\right) \tag{4.12}$$

where $\boldsymbol{x}^{(0)} = 0$ and $n = 0,1,2,\ldots$ is iteration step.

Van Cittert Algorithm. The Van Cittert based algorithms [10], [12] of deconvolution are widely applied in different areas, for example in spectroscopy, in image processing and others. The basic form of Van Cittert algorithm for discrete convolution system is

$$\boldsymbol{x}^{(n+1)} = \boldsymbol{x}^{(n)} + \mu\left(H^T HH^T \boldsymbol{y} - H^T HH^T H\boldsymbol{x}^{(n)}\right) = \boldsymbol{x}^{(n)} + \mu\left(\boldsymbol{y}' - A\boldsymbol{x}^{(n)}\right), \tag{4.13}$$

where A is system Toeplitz matrix, n represents the number iterations and μ is the relaxation factor. The convergence condition of Eq. 4.13 is that the diagonal elements of the matrix A satisfy

$$A_{ii} > \sum_{j=0, j\neq i}^{N-1} A_{ij} \ , \ i = 0,1,\ldots,N-1.$$

The relation between relaxation factor μ and Riley algorithm of deconvolution is apparent. The coefficient μ should satisfy the condition

$$0 < \mu < 2/\lambda_{max} ,$$

where λ_{max} is the greatest eigenvalue of A.

Janson Algorithm. If we introduce local variable relaxation factor μ_i then Eq. 4.13 becomes

$$x^{(n+1)}(i) = x^{(n)}(i) + \mu_i \left[y'(i) - \sum_{m=0}^{M-1} A_{im} x^{(n)}(k) \right] \tag{4.14}$$

If we choose

$$\mu_i = r \left(1 - \frac{2}{b-a} \right) \left| x^{(n)}(i) - \frac{a+b}{2} \right| \tag{4.15}$$

where r is usually about 0.2, a is usually zero and b must be greater that the ultimate height of the highest peak in the data, we get Janson algorithm of deconvolution [65].

Gold Algorithm. Further, if we choose the local variable relaxation factor

$$\mu_i = \frac{x^{(n)}(i)}{\sum_{m=0}^{M-1} A_{im} x^{(n)}(m)} \tag{4.16}$$

and we substitute it into Eq. 4.13 we get

$$x^{(n+1)}(i) = \frac{y'(i)}{\sum_{m=0}^{M-1} A_{im} x^{(n)}(m)} x^{(n)}(i). \tag{4.17}$$

This is the Gold deconvolution algorithm [66], [67]. Its solution is always positive when the input data are positive, which makes the algorithm suitable for the use for naturally positive definite data, i.e., spectroscopic data. In the rest of the chapter let us call it classic Gold deconvolution.

In practice, we have revealed that the second multiplication by the matrix $H^T H$ in Eq. 4.13, for Gold deconvolution, is redundant. On the contrary, it's omitting gives result with better resolution. Then in Eq. 4.13 we set

$$A = H^T H, \quad \boldsymbol{y'} = H^T \boldsymbol{y}.$$

Hereafter we shall call this algorithm one-fold Gold deconvolution [68]. If we take the initial solution

$$\boldsymbol{x}^{(0)} = \begin{bmatrix} 1,1,\ldots,1 \end{bmatrix}^T$$

and if all elements in the vectors $\boldsymbol{h}$, $\boldsymbol{y}$ are positive (this requirement is fulfilled for spectroscopic data), this estimate is always positive. It converges to the least square estimate in the constrained subspace of positive solutions.

Richardson-Lucy Algorithm. Richardson-Lucy like algorithms [69], [70] use a statistical model for data formation and are based on the Bayes formula [71]. The Bayesian approach consists of constructing the conditional probability density relationship

$$p(x \mid y) = \frac{p(y \mid x)p(x)}{p(y)},$$

where $p(y)$ is the probability of the output data and $p(x)$ is the probability of the input data, over all possible data realizations. The Bayes solution is found by maximizing the right part of the equation. The maximum likelihood (ML) solution maximizes only the density $p(y \mid x)$ over x [72]. For discrete data the algorithm has the form

$$x^{(n+1)}(i) = x^{(n)}(i) \sum_{j=0}^{N-1} h(j,i) \frac{y(j)}{\sum_{k=0}^{M-1} h(j,k) x^{(n)}(k)} \tag{4.18}$$

$i \in \langle 0, M-1 \rangle$. This iterative method forces the deconvolved spectra to be non-negative. The Richardson-Lucy iteration converges to the maximum likelihood solution for Poisson statistics in the data. It is also sometimes called the expectation maximization (EM) method.

Muller Algorithm. Setting the probability to have Gauss statistics, another deconvolution algorithm based on Bayes formula was derived by Muller (1997)

$$x^{(n+1)}(i) = x^{(n)}(i) \frac{\sum_{j=0}^{N-1} h(j,i) y(j)}{\sum_{j=0}^{N-1} h(j,i) \sum_{k=0}^{M-1} h(j,k) x^{(n)}(k)} \tag{4.19}$$

$i \in \langle 0, M-1 \rangle$. This algorithm again applies the positivity constraint to the decovolved data.

Maximum A Posteriori Deconvolution Algorithm. The maximum a posteriori (MAP) solution maximizes over x the product $p(y \mid x)p(x)$. For discrete data the algorithm has the form

$$x^{(n+1)}(i) = x^{(n)}(i) \exp \left\{ \sum_{j=0}^{N-1} h(j,i) \left[\frac{y(j)}{\sum_{k=0}^{M-1} h(j,k)x^{(n)}(k)} - 1 \right] \right\} \qquad (4.20)$$

Positivity of the solution is assured by the exponential function. Moreover, the non-linearity in Eq. 4.20 permits superresolution [73].

Blind Deconvolution. Up to this moment we assumed that we know the response function of the system $h(t)$. However sometimes we only suppose its shape or, respectively, we know theoretically its analytical form, which can substantially differ from the reality. In Eq. 4.1 one can observe that the operation of the convolution is commutative, which gives possibility of implementation of iterative algorithm of blind deconvolution scheme [52], [72] and [74].

a. Estimate an initial response function $h^{(0)}(i)$ from $y(i)$. This can be accomplished in different ways. For example, we choose the narrowest line lobe of $y(i)$. By Gaussian fitting of the selected part, we can get an initial estimate of the response function.
b. Accomplish the deconvolution using the estimated response function $h^{(0)}(i)$ and given $y(i)$

$$x^{(n)}(i) = y(i) * *h^{(n)}(i)$$

where $* *$ denotes the operation of deconvolution
c. Accomplish the deconvolution using the estimated spectral function for finding an improved estimation of the response function

$$h^{(n+1)}(i) = y(i) * *x^{(n)}(i)$$

d. Repeat steps b, c for improving the estimation of $x(i)$, $h(i)$.

In the steps b, c one can use any of the above-presented deconvolution algorithms.

Finally we would like to emphasize that there exists immense number of variations of deconvolution algorithms applied in various scientific fields, e.g. astronomy, image processing, geophysics, optics etc. In this section, we have introduced only basic classes of deconvolution algorithms that can be easily employed and implemented for the processing of spectrometric data.

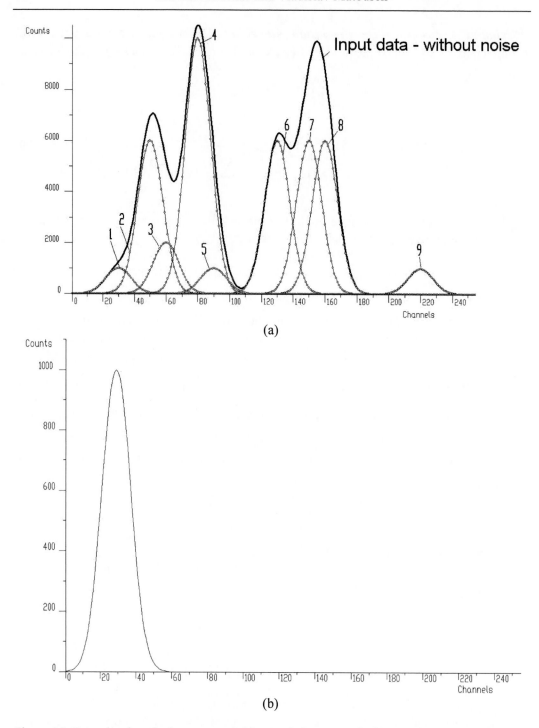

Figure 4.1. Example of synthetic spectrum (without noise) composed of 9 Gaussians (a), and response function (b).

4.3. Study of Deconvolution Algorithms

In what follows, we shall study the properties of deconvolution methods and regularization techniques as well. We shall compare their decomposition capabilities. To make these investigations we have chosen a synthetic data (spectrum, 256 channels) consisting of 9 very closely positioned, relatively wide peaks (σ=5), without noise (Fig. 4.1(a)) and corresponding response (Fig. 4.1(b)), respectively. Thin lines represent pure Gaussians. Thick line is the resulting spectrum. One can notice that some smaller peaks are not observable by eye. It should be emphasized that this is an extremely difficult task. The positions, amplitudes and areas are given in Tab. 4.1. In ideal case, we should obtain the result given in Fig. 4.2. The areas of the Gaussian components of the spectrum are concentrated completely to δ - functions.

Now let us add Gaussian noise with the amplitude of 1% of the heights of small peaks (#1, #5 and #9). When solving such an overdetermined system of linear equations in the sense of minimum least squares criterion without any regularization or constraints using algorithm given by Eq. 4.6 we obtain the result with large oscillations (Fig. 4.3).

From mathematical point of view, it is the optimal solution in the unconstrained space of independent variables. From physical point of view we are interested only in a meaningful solution, i.e., the contents of channels should be non-negative. Therefore, we have to employ regularization techniques and/or to confine the space of allowed solutions to subspace of positive solutions.

Table 4.1. Positions, heights and areas of peaks in the spectrum shown in Fig. 4.1

Peak #	Position	Height	Area
1	30	500	20022
2	50	3000	120132
3	60	1000	40044
4	80	5000	200220
5	90	500	20022
6	130	3000	120132
7	150	3000	120132
8	160	3000	120132
9	220	500	20022

Let start our considerations with classic Tikhonov method of regularization described in the previous section (algorithm given in Eq. 4.9). Let us apply this algorithm to our original data from Fig. 4.3. In Fig. 4.4 we see the original spectrum (thick line), the result of the deconvolutions for α=100 (thin line) and α=500 (thin dotted line). Some outlines of the present peaks are visible but one can observe oscillations even outside of the peak regions of the spectrum. With increasing α the oscillations are softened, but on the other hand, the outlines of some peaks disappear. The deconvolved spectra contain non-realistic negative values.

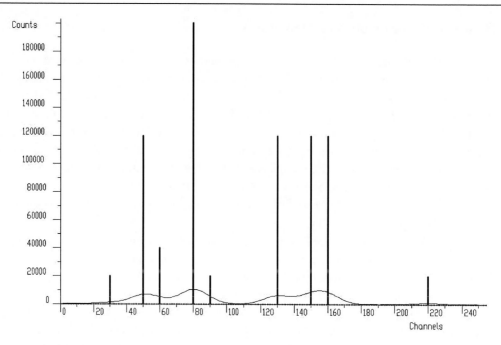

Figure 4.2. Original (thin line) and deconvolved spectrum (bars) calculated as least squares solution of the convolution system of linear equations.

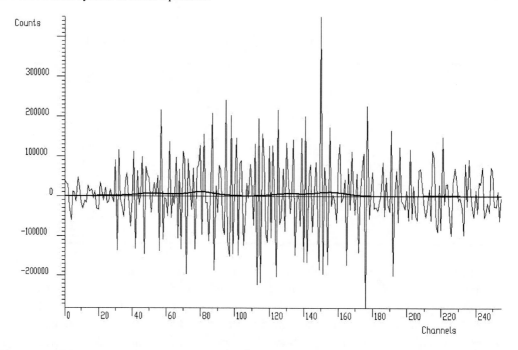

Figure 4.3. Original spectrum from Fig. 4.1 + 1% of noise of the amplitude of small peaks (#1, #5 and #9 - thick line) and deconvolved spectrum (strong oscillations) using unconstrained solution (Eq. 4.6) of the Toeplitz system of linear equations (thin line).

Let us carry out singular value decomposition of the matrix composed of shifted response from Fig. 4.1(b) according Eq. 4.3. The weights arranged according to their magnitudes are shown in Fig 4.5.

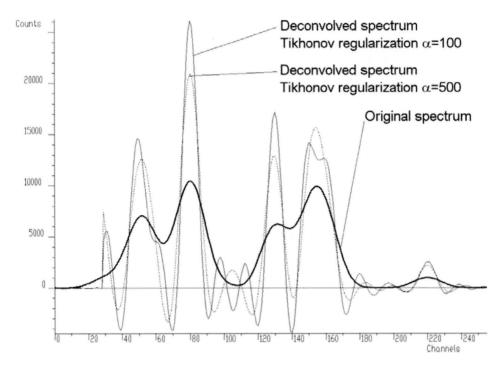

Figure 4.4. Solution obtained using Tikhonov method of regularization.

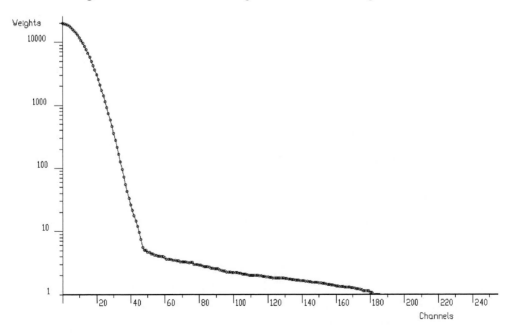

Figure 4.5. Weights (singular values) of the matrix composed of shifted responses from Fig. 4.1(a).

In Fig. 4.6(a) and (b) we show the result of SVD deconvolution after truncation of the inverses of weights to 80 and 40 elements, respectively. The rest of the inverses of weights are zeroed. In Fig. 4.6(a) one can observe strong oscillations. In Fig. 4.6(b) we obtain smoother solution, but the result contains non-realistic negative values. We can conclude that the method is not suitable for decomposition of spectrometric data.

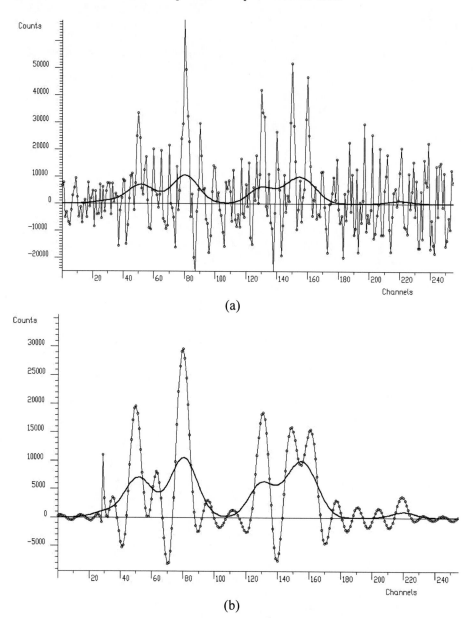

(a)

(b)

Figure 4.6. Original spectrum (thick line) and deconvolved spectrum using SVD algorithm (thin line) with 80 preserved inverses weights (the rest of inverses of weights are zeroed) (a), with 40 preserved inverses of weights (b). Channels are shown as small circles.

The Riley algorithm is sometimes called iterated Tikhonov method and is described in the previous section. Non-regularized is of little importance as it leads practically always to oscillating solution. When regularized we get the result similar to the classic Tikhonov algorithm. An example of application of Riley algorithm for our testing spectrum is given in Fig. 4.7 (1000 iterations, α =3 000).

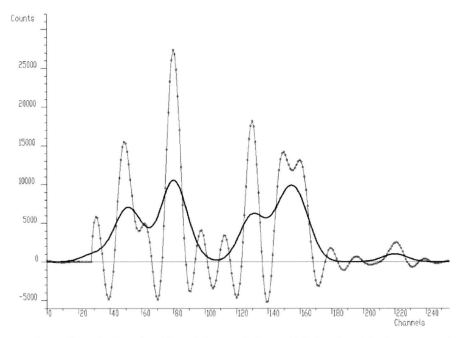

Figure 4.7. Illustration of Riley algorithm of deconvolution - thick line is original spectrum, thin line represents spectrum after deconvolution.

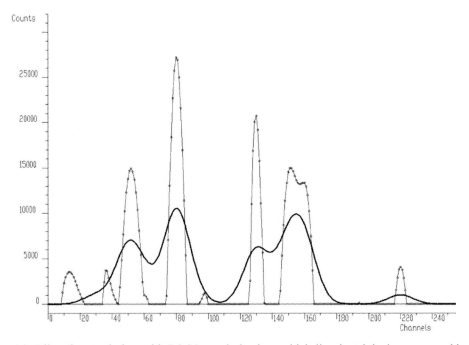

Figure 4.8. Riley deconvolution with POCS regularization - thick line is original spectrum, thin line represents spectrum after deconvolution.

However, because of its iterative nature, the Riley algorithm lends itself to another type of regularization, so called Projections On Convex Sets – POCS. There exist many regularization methods based on this kind of regularization [75-77]. A projection onto the positive $\hat{x}$'s means to set all negative elements to zero after each iteration. When applying it to our example we get the result given in Fig. 4.8 (1 000 iterations, α =9 000). Due to POCS regularization negative values disappeared from the solution.

When we employ Tikhonov-Miller regularization of squares of negative values with $\alpha = 1000$, we get the result given in Fig. 4.9. The method does not oscillate and is able to decompose the peaks in the spectrum practically to δ functions. The area of each peak is concentrated to one channel. However, from computational point of view there is a problem with the inversion in Eq. 4.10. The matrix defined according Eqs. 4.10 and 4.11 is not more Toeplitz matrix and we cannot utilize any of the fast algorithms for the inversion. So for large convolution systems or multidimensional convolution systems the method is extremely time consuming, which makes the method not applicable in practice.

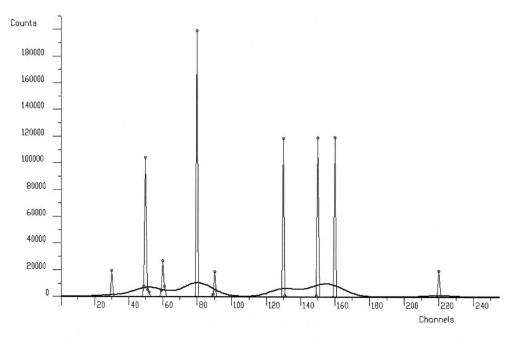

Figure 4.9. Original spectrum (thick line) and deconvolved spectrum using Tikhonov-Miller regularization of squares of negative values (thin line), $\alpha = 1000$.

Van Cittert algorithm represents the basic method of the algorithms based on iterative solution of system of linear equations. Let us continue in our previous example in the study of the properties of various deconvolution algorithms. The result achieved by the classic Van Cittert algorithm does not differ too much from that obtained by the Tikhonov algorithm. Again, it oscillates and gives negative values in clusters of channels. However, when applying above-mentioned POCS regularization after every iteration step the oscillations as well as negative values disappear. In Fig. 4.10, we present the original spectrum (thick line), spectrum deconvolved through the use of classic Van Cittert algorithm and regularized (via POCS method) deconvolved spectrum (1000 iterations). Though better, nevertheless, the small peak at the position 90, the peak between two strong peaks (position 60) as well as doublet of peaks (positions 150 and 160) could not be disclosed even by POCS regularization of the Van Cittert algorithm.

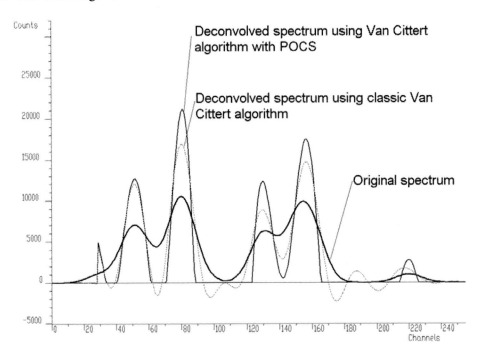

Figure 4.10. Original spectrum (thick line), deconvolved spectrum using Van Cittert algorithm (without regularization) and deconvolved spectrum using Van Cittert algorithm and regularized via POCS method.

In contradiction to Van Cittert algorithm, the Gold deconvolution (for positive response and output data) gives always positive result (see Eq. 4.17). The result of Gold deconvolution is intrinsically constrained to the subspace of positive solutions. Let us apply it to our example. In Fig. 4.11(a) we present the original spectrum and the deconvolved spectrum using classic Gold algorithm (10 000 iterations – Fig. 4.11(a)). The algorithm is unable to resolve peaks #3 and #5. Moreover, it cannot decompose peaks #7 and #8. Even if we increase number of iterations to 50 000 the result remains practically the same (Fig. 4.11(b)).

When using one-fold Gold deconvolution we can achieve much better resolution (10 000 iterations – Fig. 4.12(a)). Nevertheless, we see that the peak #3 at the position 60 cannot be resolved and that there is only an indication of the right-side small peak #5 at the position 90. The positive definite algorithms converge very slowly. In Fig. 4.12(b), after 50 000 iterations, one can observe the outline of both most critical peaks #3 and #5.

Next, let us illustrate Richardson-Lucy algorithm (Eq. 4.18). It is based on Bayes theorem of maximum probability. Like in the case of Gold deconvolution, the algorithm is positive definite. In Figs. 4.13(a), (b) we present the results after 10 000 and 50 000 iterations, respectively. Apparently, Richardson-Lucy algorithm is more efficient than the Gold one. One can clearly identify all peaks in the deconvolved spectrum. On the other hand in the channel 225 a new small fake peak appears.

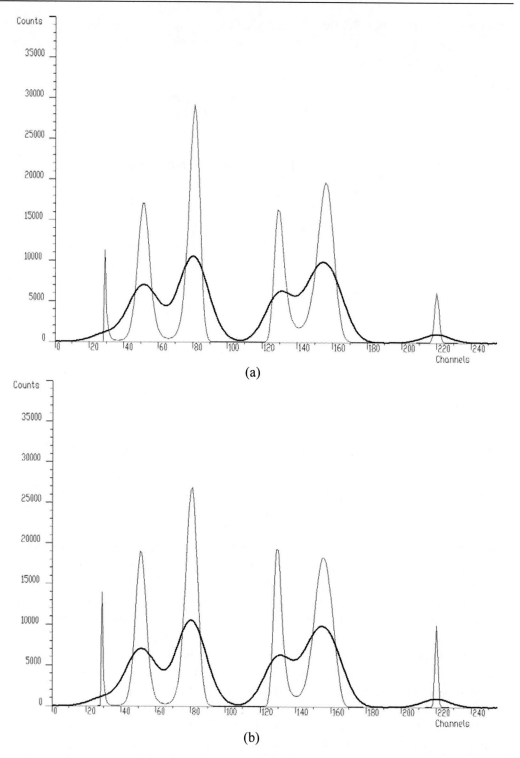

(a)

(b)

Figure 4.11. Original spectrum (thick line) and deconvolved spectrum using classic Gold algorithm (thin line) after 10000 iterations (a), 50000 iterations (b).

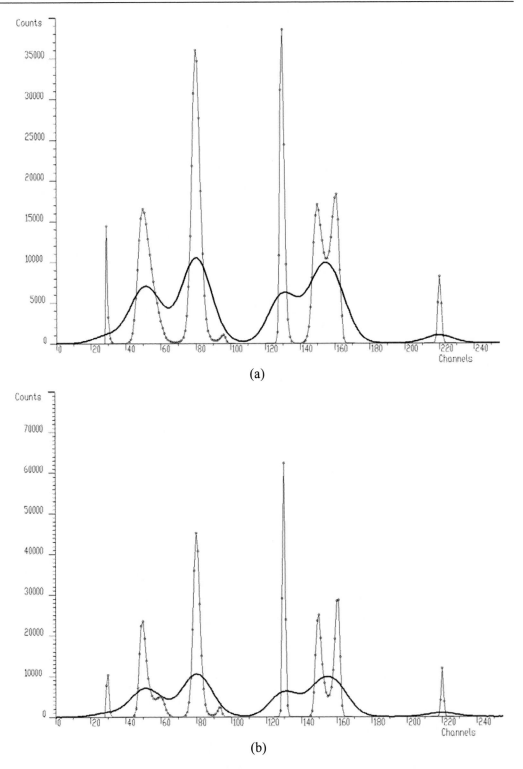

Figure 4.12. Original spectrum (thick line) and deconvolved spectrum using one-fold Gold algorithm (thin line) after 10000 iterations (a), 50000 iterations (b). Channels are shown as small circles.

For completeness' sake, we briefly present also the results obtained by the application of Muller algorithm of deconvolution (Eq. 4.19). Its behavior is very similar to the Gold algorithm. In Figs. 4.14(a), (b) we illustrate the result of the deconvolution after 10 000 and 50 000 iterations, respectively.

Let us proceed to MAP algorithm of deconvolution (Eq. 4.20). It exhibits very good properties. The results of the deconvolutions (Figs. 4.15(a), (b)) are very similar to the results achieved by Richardson-Lucy algorithm. However, it does not generate the fake peak at the position 225.

The four abovementioned algorithms are positive definite, i.e., from positive data (response and output data) one obtains always positive result. However, they converge rather slowly to the final solution.

Finally, let us study two blind deconvolutions. In the first example we employed Tikhonov blind deconvolution. In both deconvolutions, estimating input data and response function, respectively, we used regularization parameter $\alpha=50$. We have tried also other combinations, however without any relevant improvement of the results. In Fig. 4.16 we can observe oscillations in both, estimated input data and response function, as well as, non-realistic negative values. Also, the resolution of the deconvolution is rather poor.

When using classic Gold algorithm in blind deconvolution scheme, the oscillations and negative values disappear from the solutions (Fig. 4.17). On the other hand, the resolution is practically the same like in the above-mentioned forward classic Gold deconvolution. In Fig. 4.17(b) apparently the deconvolution deforms the shape of the response function. For the purposes of the deconvolution of spectroscopic data we can determine the response function with good fidelity. Therefore, the deformation of the response function is not acceptable.

From the study carried out in this section one can conclude that for the deconvolution of spectrometric data the algorithms should be positive definite. It should give results with good resolution. In the rest of the chapter we shall consider only forward deconvolution schemes. From this point of view from the examples given above it is apparent that the best results were achieved by one-fold Gold, Richardson-Lucy and MAP algorithms. Nevertheless, none of the algorithms is able to decompose the testing data to delta functions given in Fig. 4.2.

4.4. Boosted Deconvolution

The basic aim of the deconvolution (decomposition) is resolution improvement of the processed data by mathematically removing the smearing effects of an imperfect instrument using its known response function. In the previous section, we introduced a survey of several known deconvolution methods with regard to various aspects and approaches. There exist many problems inherent to the solution of the problem of deconvolution and decomposition in general:

- computational complexity,
- influence of the errors due to the measured noise,
- ill conditionality of systems,
- regularization of the solution.

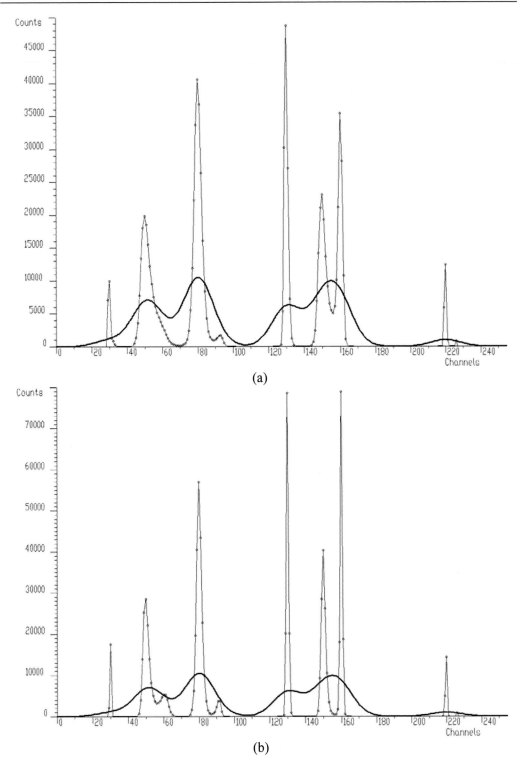

Figure 4.13. Original spectrum (thick line) and deconvolved spectrum using Richardson-Lucy algorithm (thin line) after 10 000 iterations (a), 50 000 iterations (b).

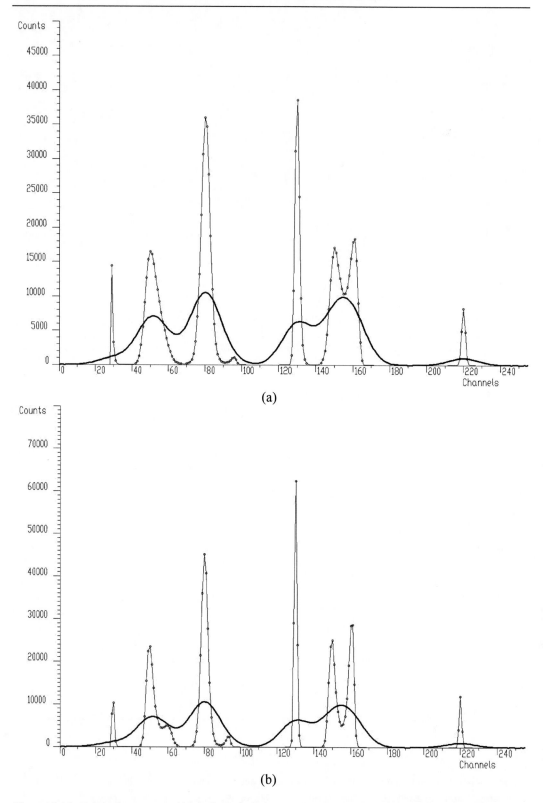

Figure 4.14. Original spectrum (thick line) and deconvolved spectrum using Muller algorithm (thin line) after 10000 iterations (a), 50000 iterations (b).

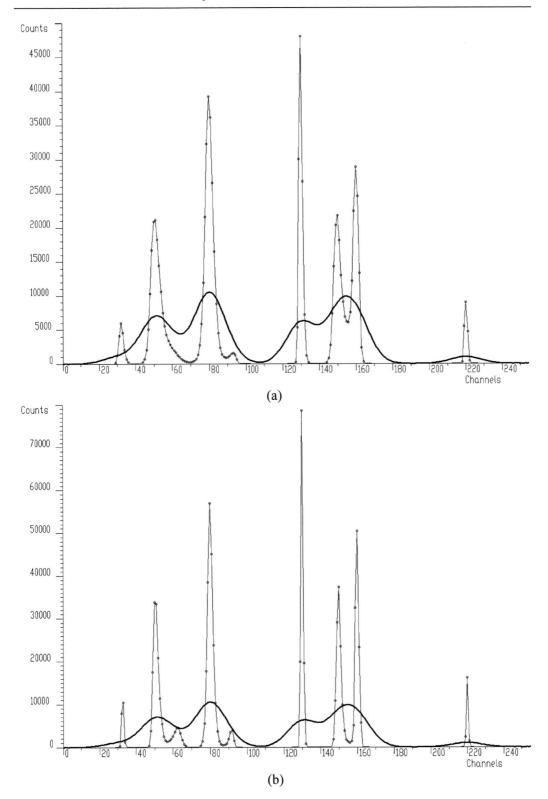

(a)

(b)

Figure 4.15. Original spectrum (thick line) and deconvolved spectrum using MAP algorithm (thin line) after 10000 iterations (a), 50000 iterations (b).

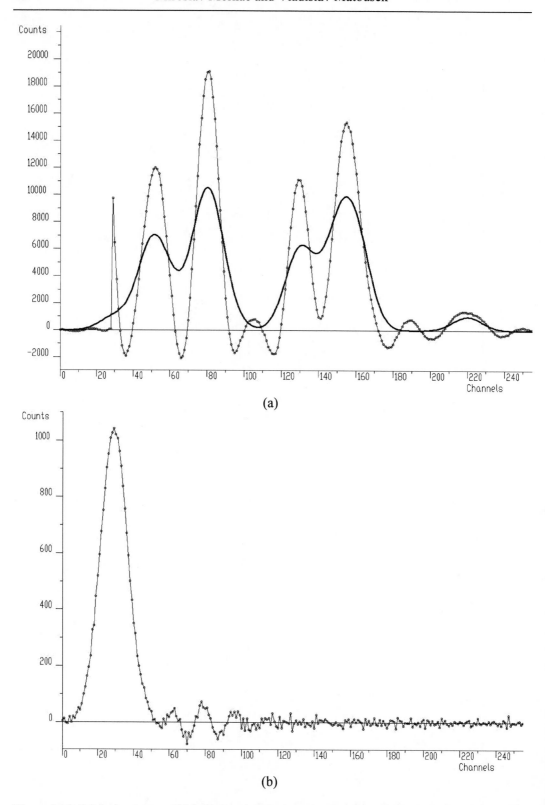

Figure 4.16. Original spectrum (thick line) and deconvolved spectrum using blind Tikhonov algorithm (thin line) after 1000 iterations (α=50) (a), deconvolved response function (α=50) (b).

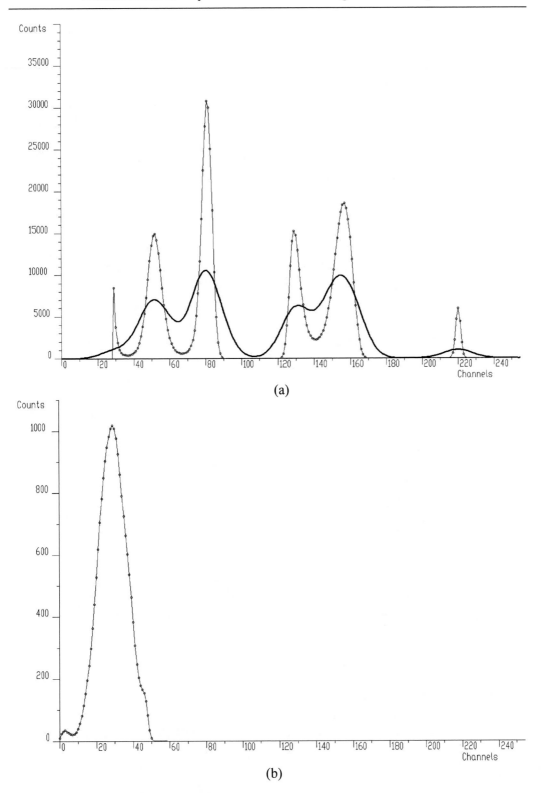

Figure 4.17. Original spectrum (thick line) and deconvolved spectrum using blind Gold algorithm (thin line) after 1000 iterations and 50 repetitions (a), deconvolved response function (b).

Before starting to employ deconvolution operation for the analysis of spectrometric data, one has to realize their specific features

- spectrometric data are composed of useless components of information (background and noise) and useful information components - peaks. Before the deconvolution can be carried out, it is necessary to remove background from the data [43], [44]. One has to pay great attention to this operation. If the estimation of the background is incorrect, all deconvolution algorithms give always false results.
- closely positioned peaks can be overlapping to such an extent that sometimes from visual observation it is impossible to say anything about their existence.
- the solution must be positive. It is absolutely obvious that counts cannot be negative. Therefore the deconvolution algorithm must be positive definite. This guarantees positivity of the solution.
- the deconvolved spectrum should have the form of spikes or δ - functions. The peak is either present or missing at a given position. The position of a δ-function says about the energy of the peak and its height says about its intensity. The area of the peak should be as much as possible concentrated into one channel (point). All the algorithms known so far give results with non-discrete quasi-continuous distribution.

The algorithm proposed in this section should address these requirements.

Iterative positive definite deconvolutions (Gold, Richardson-Lucy, Muller and MAP) converge to stable states. It is useless to increase the number of iterations; the result obtained practically does not change. To see and judge the course of the deconvolution procedure (Gold algorithm) we recorded the sum of weighted squares of errors per channel (denoted by letter Q) after each iteration step in the form of graph (Fig. 4.18).

Obviously, after initial decrease of Q at the beginning of the deconvolution operation during the following iteration steps it remains constant. Therefore, it is useless to continue in the iterations as it does not lead to a sufficient improvement of the resolution. Instead of it we can stop iterations, apply a boosting operation and repeat this procedure. Boosting operation should decrease sigma of peaks. In other words, when the solution reaches its stable state it is necessary to change the particular solution $\boldsymbol{x}^{(L)}$ in a way and repeat again the deconvolution. To change the relations among elements of the particular solution we need to apply non-linear boosting function to it. The power function proved to work satisfactorily. Then, the algorithm of boosted Gold, Richardson-Lucy or MAP deconvolution is as follows:

1. Set the initial solution $\boldsymbol{x}^{(0)} = \left[1,1,\ldots,1\right]^{T}$.
2. Set required number of repetitions R and iterations L
3. Set the number of repetitions $r = 1$.
4. According to either Eq. 4.17 (Gold), or Eq. 4.18 (Richardson-Lucy) or Eq. 4.20 (MAP) for $n = 0,1,\ldots,L-1$ find solution $\boldsymbol{x}^{(L)}$.

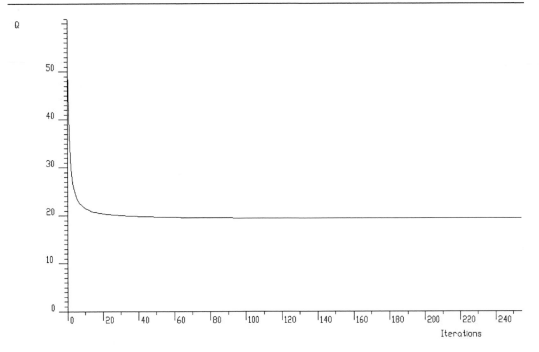

Figure 4.18. Sum of weighted squares of errors per one channel in dependence on number of iterations.

5.   If $r = R$ stop the calculation, else
    a.   apply boosting operation, i.e., set

$$x^{(0)}(i) = \left[x^{(L)}(i) \right]^{p} , \text{ where } i = 0, 1, \dots, N-1 \qquad (4.21)$$

        and p is boosting coefficient >0,

    b. $r = r+1$,
    c.   continue in 4.

The question is the choice of the boosting coefficient p in 4.21. Let us imagine the peak can be approximately described by a well-known Gaussian function

$$A \cdot e^{-\frac{(i-i_0)^2}{2\sigma^2}}, \qquad (4.22)$$

where A is amplitude, i_0 is position and σ represents width of the peak, respectively. By the application of the power function with the parameter p we get

$$A^p \cdot e^{-\frac{(i-i_0)^2}{2\left(\frac{\sigma}{\sqrt{p}}\right)^2}}. \qquad (4.23)$$

Clearly, if the coefficient p is greater than 1 the boosting operation increases the heights in favor of bigger peaks and decreases widths of peaks. However, if the coefficient is too big it relatively suppresses small peaks, which can later completely disappear from the deconvolved spectrum. Moreover, if two or more peaks are positioned closely to each other the exponential operation can produce new artificial peaks. For example let us imagine we have two peaks and $p = 2$. Then we get

$$\left(A_1 \cdot e^{-\frac{(i-i_1)^2}{2\sigma^2}} + A_2 \cdot e^{-\frac{(i-i_2)^2}{2\sigma^2}} \right)^2 = \tag{4.24}$$

$$= A_1^2 \cdot e^{-\frac{(i-i_1)^2}{2\left(\frac{\sigma}{\sqrt{2}}\right)^2}} + A_2^2 \cdot e^{-\frac{(i-i_2)^2}{2\left(\frac{\sigma}{\sqrt{2}}\right)^2}} + 2 \cdot A_1 \cdot A_2 \cdot e^{-\frac{(i-i_1)^2+(i-i_2)^2}{2\sigma^2}}$$

The third term on the right side of Eq. 4.24 represents fake peak with position between two original peaks. This can negatively influence the result of the boosted deconvolution.

Apparently, when we want to boost peaks in the spectrum the parameter p should be greater than 1. Otherwise, the peaks would be suppressed (smoothed) and consequently the resolution would decrease. On the other hand, according to our experience, it should not be too big (greater than 2). When choosing it too big, small peaks will disappear from spectrum at the expense of big ones. One can conclude that the parameter p should be chosen from the range $p \in \langle 1, 2 \rangle$. Empirically we have found that reasonable results can be obtained with boosting coefficients $p \approx 1.1 - 1.2$.

Let us illustrate and compare the classic and boosting Gold deconvolution algorithms through the use of simple example. Let us have a convolution system with two unknowns

$$\mathbf{y} = H\mathbf{x} + \mathbf{n} = \begin{bmatrix} 3 \\ 17 \\ 4 \\ 0 \end{bmatrix} = \begin{bmatrix} 1 & 0 \\ 5 & 1 \\ 2 & 5 \\ 0 & 2 \end{bmatrix} \cdot \begin{bmatrix} x(0) \\ x(1) \end{bmatrix} \tag{4.25}$$

We have to minimize

$$\left\| \begin{bmatrix} 1 & 0 \\ 5 & 1 \\ 2 & 5 \\ 0 & 2 \end{bmatrix} \cdot \begin{bmatrix} x(0) \\ x(1) \end{bmatrix} - \begin{bmatrix} 3 \\ 17 \\ 4 \\ 0 \end{bmatrix} \right\|^2 = 30x(0)^2 + 30x(0)x(1) + 27x(1)^2 \tag{4.26}$$

$$-192x(0) - 74x(1) + 314$$

which is called loss function. Least squares solution of the convolution system given by Eq. 4.25 according to Eq. 4.6 is

$$[3.444, -0.448]^T \qquad\qquad (4.27)$$

It is optimal solution in the unconstrained space of two independent variables x_0, x_1. It contains negative value. However, in the applications like spectroscopy, the negative values are unreasonable. In Fig. 4.19 we show the loss function (Eq. 4.26) in two-dimensional space and both the unconstrained solution and the solution in the constrained to positive subspace. When employing the Gold deconvolution algorithm (Eq. 4.16) we get a sequence of iterations. The situation for the first 5 iterations (vectors $\boldsymbol{X_0} - \boldsymbol{X_4}$) is shown in Fig. 4.19 as well. It converges rather slowly to the optimal solution in the positive subspace.

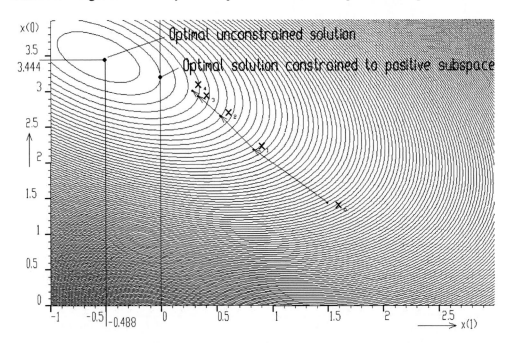

Figure 4.19. Graphical illustration of loss function, optimal solutions in both unconstrained space and constrained positive subspace and the first 5 iterative solutions of Gold deconvolution. The arrows indicate the direction of iterations.

In Fig. 4.20, we present detailed view for classic and boosted Gold deconvolution. Vectors $\boldsymbol{X_0} - \boldsymbol{X_{14}}$ represent sequence of classic Gold deconvolution iterations. Vectors $\boldsymbol{b}$ represent sequence of boosted Gold deconvolution iterations. The boosted operation was carried out after every 5 iterations, so that the solutions $\boldsymbol{X_0} - \boldsymbol{X_4}$ and $\boldsymbol{b_0} - \boldsymbol{b_4}$ coincide. However, due to the boosting, the next iterations of boosted Gold deconvolution approach the optimal solution in positive subspace much faster than iterations of classic deconvolution.

Now let us investigate the properties of boosted deconvolutions using our testing synthetic spectrum from previous section. We stopped the iterations after every 200 steps, applied boosting operations according to the above given algorithm and repeated this procedure 50 times with boosting coefficient $p = 1.2$. Entirely, it gives 10000 iteration steps. The result of the boosted one-fold Gold deconvolution is shown in Fig. 4.21.

Except for the channel 80, it concentrates areas of peaks practically into one channel, i.e. into spikes. The results in peaks (spikes) positions together with errors are presented in Tab. 4.2. Anyway, there still remains the problem in the positions of some estimated peaks. Mainly the estimate of the small peak at the position 90 is rather far from the reality.

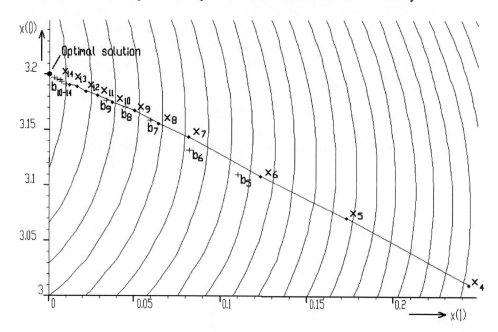

Figure 4.20. Detailed view of loss function and iterative solutions for classic Gold deconvolution (vectors x) and boosted Gold deconvolution (vectors b).

Table 4.2. Deconvolved peaks (spikes) positions, channels contents and their errors in the spectrum shown in Fig. 4.21 (boosted one-fold Gold algorithm)

Peak #	Position [channel]	Position error [channels]	Channel contents	Contents error [%]
1	29	1	17746	11
2	49	1	109870	8.5
3	59	1	53042	32
4	80	0	174890	12.6
5	96	6	12433	37
6	130	0	121600	1.2
7	151	1	138890	15.6
8	161	1	100520	16.3
9	220	0	20028	0.03

Further, in Fig. 4.22 we present the result of boosted Richardson-Lucy algorithm (again 50 iterations, repeated 20 times with boosting coefficient $p = 1.2$). It decomposes completely (to one channel) all peaks in the spectrum. Again, we present numerical results in Tab. 4.3. In peaks positions there is only one error at the position 90. Except for this channel, the errors in the channel counts in spikes positions are negligible.

Finally, in Fig. 4.23 we present the result of boosted MAP algorithm using the same parameters. Again, it decomposes completely all peaks in the spectrum.

Table 4.3. Deconvolved peaks (spikes) positions, channels contents and their errors in the spectrum shown in Fig. 4.22 (boosted Richardson-Lucy algorithm)

Peak #	Position [channel]	Position error [channels]	Channel contents	Contents error [%]
1	30	0	20187	0.82
2	50	0	117010	2.5
3	60	0	41479	3.5
4	80	0	206740	3.2
5	92	2	15158	24
6	130	0	119960	0.1
7	150	0	120160	0.02
8	160	0	120120	0.009
9	220	0	19607	2

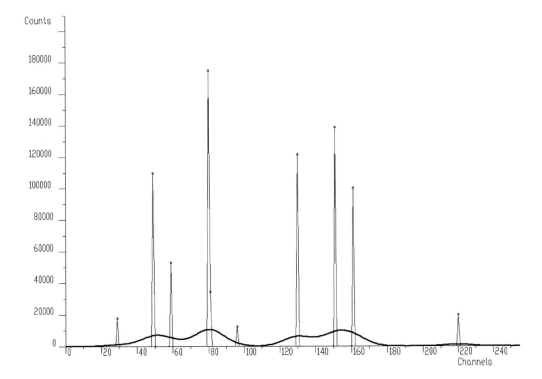

Figure 4.21. Original spectrum (thick line) and deconvolved spectrum using boosted one-fold Gold algorithm (thin line) after 200 iterations and 50 repetitions ($p = 1.2$).

Analogously to the previous examples, numerical results are shown in Tab. 4.4. Now there are two errors in the peak positions (channels 30 and 90, respectively). Logically, the errors of channels contents at these two positions are higher than at other positions.

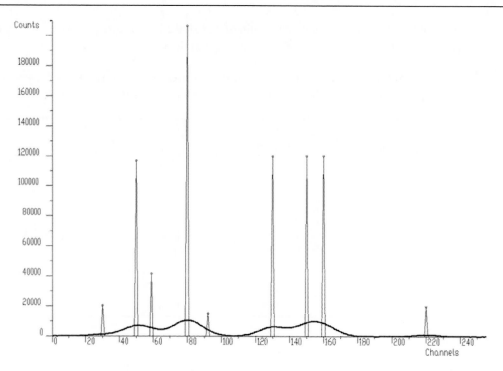

Figure 4.22. Original spectrum (thick line) and deconvolved spectrum using boosted Richardson-Lucy algorithm (thin line) after 200 iterations and 50 repetitions ($p = 1.2$).

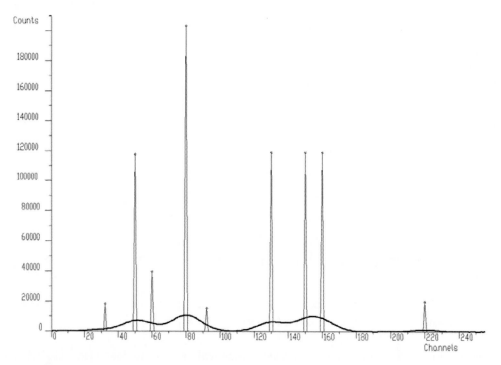

Figure 4.23. Original spectrum (thick line) and deconvolved spectrum using boosted MAP algorithm (thin line) after 200 iterations and 50 repetitions ($p = 1.2$).

Table 4.4. Deconvolved peaks (spikes) positions, channels contents and their errors in the spectrum shown in Fig. 4.23 (boosted MAP algorithm).

Peak #	Position [channel]	Position error [channels]	Channel contents	Contents error [%]
1	32	2	18418	8
2	50	0	117780	2
3	60	0	39604	1
4	80	0	203620	1.6
5	92	2	15341	23
6	130	0	118980	0.9
7	150	0	119200	0.7
8	160	0	119140	0.8
9	220	0	19848	0.8

Further, we have applied the sequence of above-mentioned methods also to experimental γ-ray spectra. However, before application of the deconvolution procedure we removed background from the spectrum using the background elimination algorithm with decreasing clipping interval presented in [43], [44]. In Fig. 4.24 and Fig. 4.25 we give results after application of classic and boosted Gold deconvolution algorithms, respectively. Again, boosted Gold deconvolution decomposes the spectrum practically to δ functions. Application of boosted Richardson-Lucy and MAP deconvolution algorithms gives results practically identical to that given in Fig. 4.25.

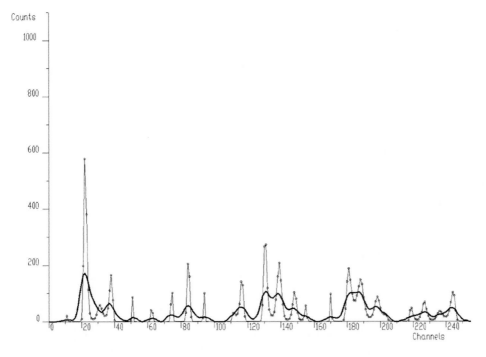

Figure 4.24. Original γ-ray spectrum (thick line) and deconvolved spectrum using classic Gold algorithm (thin line) after 10 000 iterations.

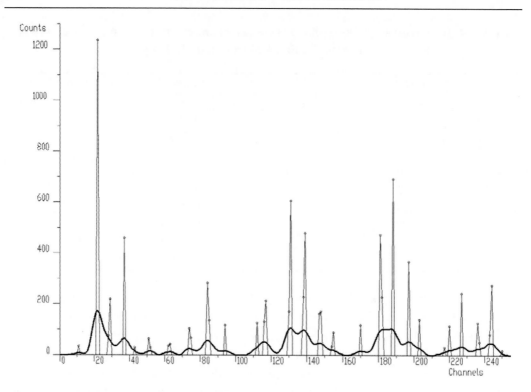

Figure 4.25. Original γ-ray spectrum (thick line) and deconvolved spectrum using boosted Gold algorithm (thin line) after 200 iterations and 50 repetitions ($p = 1.2$).

4.5. Robustness

Until now, in the examples given above, we studied various deconvolution algorithms using testing synthetic spectrum from Fig. 4.3. In this section, we want to demonstrate robustness of the deconvolution methods in respect to increasing level of noise. We shall study only positive definite deconvolution algorithms. In Fig. 4.26(a), we present a matrix, which is composed of the original spectrum from Fig. 4.1 and noise vector with increasing amplitude ranging from 0% up to 120% of the amplitude of small peaks (#1, #5, and #9). Matrix composed of deconvolved spectra (rows of the matrix shown in Fig. 4.26(a)) using classic Gold deconvolution algorithm and 10 000 iteration steps is presented in Fig. 4.26(b). One can observe that classic Gold deconvolution algorithm is robust to noise. Peaks do not change their positions and deconvolved spectra are smooth even for high level of noise. On the other hand, as mentioned above, the method is not able to resolve peaks #3, #5 and to decompose peaks #7 and #8 even for noiseless data.

In the following experiments in non-boosted deconvolutions, we have carried out 10000 iterations and in boosted ones 200 iterations repeated 50 times with boosting coefficient $p = 1.2$. Taking together it gives again 10000 iterations. Let us now compare non-boosted and boosted one-fold Gold deconvolution algorithms, respectively (Fig. 4.27). In Fig. 4.27(a) again one can see that the deconvolution is relatively robust to increasing level of noise.

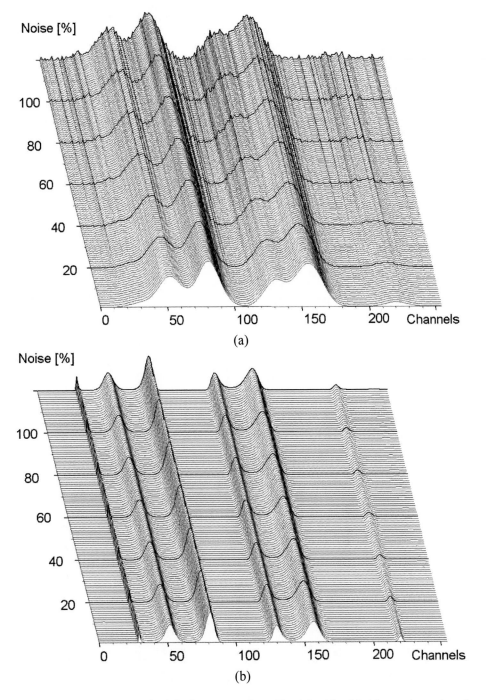

Figure 4.26. Matrix composed of original spectrum from Fig. 4.1 with added increasing noise given in percentage of the amplitude of small peaks (#1, #5 and #9) (a). Matrix composed of deconvolved spectra from (a) using classic Gold deconvolution algorithm (b).

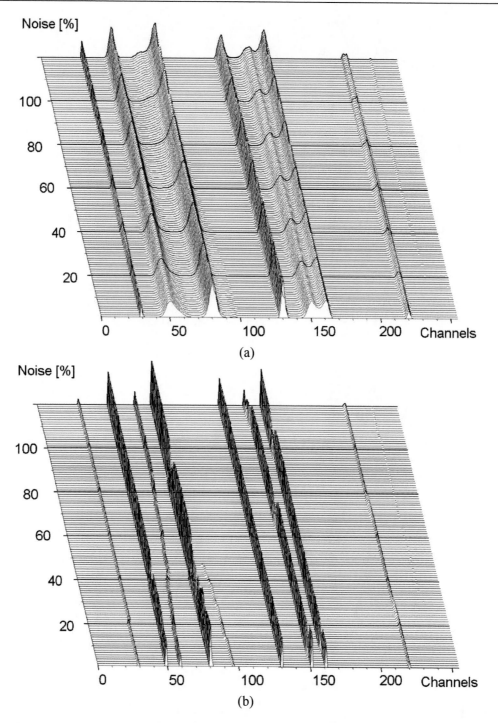

Figure 4.27. Matrix composed of deconvolved spectra from Fig. 4.26(a) using one-fold Gold deconvolution algorithm (a) and boosted one-fold Gold deconvolution algorithm (b), respectively.

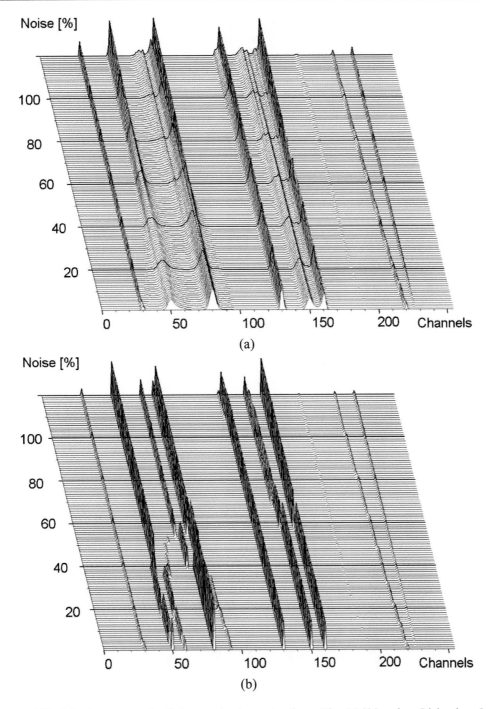

Figure 4.28. Matrix composed of deconvolved spectra from Fig. 4.26(a) using Richardson-Lucy deconvolution algorithm (a) and boosted Richardson-Lucy deconvolution algorithm (b), respectively.

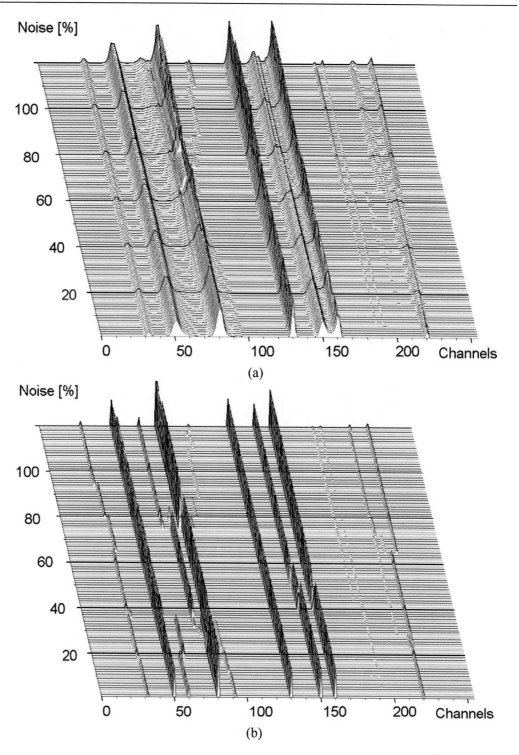

Figure 4.29. Matrix composed of deconvolved spectra from Fig. 4.26(a) using MAP deconvolution algorithm (a) and boosted MAP deconvolution algorithm (b), respectively.

It resolves the doublet composed of peaks #7 and #8, respectively. Nevertheless, the resolution capabilities are quite limited. On the other hand, boosted one-fold deconvolution decomposes all peaks practically to one channel. However, for higher levels of noise the peak #3 changes its position. The peak #5 for more than 40% of noise disappears. Other peaks practically do not change their positions.

Further, let us compare non-boosted and boosted Richardson-Lucy deconvolution (Fig. 4.28).

For noiseless data, the behavior of standard non-boosted Richardson-Lucy algorithm is similar to that of one-fold Gold algorithm. However, for noisy data it splits the peak #9 to two peaks, naturally the second one being fake. The boosted Richardson-Lucy algorithm improves resolution, but it also generates false twin peak to the peak #9. The positions of other peaks change more dramatically than in one-fold Gold deconvolution.

Finally, let us analyze the robustness of non-boosted and boosted MAP deconvolution algorithms, respectively (Fig. 4.29). Again, for noiseless data the behavior of non-boosted MAP algorithm is similar to that of one-fold Gold algorithm. However, for the noise level higher than 40% it starts to generate false peaks. Boosted MAP algorithm gives good results up to 20% of noise. After that, one can observe changing positions of the peaks. There appear fake peaks in deconvolved data as well.

4.6. Two-Dimensional Spectra

Analogously to Eq. 4.1 for two-dimensional convolution system one can write

$$
\begin{aligned}
y(i_1,i_2) &= \sum_{k_1=0}^{i_1}\sum_{k_2=0}^{i_2} x(k_1,k_2)h(i_1-k_1,i_2-k_2)+n(i_1,i_2) \\
&= x(i_1,i_2)*h(i_1,i_2)x(i)*h(i)+n(i_1,i_2),
\end{aligned}
\tag{4.28}
$$

where $i_1 = 0,1,...,N_1-1,$ $i_2 = 0,1,...,N_2-1$. Obviously, with increasing number of dimensions the number of operations grows exponentially. Therefore, the question of computational time becomes striking. When comparing Eqs. 4.17, 4.18 and 4.20 for Gold, Richardson-Lucy and MAP deconvolution algorithms, respectively, one can observe that the Gold algorithm is the simplest one. For each iteration step, it requires one matrix vector multiplication whereas in the Richardson-Lucy and MAP deconvolution two matrix vector multiplications must be carried out. Moreover, the one-fold Gold deconvolution can be optimized even for higher-dimensional data [78] thus enabling to decrease the number of operations to the minimum. Also, the results obtained in the previous section, dealing with the study of robustness, say in the favor of the one-fold Gold deconvolution algorithm. Therefore, in the rest of the section we shall analyze only the properties of Gold deconvolution.

To study the properties of two-dimensional Gold algorithms let us take two-dimensional synthetic spectrum containing 17 peaks and added noise with the amplitude 5% of the amplitude of smallest peak (Fig. 4.30). There are 5 overlapped peaks concentrated in a cluster (positions x - y, 14 - 40, 17 - 37, 19 - 39, 19 - 46, 22 - 43) and 2 overlapped peaks in a doublet (44 - 46, 48 - 46).

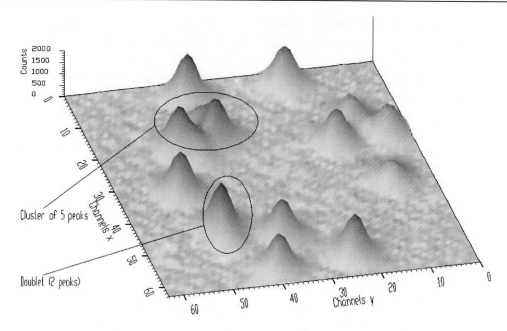

Figure 4.30. Two-dimensional original synthetic spectrum.

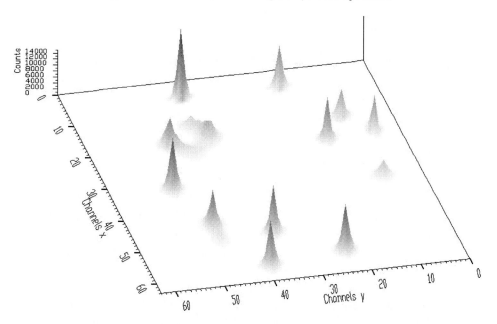

Figure 4.31. Spectrum from Fig. 4.30 after classic Gold deconvolution (1 000 iterations).

After two-fold Gold deconvolution (1 000 iteration steps), we get the result given in Fig. 4.31. Analogously to one-dimensional spectrum the resolution is better than in the original data, however the overlapped peaks remained unresolved.

The result is slightly better after the application of one-fold Gold deconvolution (1000 iterations) given in Fig. 4.32. Some of peaks in the cluster are separated but some of them are still overlapping.

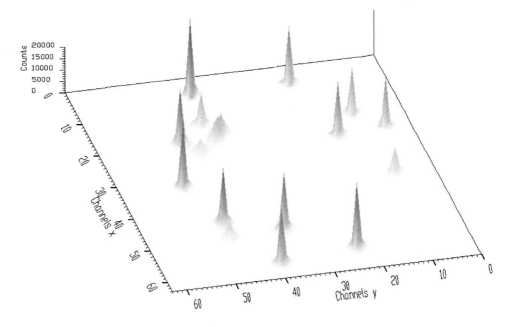

Figure 4.32. Spectrum from Fig. 4.30 after one-fold Gold deconvolution (1 000 iterations).

When applying one-fold boosted Gold deconvolution (50 iterations, repeated 20 times with boosting coefficient $p = 1.2$) we get the result presented in Fig. 4.33. Even the close positioned peaks from the cluster and doublet are practically completely decomposed to one-channel "peaks". The positions in the deconvolved spectrum coincide with the positions of peaks in the original data.

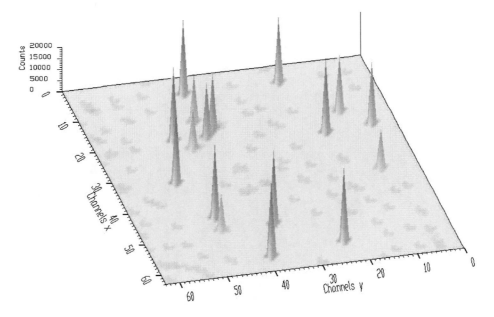

Figure 4.33. Spectrum from Fig. 4.30 after boosted one-fold Gold deconvolution (50 iterations repeated 20 times).

Before application of the deconvolution procedure to experimental γ - γ ray spectrum we removed background from the spectrum using the background elimination algorithm described in Refs. 43 and 44 (Fig. 4.34). In Fig. 4.35 we present the spectrum after application of boosted Gold deconvolution algorithms (50 iterations, repeated 20 times with boosting coefficient $p = 1.2$). It decomposes the multiplets in the experimental γ - γ ray spectrum practically to δ functions.

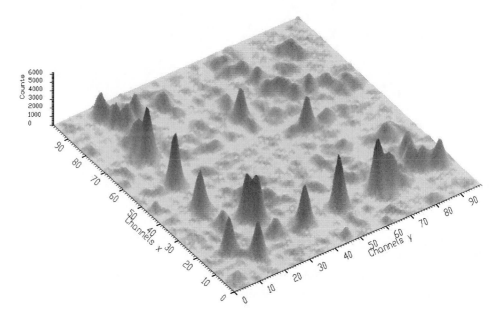

Figure 4.34. Experimental γ-γ ray spectrum (after background elimination).

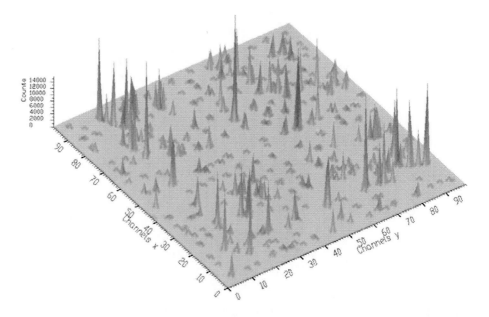

Figure 4.35. Spectrum from Fig. 4.34 after boosted one-fold Gold deconvolution (50 iterations repeated 20 times).

4.6. Three-Dimensional Spectra

The extension of Eqs. 4.1 and 4.28 to three-dimensional convolution system is straightforward. The algorithm of boosted one-fold Gold deconvolution lends itself for extension to and optimization of three-dimensional system [78]. We have implemented the above-mentioned one-fold Gold deconvolution algorithms also for three-dimensional data. In the next experiment we have applied the algorithm to three-dimensional synthetic spectrum consisting of 4 peaks positioned close to each other (positions x - y - z, $15 - 15 - 15$ with amplitude A = 100, $12 - 12 - 18$, A = 250, $14 - 19 - 12$, A = 150, $20 - 20 - 13$ A = 50) shown in Fig. 4.36. Channels are shown as spheres with sizes proportional to the counts they contain.

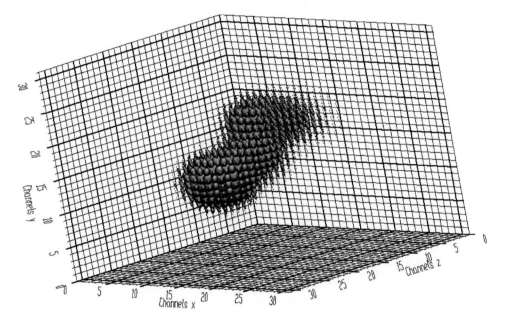

Figure 4.36. Three-dimensional synthetic spectrum (sizes of spheres are proportional to counts the channels contain).

In Fig. 4.37 we present the same spectrum after boosted one-fold Gold deconvolution (50 iterations, with boosting coefficient $p = 1.2$, repeated 20 times). The volumes of peaks are concentrated separately practically into one channel each. The positions of peaks coincide with the peaks positions in the original spectrum. The relations among amplitudes of peaks are also approximately preserved. The counts scales in both figures are different (1 - 250 in Fig. 4.36, $1 - 20\,000$ in Fig. 4.37).

Analogously to two-dimensional spectra in the last example, we introduce experimental γ-γ-γ ray spectrum (after background elimination). In Fig. 4.38, we present the spectrum before deconvolution and in Fig. 4.39 the same spectrum after boosted one-fold Gold deconvolution (50 iterations, repeated 20 times with boosting coefficient $p = 1.2$). Now the counts scales in both figures were the same. The concentration of the volume of experimental peaks in deconvolved data is apparent.

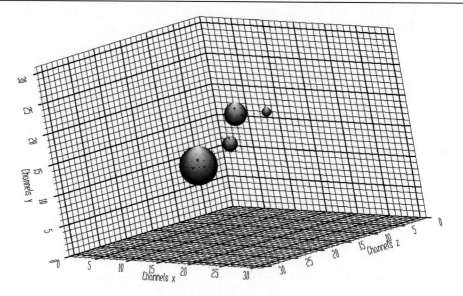

Figure 4.37. Spectrum from Fig. 4.36 after boosted one-fold Gold deconvolution (50 iterations repeated 20 times).

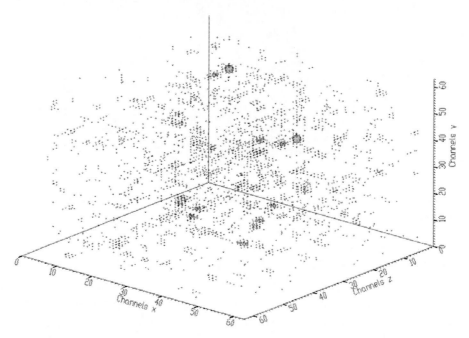

Figure 4.38. Experimental γ-γ-γ ray spectrum (after background elimination).

4.7. Linear Time Dependent Systems

For time dependent continuous linear system one can write

$$y(t) = \int_{-\infty}^{t} h(t,\tau) x(\tau) d\tau \qquad (4.29)$$

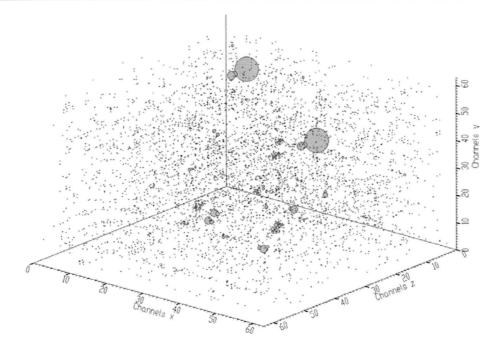

Figure 4.39. Spectrum from Fig. 4.38 after boosted one-fold Gold deconvolution (50 iterations repeated 20 times).

Analogously, for discrete system it holds

$$y(i) = \sum_{k=0}^{i} h(i,k) x(k).$$
(4.30)

The relation (4.30) represents general system of linear equations. In this case, the columns of the system matrix are not simply shifted as it was in the above-examples of the deconvolution of γ-ray spectra. Naturally, the Fourier transform does not factorize these systems of equations to the product of coefficients and the algorithms based on this approach are not applicable. However, all the above-given deconvolution algorithms (4.8-4.20) are applicable also for linear systems with changing response. The main problem consists in determination of system matrix H.

In these systems, the shape of response changes with independent variable (time, in the case of nuclear spectra energy, etc.). Basically, these systems can be divided into two groups:

− general discrete linear systems that can be described by a system matrix. The dependence (e.g. the change of the shape of columns in the matrix) cannot be formalized, e.g. by formula
− special systems - the change of the response can be reflected in mathematical formula, algorithm, table, i.e., the system matrix has special form.

If the change of the shape of the response function could be expressed analytically, we could make a benefit of the known formula in generation of the system matrix. We have employed this technique in the decomposition of the electron spectra. In the 4.40 we present the processed spectrum. The first few responses at the beginning of the coordinate system are shown in 4.41. The shape of the responses is changing with increasing energy (number of the response). It should reflect the left-side tail typical for this kind of data.

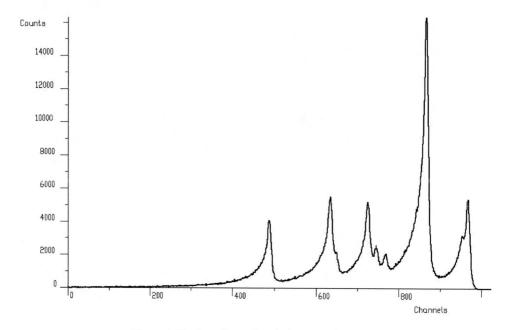

Figure 4.40. One-dimensional electron spectrum.

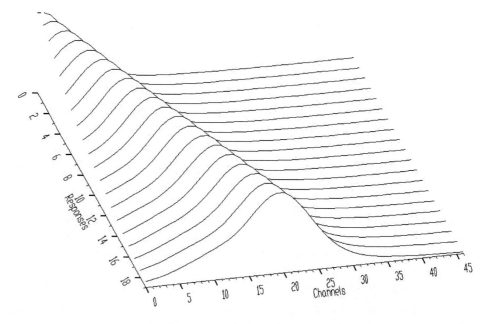

Figure 4.41. Part of response matrix composed of response electron spectra.

Using this system matrix, we applied the Gold deconvolution algorithm to the original electron spectrum (with boosting). The result of the deconvolution (thin line) together with the original data (thick line) in log scale is given in Fig. 4.42.

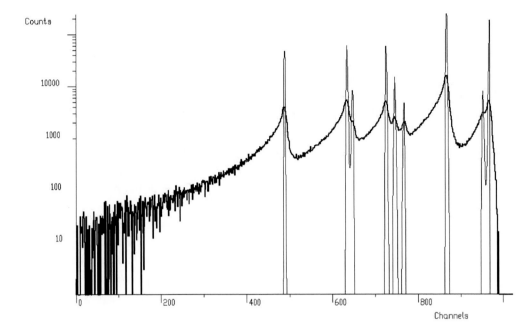

Figure 4.42. Original electron spectrum (thick line) and deconvolved (boosted one-fold deconvolution - thin line).

To see details close to the baseline we introduce the same spectra in linear scale in Fig. 4.43.

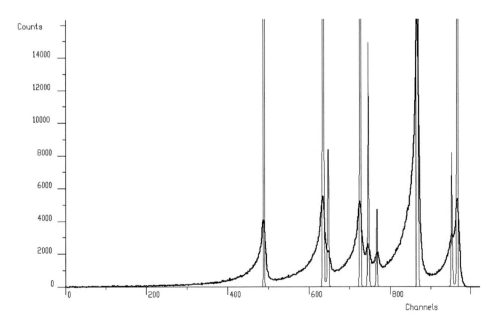

Figure 4.43. Detail of the spectrum from Fig. 4.42.

The results given in Fig. 4.42 and Fig. 4.43 prove in favor of the proposed method. The method finds correctly also overlapped small peaks positioned close to the big ones.

There exist various ways how to determine the response matrix. It strongly depends on the application. In the next example, the response matrix was composed of the neutron spectra of pure chemical elements (Fig. 4.44). These spectra define the columns of the system matrix, which is then used in the deconvolution procedure. The method was used to determine chemical composition of the unknown material. The original neutron spectrum of the material (Fig. 4.45) is unfolded according to the system matrix.

An example of the decomposition of continuum γ-ray spectra was described in detail in [79]. The system matrix (Figs. 4.47 and 4.48) was synthesized from simulated responses employing a new developed interpolation algorithm. Then the method is able to move the background area belonging to appropriate peak to its photopeak position. We get the spectrum consisting of deconvolved narrow photopeaks. We extended the boosting deconvolution also to this kind of data and finally we obtained δ-like lines placed in the photopeak positions with heights proportional to the peak areas.

In Fig. 4.49 we present measured spectrum of ^{56}Co. The spectrum of ^{56}Co after Gold deconvolution using above displayed response is given in Fig. 4.50. The detail of the original γ-ray spectrum before deconvolution (dotted line), after classic (thin line) and boosted Gold deconvolution (bars) are shown in Fig. 4.51. Again, the boosted deconvolution algorithm is able to decompose the peaks practically to δ-functions.

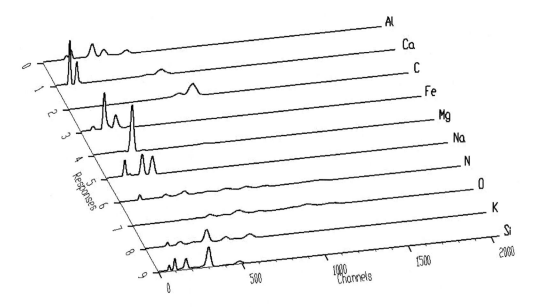

Figure 4.44. Response matrix composed of neutron spectra of pure chemical elements.

After unfolding the obtained coefficients, i.e., the contents of the responses in the original spectrum are presented in Fig. 4.46. They determine the contents of the chemical elements in the investigated material.

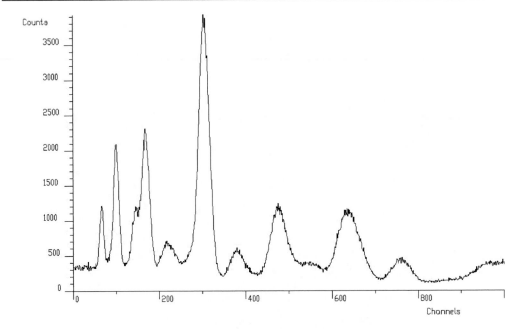

Figure 4.45. Original neutron spectrum to be deconvolved.

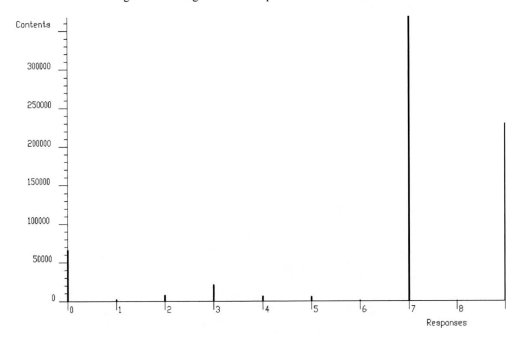

Figure 4.46. Unfolded spectrum (coefficients correspond to the contents of appropriate elements).

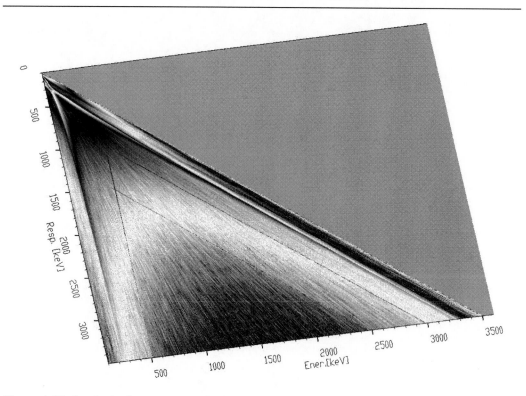

Figure 4.47. Synthesized response matrix for continuum decomposition for experimental data from Gammasphere.

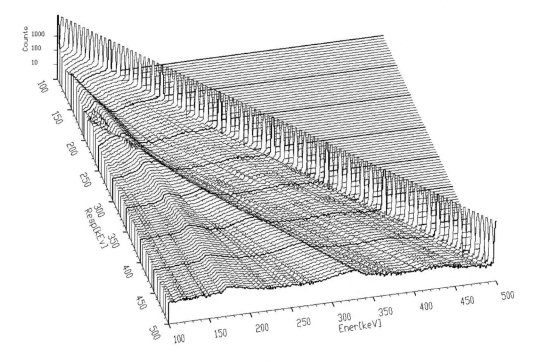

Figure 4.48. Detail of the response matrix.

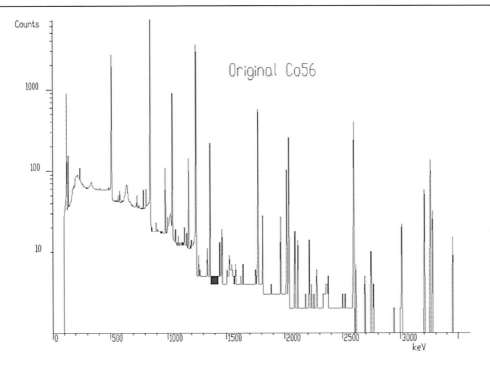

Figure 4.49. Measured spectrum of ^{56}Co.

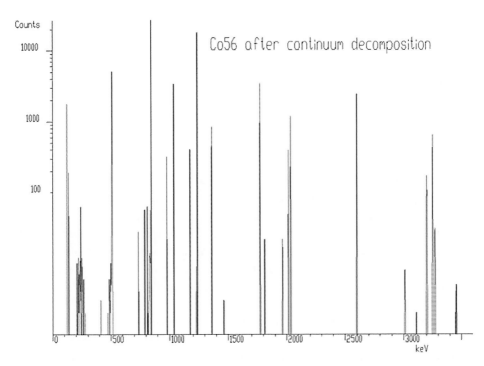

Figure 4.50. Spectrum of ^{56}Co after Gold deconvolution using above displayed response matrix shown in Figs. 4.47 and 4.48.

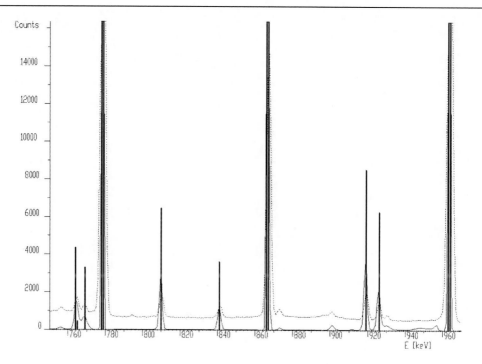

Figure 4.51. Spectrum before deconvolution (dotted line), after classic (thin line) and boosted Gold deconvolution (bars).

5. Identification of Spectroscopic Information Carrier Objects

5.1. Peak Searching

5.1.1. Peak Searching Algorithm Based on Smoothed Second Differences

a. One-dimensional Spectra

The essential peak searching algorithm is based on smoothed second differences (SSD) that are compared to its standard deviations [6]. The SSD and its standard deviations in the vicinity of a peak and the principle of the algorithm are outlined in Fig. 5.1.

In Fig. 5.2 we show entire synthetic one-dimensional spectrum together with SSD in its bottom part. For synthetic spectra, we know in advance the result that should be obtained and thus we can verify the reliability of the proposed algorithm. One can observe that SSD is independent of background, noise and its negative amplitudes are proportional to the amplitudes of original peaks.

We have modified the algorithm so that it was able to recognize peaks positioned relatively close to each other and also the peaks positioned very close to the edges of the spectrum. The algorithm is selective to the peaks with a given σ that should be passed to the peak searching function as parameter. An example of one-dimensional γ-ray experimental noisy spectrum with lot of peaks is given in Fig. 5.3 (σ =2). Identified peaks are denoted by markers with peak position. Inverted positive SSD spectrum is shown in the bottom of the figure. The method is able to find peaks also at the edges of the spectrum region and also peaks positioned relatively close to each other.

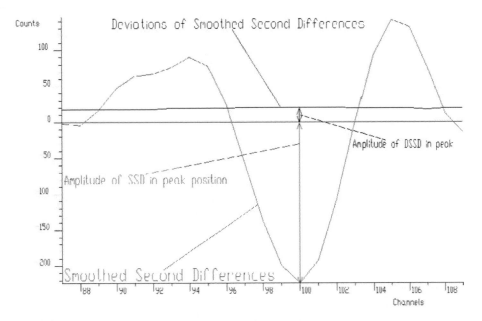

Figure 5.1. Smoothed Second Differences (SSD) and its standard deviation in the vicinity of peak.

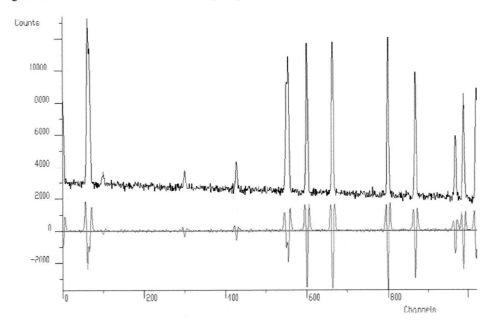

Figure 5.2. One-dimensional synthetic spectrum and its SSD spectrum.

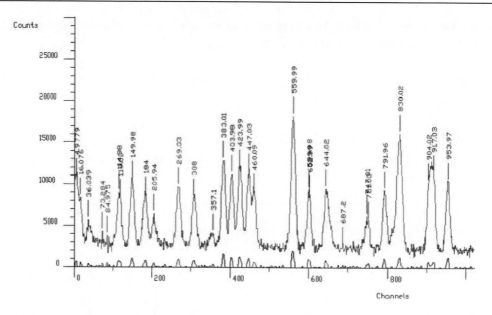

Figure 5.3. An example of one-dimensional γ-ray experimental spectrum with found peaks and its inverted positive SSD spectrum.

The algorithm is able to recognize also Compton edges from Gaussians, see Fig. 5.4.

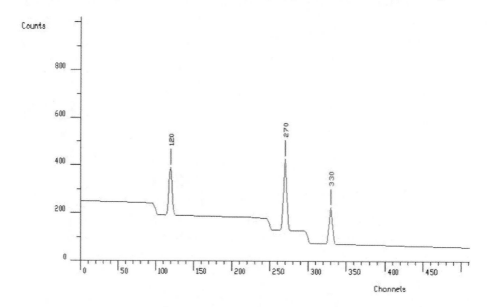

Figure 5.4. Illustration of the algorithm for synthetic spectrum with Compton edges.

The algorithm can be used to discover non-Gaussian peaks as well. An example of the data with found peaks is shown in Fig. 5.5 (electron spectrum). Again, inverted positive SSD spectrum is shown in the bottom part of the figure. The threshold of heights of identified peaks can be defined by threshold value given in percentage of amplitude of inverted positive SSD of the highest peak. Employing this technique one can influence and control the sensitivity of the procedure.

An illustration of the capabilities of the peak search algorithm is demonstrated in Fig. 5.6 (σ =2, threshold=5%). In the synthetic spectrum, we generated 15 peaks. All of them were identified by the presented algorithm. Again, the method finds the peaks at the edges (channels 2, 1020). It decomposes and finds also peaks in the doublets (61, 66) and (551, 556).

Let us now investigate the ability of the algorithm to work with peaks of different widths. In Fig. 5.7 we show an example of such a spectrum. The width of generated peaks varies from σ =1 up to σ =10. In Fig. 5.7 for σ =10 and threshold=5% the algorithm identifies correctly all peaks in the spectrum. However, the resolution of the algorithm for this kind of data is relatively small.

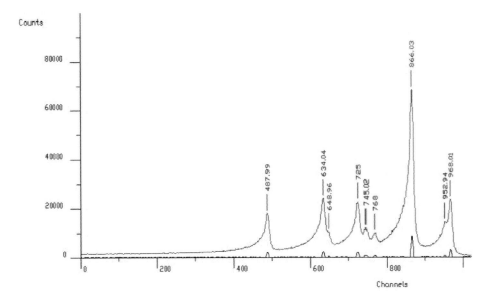

Figure 5.5. Example of application of the peak searching algorithm to the electron spectrum.

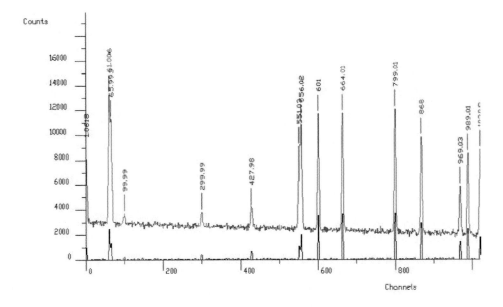

Figure 5.6. Synthetic spectrum with doublet and peaks positioned very close to the edges.

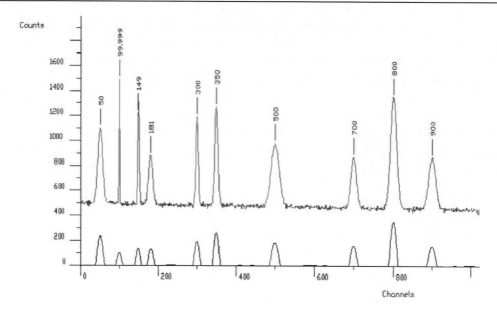

Figure 5.7. Illustration of the application of the peak search algorithm to peaks with different widths.

b. Two-dimensional Spectra [80]

We suppose that the number of counts in a two-dimensional γ-ray spectrum in the channels x, y can be approximated by

$$N(x,y) = G_1(x,y) + G_2(x) + G_3(y) + B + Cx + Dy. \qquad (5.1)$$

The two-dimensional Gaussian is defined as

$$G_1(x,y) = A_1 \exp\{-\frac{1}{2(1-\rho^2)}[\frac{(x-x_1)^2}{\sigma_{x_1}^2} - \frac{2\rho(x-x_1)(y-y_1)}{\sigma_{x_1}\sigma_{y_1}} + \frac{(y-y_1)^2}{\sigma_{y_1}^2}]\}, \qquad (5.2)$$

where x_1, y_1 give its position, σ_{x_1} and σ_{y_1} are standard deviations of normal distributions in the directions x and y, respectively, and ρ is a correlation coefficient. Analogously one-dimensional ridges parallel with axes y and x can be described by

$$G_2(x) = A_2 \exp\{-\frac{(x-x_2)^2}{2\sigma_{x_2}^2}\} \qquad (5.3)$$

and

$$G_3(y) = A_3 \exp\{-\frac{(y-y_3)^2}{2\sigma_{y_3}^2}\}, \qquad (5.4)$$

respectively. For a short interval we assume that the background in (5.1) can be approximated by the linear function

$$B + Cx + Dy. \tag{5.5}$$

Obviously the mixed partial second derivative of (5.1) in both dimensions $\dfrac{\partial^4 N(x,y)}{\partial x^2 \partial y^2}$ does not depend on $G_2(x), G_3(y)$ and the linear background.

Due to the discrete nature of channel counts, we shall replace the mixed partial second derivatives by the mixed partial second differences in both dimensions, thus

$$\begin{aligned}
S(i,j) = {}& N(i+1,j+1) - 2N(i,j+1) + N(i-1,j+1) \\
& - 2N(i+1,j) + 4N(i,j) - 2N(i-1,j) \\
& + N(i+1,j-1) - 2N(i,j-1) + N(i-1,j-1).
\end{aligned} \tag{5.6}$$

In order to eliminate statistical fluctuations we shall employ smoothed mixed partial second differences in both dimensions. The smoothing is achieved by summing $S(i,j)$ in a given window

$$S_w(i,j) = \sum_{i_1=i-r}^{i+r} \sum_{j_1=j-r}^{j+r} S(i_1,j_1), \tag{5.7}$$

where the window is $w = 2r + 1$. In accordance with [6] the repetitive application of (5.7) yields the general smoothed mixed partial second differences in both dimensions, $S_{z,w}(i,j)$, where z is the number of repetitions. For the sake of simplicity, from now on in the paper we shall call the difference spectrum $S_{z,w}(i,j)$ the two-parameter smoothed second differences or two-parameter SSD.

The contents of the channel i, j of this difference spectrum can then be expressed as

$$S_{z,w}(i,j) = \sum_{i_1=i-r}^{i+r} \sum_{j_1=j-r}^{j+r} C_{z,w}(i_1-i, j_1-j) N(i_1, j_1), \tag{5.8}$$

where $r = z \cdot m + 1, m = (w-1)/2$. Eq. (5.6) and (5.7) imply that the elements of the filter matrix $C_{z,w}(i,j)$ can be factorized, i.e.

$$C_{z,w}(i,j) = C_{z,w}(i) \cdot C_{z,w}(j). \tag{5.9}$$

The vector elements $C_{z,w}(i), i \in <-r, r>$, can be calculated using the recursion formula

$$C_{z,w}(i) = \sum_{i_1=i-r}^{i+r} C_{z-1,w}(i_1), \tag{5.10}$$

where $i \in <-z \cdot r - 1, z \cdot r + 1>$ and

$$C_{0,w} = \begin{cases} -2 & \textit{if } i = 0 \\ 1 & \textit{if } |i| = 1 \\ 0 & \textit{otherwise.} \end{cases}$$

In analogy with the algorithm in [6] and in view of (5.9), we define the standard deviation of a two-parameter SSD as

$$F_{z,w}(i,j) = \sqrt{\sum_{i_1=i-r}^{i+r} \sum_{j_1=j-r}^{j+r} C_{z,w}^2(i_1 - i)C_{z,w}^2(j_1 - j)N(i_1, j_1)}. \tag{5.11}$$

Another way to compute the coefficients of the filter vector C is used in [81]. The vector C is defined as the second derivative of the Gaussian

$$C_\sigma(i) = \frac{d^2}{dx^2} \exp(-\frac{x^2}{2\sigma^2})\Big|x = i \quad , \tag{5.12}$$

where σ is the standard deviation of the searched peaks. The replacement of the filter coefficients $C(i)$ from (5.12) to (5.8), (5.9), (5.11) is straightforward. The choice of method employed in a peak-searching procedure is optional and depends on the application. While in the first case the free parameters are $z, w,$ in the second method the only free parameter is σ. In general, independently of the calculation method of the SSD filter, the relations (5.8), (5.11) can be expressed as

$$S(i,j) = \sum_{i_1=i-r}^{i+r} \sum_{j_1=j-r}^{j+r} C(i_1 - 1)C(j_1 - j)N(i_1, j_1), \tag{5.13}$$

$$F(i,j) = \sqrt{\sum_{i_1=i-r}^{i+r} \sum_{j_1=j-r}^{j+r} C^2(i_1 - i)C^2(j_1 - j)N(i_1, j_1)}. \tag{5.14}$$

The two-parameter SSD spectrum calculated using the algorithm presented above is insensitive to the intersection of two ridges representing coincidence peak - background. For illustration, using (5.1), (5.2), (5.3), (5.4), (5.5) we have generated a (synthetic) spectrum (Fig. 5.8) with two two-dimensional Gaussians

$$G_{1A} \quad (A_{1A} = 100, \quad x_{1A} = 60, \quad y_{1A} = 8)$$
$$G_{1B} \quad (A_{1B} = 50, \quad x_{1B} = 180, \quad y_{1B} = 200),$$

(5.15)

two ridges parallel with the y axis

$$G_{2A} \quad (A_{2A} = 200, \quad x_{2A} = 60)$$
$$G_{2B} \quad (A_{2B} = 180, \quad x_{2B} = 180),$$

(5.16)

two ridges parallel with the x axis

$$G_{3A} \quad (A_{3A} = 200, \quad x_{3A} = 80)$$
$$G_{3B} \quad (A_{3B} = 500, \quad x_{3B} = 200)$$

(5.17)

and linear background

$$800 - 1,176.x - 1,96.y.$$

The parameters σ were set to 6 and $\rho = 0$.

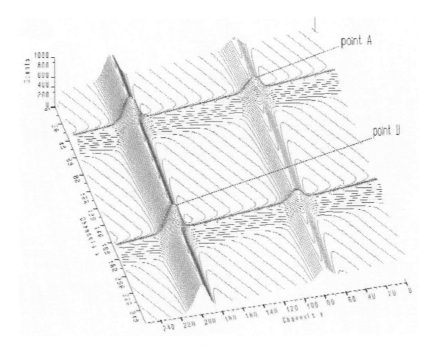

Figure 5.8. Synthetic spectrum with two two-dimensional Gaussians located at the points A, B.

At a first glance it is very difficult to estimate the positions of two-dimensional Gaussians. Their positions are denoted in Fig. 5.8 by the points A, B. The difference spectrum calculated according to (5.8), (5.9), (5.10) with $w = 7, z = 7$ is shown in Fig. 5.9.

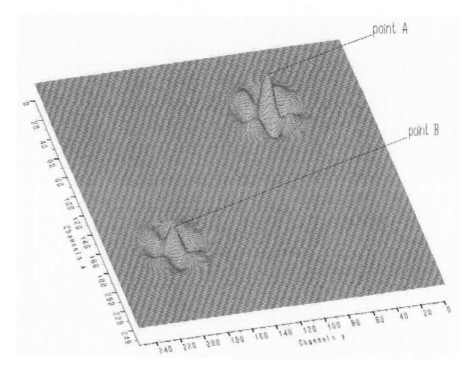

Figure 5.9. The difference spectrum for the spectrum from Fig. 5.8.

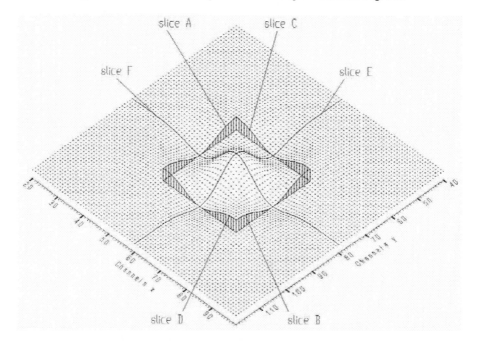

Figure 5.10. The shape of the difference spectrum of the two-dimensional Gaussian.

Apparently the difference spectrum reflects only the presence of two-dimensional peaks G_{1A}, G_{1B}. It ignores linear background, ridges $G_{2A}, G_{2B}, G_{3A}, G_{3B}$ and their intersections.

Now let us have a closer look at the shape of the difference spectrum of a two-dimensional peak (Fig. 5.10).

We can observe that in slices A, B, C, D the courses of the curves correspond to the smoothed second difference spectrum of a one-dimensional Gaussian. In slices E, F they correspond to negation of the smoothed second difference spectrum of a one-dimensional Gaussian. The difference spectrum and its standard deviation for one-dimensional Gaussian are given in Fig. 5.11.

According to Fig. 5.11, to decide whether an object found is a Gaussian peak or not, we have to define the points $i_1 - i_6$. Subsequently we can apply evaluation criteria for a one-dimensional Gaussian described in detail in [6]. We repeat this procedure for each dimension (in our example slices E, F).

When looking at Figs. 5.9 and 5.10 more thoroughly, we can observe that one two-dimensional peak gives rise to 5 local maxima in the difference spectrum. The position of one of the local maxima located in the middle of the surrounding ones corresponds to the position of the two-dimensional peak. Let us call this local maximum true and the remaining four local maxima false (Fig. 5.12).

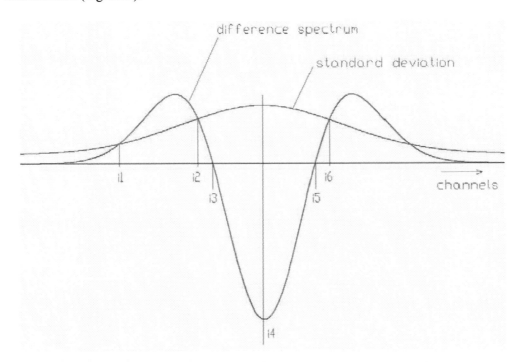

Figure 5.11. The difference spectrum and its standard deviation for one-dimensional Gaussian.

Once we have calculated two-parameter SSD (difference spectrum), the problem is how to distinguish between true and false maxima. False maxima should be ignored by the searching algorithm. In simple instances, when we have just a few peaks in the spectrum, it may seem trivial. It can be solved by finding groups of corresponding five maxima. Such an example can be illustrated by the spectrum with one peak from an experiment of positron annihilation (Fig. 5.13) [82].

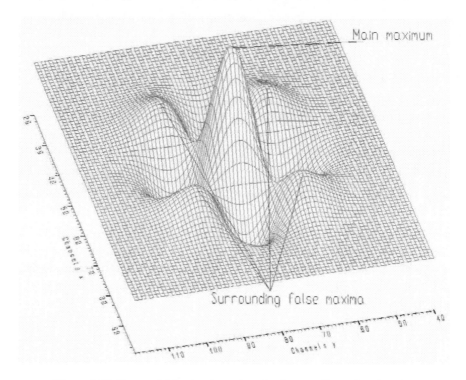

Figure 5.12. True and false local maxima for the two-dimensional peak.

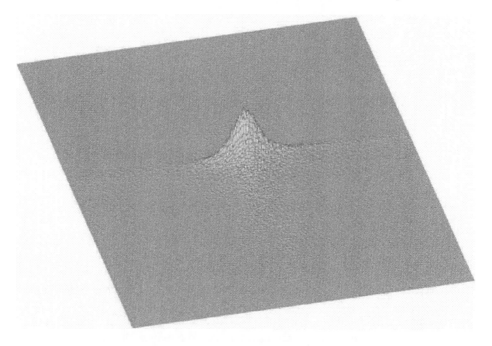

Figure 5.13. Spectrum with one peak from experiment of positron annihilation.

Its two-parameter SSD is shown in Fig. 5.14.

Figure 5.14. Two-parameter SSD spectrum from Fig. 5.13.

However, when a large number of peaks has to be evaluated in a spectrum, as is usually the case for γ-ray spectra, this task is very difficult. Figs. 5.15 and 5.16 give an example of such a two-dimensional γ-ray spectrum and its two-parameter SSD, respectively.

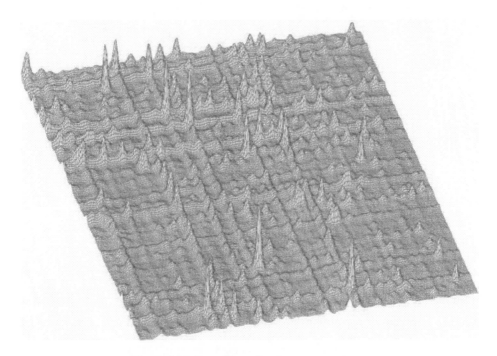

Figure 5.15. Two-dimensional γ-ray spectrum.

From Fig. 5.16 it is obvious that it is practically impossible to find groups of corresponding five local maxima. The true and false local maxima of two-parameter SSD of neighbouring peaks in an original spectrum can either be mixed or may coincide. Therefore it is necessary to suppress in a way the false maxima in the two-parameter SSD.

In keeping with (5.7), (5.8), and (5.13), we shall define spectra of the one-parameter smoothed second derivatives (one-parameter SSD) in both x and y independent variables, i.e.

$$X(i,j) = \sum_{i_1=i-r}^{i+r} C(i_1 - i)N(i_1, j) \qquad (5.18)$$

and

$$Y(i,j) = \sum_{j_1=j-r}^{j+r} C(j_1 - j)N(i, j_1), \qquad (5.19)$$

where the coefficients of the filter C are defined either according to (5.10) or (5.12).

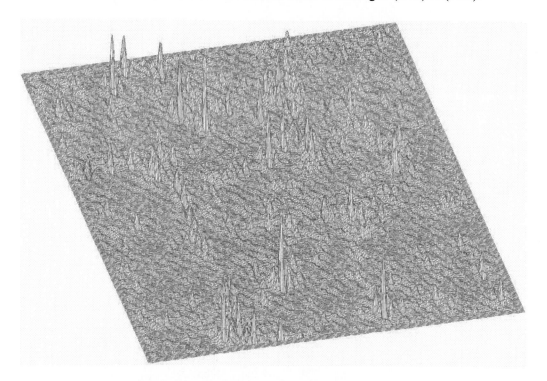

Figure 5.16. Two-parameter SSD spectrum from Fig. 5.15.

From the filter defined as $C_s(i, j)$ for $z = w = 0$ according to (5.10) for two-parameter SSD as well as filters $C_x(i), C_y(j)$ for one-parameter SSDs in a two-dimensional space in this way we can observe that, at the expected position of a two-dimensional peak $i_1 = i, j_1 = j$, we have

$$C_s(i, j) > 0, \quad C_x(i) < 0, \quad C_y(j) < 0.$$

Likewise from the definitions of a two-parameter SSD given by (5.7), (5.8), (5.9), (5.10), (5.13) and a one-parameter SSD given by (5.18), (5.19) as well as from the Fig. 5.11 and presented examples, it is apparent that in the position of a true local maximum of a two-parameter SSD (i_t, j_t), the following conditions are satisfied

$$S(i_t, j_t) > 0, \qquad X(i_t, j_t) < 0 \quad and \quad Y(i_t, j_t) < 0. \qquad (5.20)$$

Together with them also the following conditions, saying that in the point (i_t, j_t) one-parameter SSDs have local minima in both dimensions, must be satisfied

$$X(i_{t-1}, j_t) \geq X(i_t, j_t) \leq X(i_{t+1}, j_t) < 0, \qquad (5.21)$$

$$Y(i_t, j_{t-1}) \geq Y(i_t, j_t) \leq Y(i_t, j_{t+1}) < 0. \qquad (5.22)$$

The condition (5.20) is not satisfied in the point of a false local maximum in a two-parameter SSD (i_f, j_f), i.e.,

Figure 5.17. Two-parameter SSD spectrum from Fig. 5.14 after application of the condition (5.20).

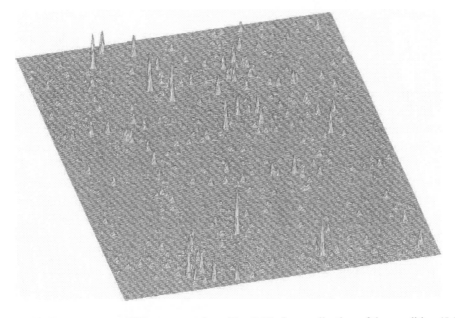

Figure 5.18. Two-parameter SSD spectrum from Fig. 5.16 after application of the condition (5.20).

$$S(i_f, j_f) > 0 \quad and \quad (X(i_f, j_f) \geq 0 \quad or \quad Y(i_f, j_f) \geq 0). \qquad (5.23)$$

By application of the condition (5.20) we can unambiguously distinguish between true and false maxima in a two-parameter SSD. The examples of two-parameter SSD spectra from Figs. 5.14, 5.16, after application of the condition (5.20), are presented in Figs. 5.17, 5.18.

Summarizing all together the algorithm to search for peaks in a two-dimensional spectrum is as follows:

- Using (5.13), we calculate the spectrum of the smoothed second differences in both dimensions (two-parameter SSD) $S(i, j)$.

- Using (5.14), we calculate the spectrum of the standard deviations of the smoothed second differences in both dimensions $F(i, j)$.

- For $i \in <0, n_1 - 1>, j \in <0, n_2 - 1>$, where n_1, n_2 are sizes of the spectrum, we search for local maxima in the spectrum of two-parameter SSD - $S(i, j)$.

- Once we have found a local maximum in the position i_l, j_l, then by using (5.18) and (5.19), we calculate spectra of one-parameter SSD $X(i_l, j_l)$ and $Y(i_l, j_l)$, respectively. Then by applying condition (5.20) we decide whether the found local maximum is true or false.

- If condition (5.20) is satisfied we test the slice $S(i, j_l), F(i, j_l), i \in <0, n_1 - 1>$ (slice F in Fig. 5.10) by Mariscotti's criteria for the shape of the peak in x dimension. Likewise we test the slice $S(i_j, j), F(i_l, j), j \in <0, n_2 - 1>$ (slice E in Fig. 5.10) by the same criteria for the shape of the peak in y dimension.

- If the shape of the peak in both dimensions satisfies these criteria, we have found a two-dimensional peak in the position i_l, j_l.

- We repeat the whole procedure for another local maximum from point c onwards.

The algorithm to search for peaks in coincidence two- dimensional spectra described so far has been tested using computer-generated synthetic spectra. For this kind of spectra, we know in advance the result that should be obtained and thus we can verify the reliability of the algorithms. In Fig. 5. 19 we can see an example of such a two-dimensional spectrum. It contains two-dimensional peaks, ridges in both dimensions, and noise. The peak-searching procedure finds two-dimensional peaks exactly at the location where they were generated. The markers denote the positions of the found two-dimensional peaks. The method is insensitive to the crossing of ridges. Fig. 5.20 illustrates real coincidence two-dimensional γ-ray spectrum. Again, the found two-dimensional peaks are denoted by markers.

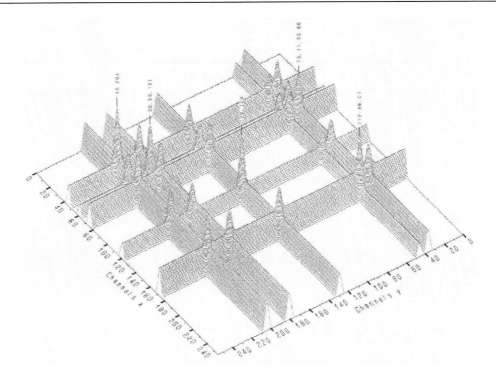

Figure 5.19. Synthetic two-dimensional spectrum with found peaks denoted by markers.

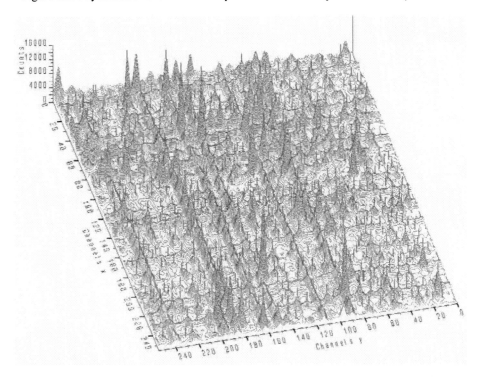

Figure 20. Coincidence of two-dimensional γ-ray spectrum with found peaks.

c. Three and *m*-dimensional Spectra [80]

The algorithm of peak searching in two-dimensional spectra described so far can be generalized to three-, four- and *m*-dimensional spectra. Analogously to the two-dimensional case, we can define for three dimensions, a spectrum of the smoothed second differences in all three dimensions, three-parameter SSD,

$$S(i,j,k) = \sum_{i_1=i-r}^{i+r} \sum_{j_1=j-r}^{j+r} \sum_{k_1=k-r}^{k+r} C(i_1-i, j_1-j, k_1-k) N(i_1, j_1, k_1) \tag{5.24}$$

and its standard deviation

$$F(i,j,k) = \sqrt{\sum_{i_1=i-r}^{i+r} \sum_{j_1=j-r}^{j+r} \sum_{k_1=k-r}^{k+r} C^2(i_1-i, j_1-j, k_1-k) N(i_1, j_1, k_1)}. \tag{5.25}$$

If we choose the filter according to (5.10), we obtain

$$C_s(i,j,k) = C(i)C(j)C(k). \tag{5.26}$$

With regard to (5.10) for $z = w = 0$ we can see that of the expected position of a three-dimensional peak $i_1 = i, j_1 = j, k_1 = k$ we have

$$C_s(i,j,k) < 0 \tag{5.27}$$

and

$$C(i) < 0, \qquad C(j) < 0, \qquad C(k) < 0. \tag{5.28}$$

Obviously, the point $i_1 = i, j_1 = j, k_1 = k$ is the only one that satisfies the conditions (5.27) and (5.28). Further in analogy to (5.18), (5.19), we define spectra of the one-parameter smoothed second differences in all three dimensions by

$$X(i,j,k) = \sum_{i_1=i-r}^{i+r} C(i_1-i) N(i_1,j,k), \tag{5.29}$$

$$Y(i,j,k) = \sum_{j_1=j-r}^{j+r} C(j_1-j) N(i,j_1,k), \tag{5.30}$$

$$Z(i,j,k) = \sum_{k_1=k-r}^{k+r} C(k_1-k) N(i,j,k_1). \tag{5.31}$$

The spectrum of a three-parameter SSD defined according to (5.24) is insensitive to continuous background, one-fold coincidences peak-background, two-fold coincidences peak-background or to any combinations of their intersections. Now in the position of the expected three-dimensional peak, the function $S(i,j,k)$ has corresponding local minima. Besides in the vicinity of a three-dimensional peak the function $S(i,j,k)$ contains additional 12 local minima. Again, the true local minimum must be distinguished from the false ones. Analogously with the condition (5.20) and in line with the example given above of filter C, one can conclude that in the position of true local minimum (i_t, j_t, k_t) it holds

$$S(i_t, j_t, k_t) < 0, \qquad\qquad X(i_t, j_t, k_t) < 0,$$
$$Y(i_t, j_t, k_t) < 0 \qquad and \qquad Z(i_t, j_t, k_t) < 0. \tag{5.32}$$

Now, we can generalize the search algorithm for m-dimensional spectra. For the spectrum of the m-parameter SSD we can write

$$S(i_1, i_2, \ldots, i_m) =$$
$$= \sum_{a_1 = i_1 - r}^{i_1 + r} \sum_{a_2 = i_2 - r}^{i_2 + r} \cdots \sum_{a_m = i_m - r}^{i_m + r} C(a_1 - i_1, a_2 - i_2, \ldots, a_m - i_m) N(a_1, a_2, \ldots, a_m). \tag{5.33}$$

Its standard deviation is

$$F(i_1, i_2, \ldots, i_m) =$$
$$= \sqrt{\sum_{a_1 = i_1 - r}^{i_1 + r} \sum_{a_2 = i_2 - r}^{i_2 + r} \cdots \sum_{a_m = i_m - r}^{i_m + r} C^2(a_1 - i_1, a_2 - i_2, \ldots, a_m - i_m) N(a_1, a_2, \ldots, a_m)}. \tag{5.34}$$

Similarly, the spectra of a one-parameter SSD are

$$X_1(i_1, i_2, \ldots, i_m) = \sum_{a_1 = i_1 - r}^{i_1 + r} C(a_1 - i_1) N(a_1, i_2, \ldots, i_m),$$

$$X_2(i_1, i_2, \ldots, i_m) = \sum_{a_2 = i_2 - r}^{i_2 + r} C(a_2 - i_2) N(i_1, a_2, \ldots, i_m), \tag{5.35}$$

$$\vdots$$

$$X_m(i_1, i_2, \ldots, i_m) = \sum_{a_m = i_m - r}^{i_m + r} C(a_m - i_m) N(i_1, i_2, \ldots, a_m).$$

Provided we have defined filter C, either according to (5.9), (5.10), or to (5.12), then for an m-dimensional peak $(i_1, i_2, \ldots, i_m)$ one can conclude that

- if m is odd $S(i_1, i_2, \ldots, i_m) < 0$,   (5.36)
- if m is even $S(i_1, i_2, \ldots, i_m) > 0$.

The spectrum of an m-parameter SSD (5.33) is insensitive to lower-fold coincidences of the peak-background. It is selective only to m-fold coincidence peaks. However, around the m-dimensional peak it generates false local extremes, if m is odd - minima, if m is even - maxima. Their number is

- if m is odd

$$\frac{3^m - 1}{2} - 1,$$

- if m is even

$$\frac{3^m - 1}{2}.$$

In analogy to two- and three-dimensional spectra, the false local extremes can be suppressed by applying conditions

$$X_j(i_1, i_2, \ldots, i_m) < 0; \qquad j \in <1, m>,$$ (5.37)

where $X_j(i_1, i_2, \ldots, i_m)$ are calculated according to (5.35).

Finally, in the general case of m-dimensional spectra the search algorithm of m-fold coincidence peaks can be expressed as follows :

- Using (5.33), we calculate the spectrum of the smoothed second differences in all dimensions (m-parameter SSD) $S(i_1, i_2, \ldots, i_m)$.
- Using (5.34), we calculate the spectrum of standard deviation of the smoothed second differences in all dimensions $F(i_1, i_2, \ldots, i_m)$.
- For $i_j \in <0, n_{j-1}>, j \in <1, m>$, where $n_1, n_2, \ldots, n_m$ are sizes of the spectrum, we search for local extremes
  - if m is odd - local minima
  - if m is even - local maxima

in $S(i_1, i_2, \ldots, i_m)$.

- Once we have found an appropriate local extreme in the point $l_1, l_2, \ldots, l_m$, using (5.35), we calculate spectra of 1-parameter SSD

$$X_1(l_1, l_2, \ldots, l_m), \ldots, X_m(l_1, l_2, \ldots, l_m).$$

Then, by applying conditions (5.37) we decide whether the found local extreme is true or false.

- If the conditions (5.37) are satisfied, we test the slices

$$(-1)^{m+1} S(i_1, l_2, \ldots, l_m), \quad F(i_1, l_2, \ldots, l_m), \quad i_1 \in < 0, n_1 >$$
$$(-1)^{m+1} S(l_1, i_2, \ldots, l_m), \quad F(l_1, i_2, \ldots, l_m), \quad i_2 \in < 0, n_2 >$$
$$\vdots$$
$$(-1)^{m+1} S(l_1, l_2, \ldots, i_m), \quad F(l_1, l_2, \ldots, i_m), \quad i_m \in < 0, n_m >$$

by Mariscotti's criteria for the shape of the peak in each dimension.

- If the shape of the peak in all dimensions satisfies these criteria, we have found the m-dimensional peak in the position $(l_1, l_2, \ldots, l_m)$
- We repeat the whole procedure for another local extreme from point c onwards.

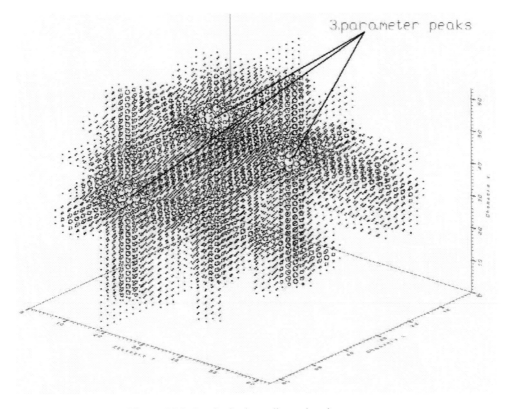

Figure 5.21. Synthetic three-dimensional spectrum.

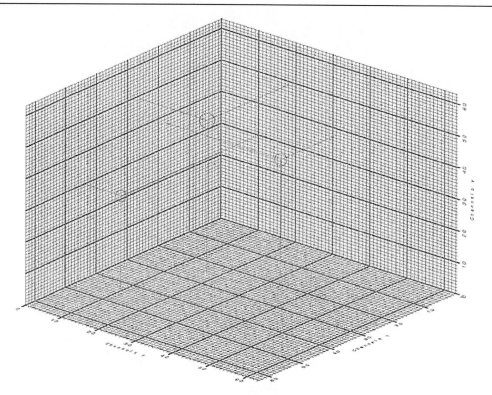

Figure 5.22. Found peaks in three-dimensional spectrum.

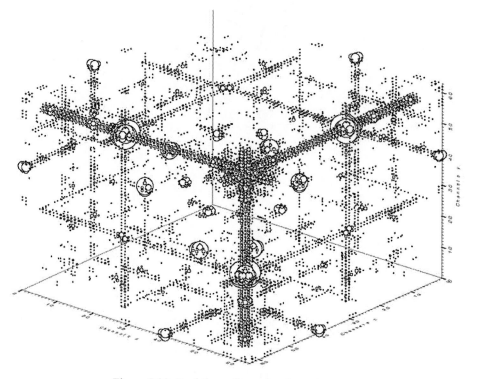

Figure 5.23. Real three-dimensional spectrum.

We have tested the peak-searching algorithm for three-dimensional spectra as well. In Fig. 5.21, we can see again the synthetic three-dimensional spectrum. The channels are represented by balls. Their sizes are proportional to the counts that the channels contain. The spectrum is composed of three-dimensional peaks, ridges of one- and two-fold coincidences and noise. Again, the peak-searching procedure determines exactly the positions of three-dimensional peaks in the locations where they were generated. In Fig. 5.22 we can see the found peaks without spectrum. The employed visualization technique allows us to read out in a simple way the positions of found peaks.

Likewise, we have used the derived algorithm to search for peaks in the real three-dimensional spectrum (Fig. 5.23). The peaks found in this spectrum are shown in Fig. 5.24. Again, this example gives evidence that the method is insensitive to crossing of lower-fold coincidences.

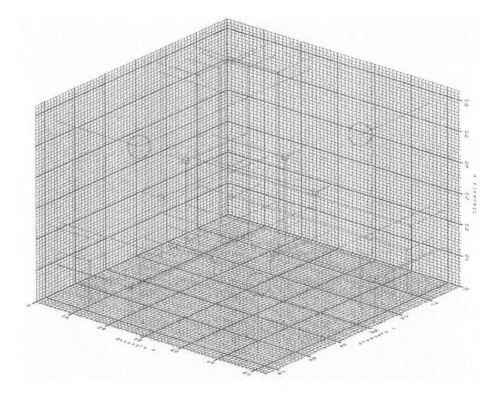

Figure 5.24. Found peaks in three-dimensional spectrum from Fig. 5.23.

5.1.2. Peak Searching Algorithm Based on Markov Chain Method

The objective of the analysis of experimental data is to extract the relevant features from data while decreasing to minimum the influence of statistical fluctuations on final result. Processing of experimental data in nuclear physics means ceaseless struggle with noise at every level of the analysis. The problem of noise eliminating can be met at background determination, deconvolution, peak searching, fitting etc. All these algorithms intrinsically contain the procedures able to cope with the elimination of the noise.

The identification of relatively narrow peaks either on smoothly varying background or after its removal is an ostensible simple task. Although there are many different algorithms and approaches to detect peaks in data, there is no one perfect method. The main problem lies in distinguishing true peaks from statistical fluctuations, Compton edges, background and other undesired spectrum features [17]. The use of the convolution method to locate the peaks is a well-established approach and is utilized in many of the available algorithms. It removes statistical fluctuations and is insensitive to linear background under Gaussians [6].

However, application of these methods cannot be directly extended to searching for peaks in multidimensional spectra. In a simplified way, one can think of a one-dimensional spectrum as a composition of background, peaks and noise. For instance, two-dimensional spectrum, besides of these components, contains also one-fold coincidences (ridges as a result of peak-background coincidences) in both directions. The algorithm to search for peaks in n-dimensional γ - ray spectra must be able to distinguish between intersections of lower-fold coincidences and n-fold coincidence. The extension of the algorithm based on smoothed second differences for multidimensional spectra [80] was described in previous section. The algorithm works satisfactorily if the peaks are well separated. It is able to discover, to some extent, even partially overlapped peaks. However, poorly resolved peaks (too close to each other, multiplets) usually require more sophisticated peak determination methods such as deconvolution [67]. Before employment of the deconvolution methods one has to determine and remove the background including lower-order coincidences [43].

Specific problems arise when trying to localize peaks in low-statistics spectra. To suppress statistical fluctuations one has to employ some additional more efficient smoothing tools. The peak searching algorithm based on Markov chain smoothing was proposed in [83]. In [84] this algorithm has been extended also for multidimensional spectra.

The smoothing algorithm based on the discrete Markov chain has a very simple invariant distribution

$$U_2 = \frac{p_{1,2}}{p_{2,1}} U_1, \quad U_3 = \frac{p_{2,3}}{p_{3,2}} U_2 U_1, \ldots, \quad U_n = \frac{p_{n-1,n}}{p_{n,n-1}} U_{n-1} \ldots U_2 U_1 \qquad (5.38)$$

U_1 being defined from the normalization condition

$$\sum_{i=1}^{n} U_i = 1$$

n is the length of the smoothed spectrum and

$$p_{i,i\pm1} = A_i \sum_{k=1}^{m} \exp\left[\frac{y(i \pm k) - y(i)}{\sqrt{y(i \pm k) + y(i)}} \right] \qquad (5.39)$$

is the probability of the change of the peak position from channel i to other channel $i + 1$. A_i is the normalization constant so that

$$p_{i,i-1} + p_{i,i+1} = 1 \qquad (5.40)$$

and m is a width of smoothing window. This invariant distribution has sharp peaks, which correspond to relevant local maxima in the original spectrum.

a. One-dimensional Spectra

Let us assume that the original one-dimensional spectrum to be smoothed is stored in the vector $y(i)$, $i \in \langle 0, N-1 \rangle$. From these data, we calculate smoothed spectrum $S(i)$, $i \in \langle 0, N-1 \rangle$ according to the algorithm outlined in Fig. 5.25.

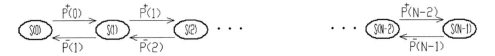

Figure 5.25. Principle of the algorithm of one-dimensional spectra smoothing based on Markov chains

The influence of the contents of the left and right neighboring channels to the smoothed values is expressed via probabilities bound by the system of equations

$$
\begin{aligned}
S(0)P^+(0) &= S(1)P^-(1) \\
S(1)P^+(1) &= S(2)P^-(2) \\
&\vdots \\
S(N-2)P^+(N-2) &= S(N-1)P^-(N-1)
\end{aligned}
\qquad (5.41)
$$

where

$$
P^\pm(i) = A(i) \sum_{k=1}^{m} \exp\left[\frac{y(i \pm k) - y(i)}{\sqrt{y(i \pm k) + y(i)}} \right]
\qquad (5.42)
$$

$A(i)$ is determined from the condition

$$
P^+(i) + P^-(i) = 1
\qquad (5.43)
$$

and

$$
\sum_{i=0}^{N-1} S(i) = 1,
\qquad (5.44)
$$

m is smoothing window. From relations (5.41) one can express the values of the smoothed spectrum

$$
S(i) = S(i-1) \cdot P^+(i-1) / P^-(i), \quad i \in \langle 1, N-1 \rangle
\qquad (5.45)
$$

where $S(0)$ can be calculated from relation (5.44).

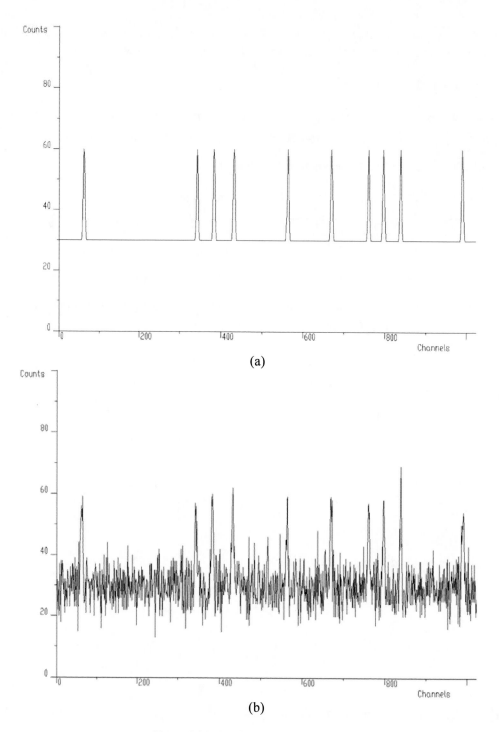

(a)

(b)

Figure 5.26. Continued on next page.

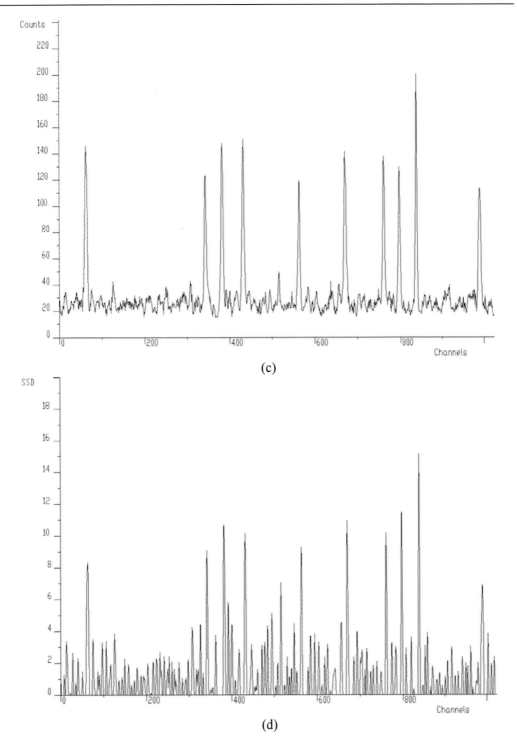

(c)

(d)

Figure 5.26. An example of synthetic spectrum with 10 peaks plus constant background (a), spectrum from Fig. 5.26(a) plus Gaussian noise (30% of peak amplitudes) (b) Markov spectrum ($m = 5$) (c) and inverted SSD spectrum ($\sigma = 2$) of the spectrum from the Fig. 5.26 (d).

The original idea of the Markov smoothing algorithm is to facilitate identification of peaks in spectra with low-statistics and very noisy spectra. An example of synthetic spectrum with 10 peaks plus constant background is given in Fig. 5.26 (a) In Fig. 5.26 (b) one can see the spectrum from Fig. 5.26 (a) plus Gaussian noise (30% of peak amplitudes). From these data, it is very difficult to distinguish the original peaks. The resulting Markov spectrum for m = 5 is presented in Fig. 5.26 (c). Obviously the peaks in Markov spectrum correspond with the peaks in original spectrum. As the operation calculating Markov spectrum is nonlinear the counts scale differs from the scale in the original spectrum given in Fig. 5.26 (b).

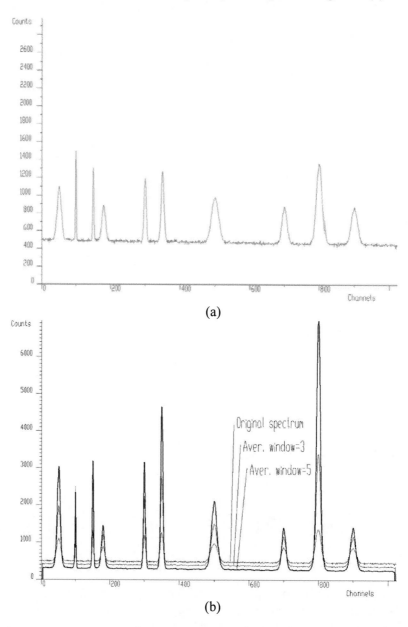

(a)

(b)

Figure 5.27. Study of the influence of the filter width m on smoothed synthetic spectrum with different widths of peaks (a). and Markov spectra for m = 3 and m = 5 (b).

Let us compare the method with the peak searching algorithm published in [80]. This method is based on calculation of smoothed second differences (SSD). Besides of other criteria (see [6, 80]), the peaks positions are identified as local maxima from inverted SSD above a given threshold value (given in percentage of the maximum amplitude). In Fig. 5.26 (d) one can see inverted positive SSD spectrum for $\sigma = 2$. Apparently, for low-statistics data or data with high level of noise, in Markov spectrum the peaks are emphasized (boosted) more evidently than in the inverted SSD spectrum. For such noisy data, when using the SSD algorithm, there exists a danger to identify fake peaks.

In the next example we have investigated the robustness of the proposed algorithm to the spectrum with the peaks with sigma changing from 1 to 10 (see Fig. 5.27 (a)). The spectra for averaging windows $m = 3, 5$ are shown in Fig. 5.27 (b). One can notice that the algorithm discovers reliably the peaks independently on their widths. Besides of smoothing with increasing m it suppresses also the level of the background. Again, due to nonlinearity of the algorithm, the scales in original and Markov spectra are different.

An example of experimental γ - ray spectrum is given in Fig. 5.28 (a). In these data one can hardly estimate the positions of true peaks. From visual estimation of the raw spectrum, one can see that two most probable peaks are positioned approximately at channels 420 and 500. The presence of other peaks is very improbable. In Fig. 5.28 (b), we present the spectrum smoothed with Markov chain based smoothing algorithm with filter width $m = 20$. In these data one can observe two dominant peaks at the above given positions. When we apply peak searching algorithm ($\sigma = 10$) to original spectrum we get SSD spectrum given in Fig. 5.28 (c). One can observe that this algorithm has a tendency to discover also fake peaks.

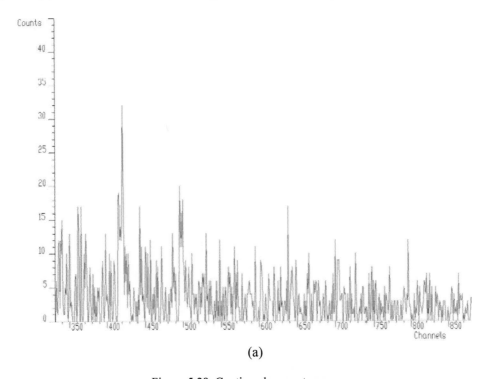

(a)

Figure 5.28. Continued on next page.

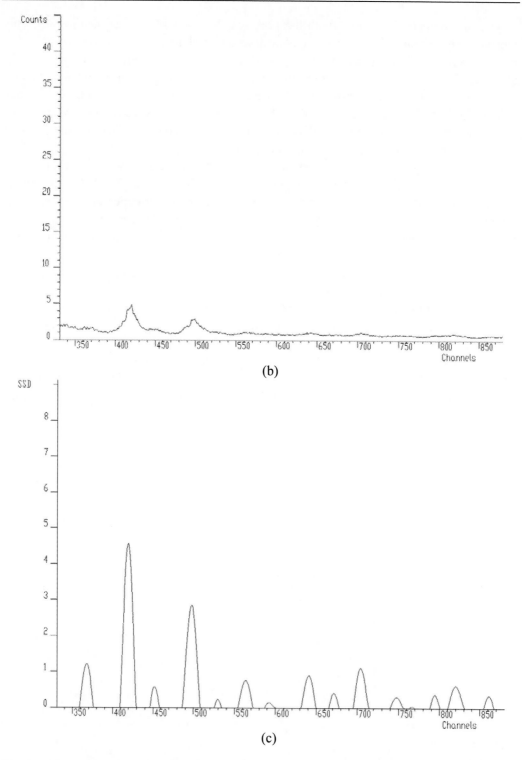

Figure 5.28. Experimental low-statistics γ - ray spectrum (a) smoothed spectrum using Markov chain algorithm (m=20) (b) and inverted SSD spectrum (σ = 10) generated from original spectrum (c).

b. Two-Dimensional Spectra

Now the situation is outlined in Fig. 5.29.

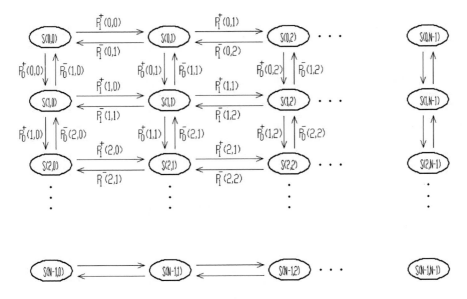

Figure 5.29. Principle of the algorithm of two-dimensional spectra smoothing based on Markov chains.

The influence of the channel contents of the smoothed spectrum $S(i_0, i_1)$

- on the value $S(i_0 + 1, i_1)$ is denoted as $P_0^+(i_0, i_1)$
- on the value $S(i_0, i_1 + 1)$ is denoted as $P_1^+(i_0, i_1)$
- on the value $S(i_0 - 1, i_1)$ is denoted as $P_0^-(i_0, i_1)$
- on the value $S(i_0, i_1 - 1)$ is denoted as $P_1^-(i_0, i_1)$.

Analogously to one-dimensional system for the first column of the spectrum one can write

$$S(0,0) P_0^+(0,0) = S(1,0) P_0^-(1,0)$$
$$S(1,0) P_0^+(1,0) = S(2,0) P_0^-(2,0)$$
$$\vdots$$
$$S(N-2,0) P_0^+(N-2,0) = S(N-1,0) P_0^-(N-1,0)$$

, (5.46)

where

$$P_0^\pm = A(i_0,0) \sum_{k=1}^{m} \exp\left[\frac{y(i_0 \pm k,0) - y(i_0,0)}{\sqrt{y(i_0 \pm k,0) + y(i_0,0)}} \right].$$

(5.47)

$A(i,0)$ is calculated from the condition

$$P_0^+ (i_0,0) + P_0^- (i_0,0) = 1, \quad i_0 \in \langle 0, N-1 \rangle. \tag{5.48}$$

Then for smoothed elements of the first column of the resulting matrix it holds

$$S(i_0,0) = S(i_0-1,0) \cdot P_0^+ (i_0-1,0)/P_0^- (i_0,0), \quad i_0 \in \langle 1, N-1 \rangle. \tag{5.49}$$

Analogously for smoothed elements of the first row of the resulting matrix it holds

$$S(0,i_1) = S(0,i_1-1) \cdot P_1^+ (0,i_1-1)/P_1^- (0,i_1), \quad i_1 \in \langle 0, N-1 \rangle. \tag{5.50}$$

The rest of smoothed elements (outside of the first column and the first row) are bound by the system of equations

$$\begin{aligned} S(i_0,i_1) P_0^- (i_0,i_1) + S(i_0,i_1) P_1^- (i_0,i_1) = \\ = S(i_0-1,i_1) P_0^+ (i_0-1,i_1) + S(i_0,i_1-1) P_1^+ (i_0,i_1-1) \end{aligned}, \tag{5.51}$$

where

$$P_0^\pm (i_0,i_1) = A(i_0,i_1) \sum_{k=1}^{m} \exp \left[\frac{y(i_0 \pm k, i_1) - y(i_0,i_1)}{\sqrt{y(i_0 \pm k, i_1) + y(i_0,i_1)}} \right], \tag{5.52}$$

$$P_1^\pm (i_0,i_1) = A(i_0,i_1) \sum_{k=1}^{m} \exp \left[\frac{y(i_0, i_1 \pm k) - y(i_0,i_1)}{\sqrt{y(i_0, i_1 \pm k) + y(i_0,i_1)}} \right] \tag{5.53}$$

and $A(i_0,i_1)$ is calculated from the condition

$$P_0^+ (i_0,i_1) + P_0^- (i_0,i_1) + P_1^+ (i_0,i_1) + P_1^- (i_0,i_1) = 1, \quad i_0,i_1 \in \langle 1, N-1 \rangle \tag{5.54}$$

Then from (15.51) one can express smoothed values of the spectrum elements outside of the first column and row

$$S(i_0,i_1) = \frac{S(i_0-1,i_1) P_0^+ (i_0-1,i_1) + S(i_0,i_1-1) P_1^+ (i_0,i_1-1)}{P_0^- (i_0,i_1) + P_1^- (i_0,i_1)}, \tag{5.55}$$

where $i_0,i_1 \in \langle 1, N-1 \rangle$. The value $S(0,0)$ can be calculated from the condition

$$\sum_{i_0=0}^{N-1}\sum_{i_1=0}^{N-1} S(i_0,i_1)=1.\tag{5.56}$$

An example of two-dimensional synthetic spectrum with 8 peaks plus constant background is shown in Fig. 5.30 (a). In Fig. 5.30 (b) we present the spectrum from Fig. 5.30 (a) plus Gaussian noise (30% of peak amplitudes). Again, from these data it is very difficult to distinguish the original peaks. In Markov spectrum ($m = 5$) in Fig. 5.30 (c) one can observe that the positions of peaks correspond with the peaks in original spectrum. Likewise in one-dimensional data due to nonlinearity of the algorithm, the scales in original and Markov spectra are different.

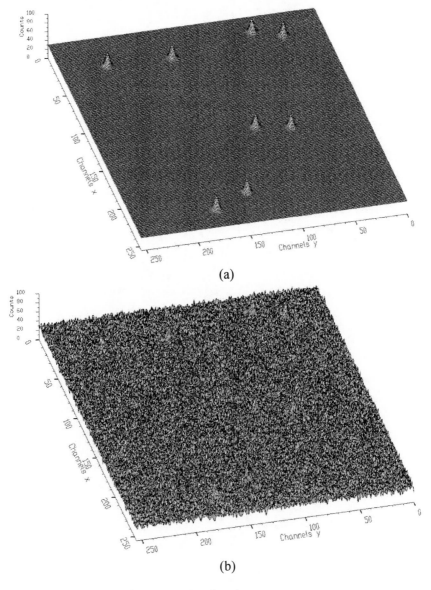

(a)

(b)

Figure 5.30. Continued on next page.

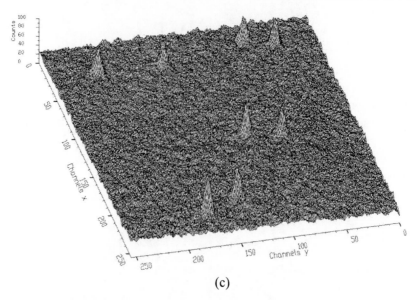

(c)

Figure 5.30. An example of two-dimensional synthetic spectrum with 8 peaks plus constant background (a), spectrum from Fig. 5.30 (a) plus Gaussian noise (30% of peak amplitudes) (b) and Markov spectrum (m = 5) of the spectrum from the Fig. 5.30 (b) - (c).

An example of two-dimensional experimental γ-γ – ray low-statistics spectrum (256x256 channels) and smoothed data using Markov algorithm (filter width m =5) is given in Fig. 5.31 (a) and Fig. 5.31 (b), respectively. Apparently, it emphasizes the relevant peaks and suppresses noise. To identify peaks in the smoothed spectrum one can employ any of the established peak searching algorithms.

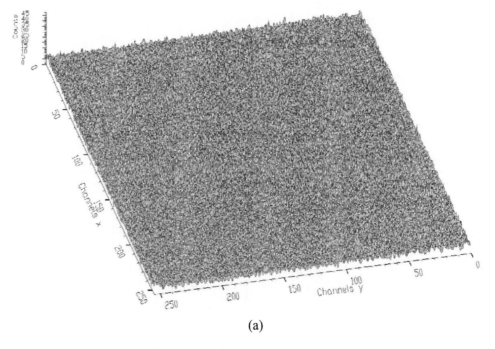

(a)

Figure 5.31. Continued on next page.

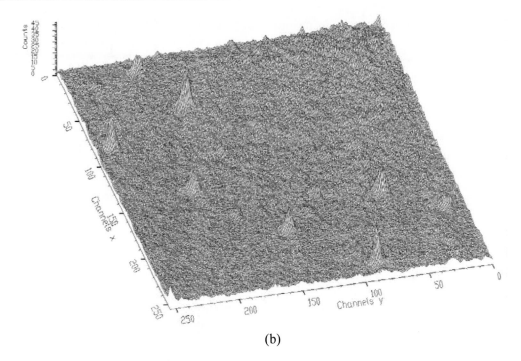

(b)

Figure 5.31. Two-dimensional experimental γ-γ - ray low-statistics spectrum (a) and Markov spectrum (m =5) (b).

c. Three-Dimensional Spectra

One element of three-dimensional array of Markov chain of smoothed spectrum for $i_0 \neq 0 \,\&\, i_1 \neq 0 \,\&\, i_2 \neq 0$ is shown in Fig.5.32.

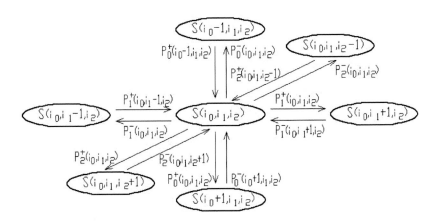

Figure 5.32. One element of three-dimensional array of Markov chain of smoothed spectrum.

In the calculation of three-dimensional smoothed spectrum, we start from the vectors (one-dimensional arrays) $S(i_0,0,0), S(0,i_1,0), S(0,0,i_2), \quad i_0,i_1,i_2 \in \langle 0, N-1 \rangle$. We employ the algorithm according to the relations (5.46) - (5.50). Its extension for the third variable is straightforward. From these vectors using the relations (5.51) – (5.55) we determine the matrices (two-dimensional arrays) $S(i_0,i_1,0), S(i_0,0,i_2), S(0,i_1,i_2)$, $i_0,i_1,i_2 \in \langle 0, N-1 \rangle$. Again, the extension for the third variable is apparent. Using the procedure analogous with two-dimensional case one can determine the elements of three-dimensional array for $i_0,i_1,i_2 \in \langle 1, N-1 \rangle$

$$S(i_0,i_1,i_2) = [S(i_0-1,i_1,i_2)P_0^+(i_0-1,i_1,i_2) + S(i_0,i_1-1,i_2)P_1^+(i_0,i_1-1,i_2) + \\ + S(i_0,i_1,i_2-1)P_2^+(i_0,i_1,i_2-1)]/[P_0^-(i_0,i_1,i_2) + P_1^-(i_0,i_1,i_2) + P_2^-(i_0,i_1,i_2)] \quad . \quad (5.57)$$

The value of the element $S(0,0,0)$ can be determined from the condition

$$\sum_{i_0=0}^{N-1}\sum_{i_1=0}^{N-1}\sum_{i_2=0}^{N-1} S(i_0,i_1,i_2) = 1. \quad (5.58)$$

An example of three-dimensional synthetic spectrum with 8 peaks is given in Fig. 5.33(a). Channels are shown as spheres with diameter proportional to the counts they contain. In Fig. 5.33(b), we present the spectrum from Fig. 5.33(a) plus Gaussian noise (30% of peak amplitudes). The resulting Markov spectrum for $m = 3$ is shown in Fig. 5.33(c). These data can be subsequently submitted for the identification of peaks. Their positions correspond with the peaks in original data.

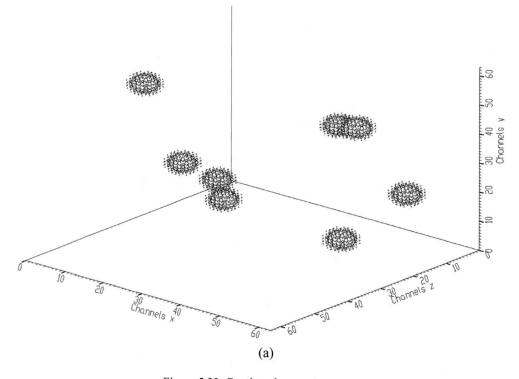

(a)

Figure 5.33. Continued on next page.

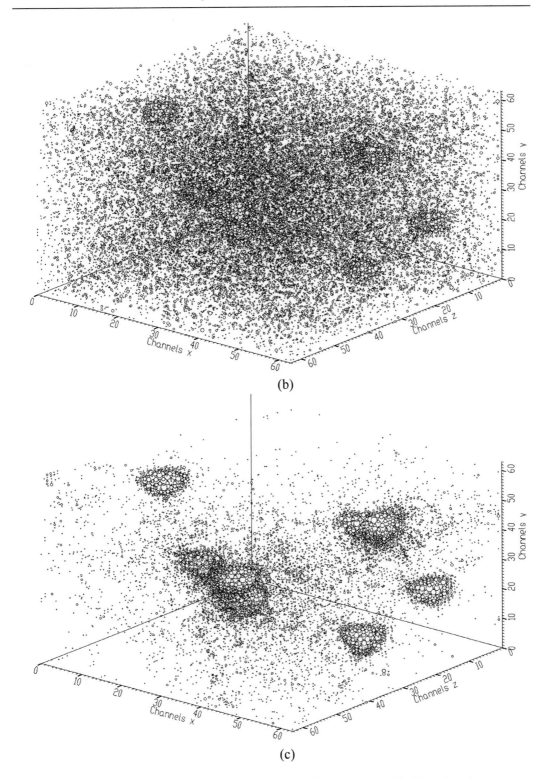

(b)

(c)

Figure 5.33. An example of three-dimensional synthetic spectrum with 8 peaks plus constant background (a), spectrum from Fig. 5.33 (a) plus Gaussian noise (30% of peak amplitudes) (b) and Markov spectrum ($m = 3$) of the spectrum from the Fig. 5.33 (b) - (c).

Finally, in Fig. 5.34 (a), we present an example of three-dimensional experimental γ-γ-γ – ray low-statistics spectrum (64x64x64 channels). In Fig. 5.34 (b) we can see smoothed data using three-dimensional Markov algorithm (filter width m =7). Again, it suppresses noise. The identification of peaks in the smoothed spectrum is much feasible than in the original one.

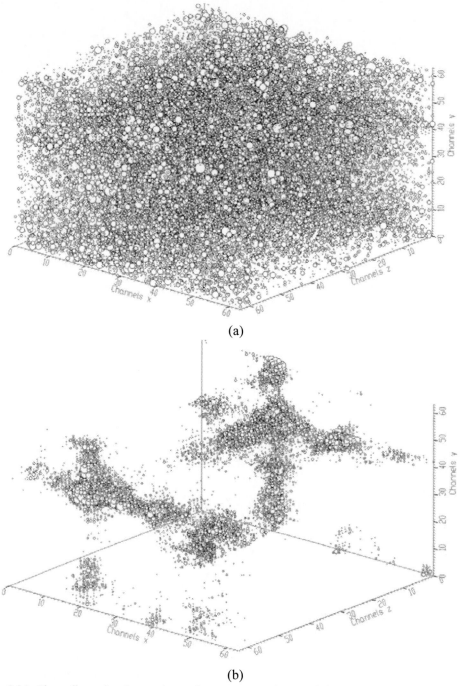

(a)

(b)

Figure 5.34. Three-dimensional experimental γ-γ- γ - ray low-statistics spectrum (a) and Markov spectrum (m =7) (b).

d. *M*-dimensional Spectra

The extension of the above-derived smoothing algorithms based on Markov chains to m-dimensional data is a matter of combinatorics. The algorithm can be broken down to the calculation of

- m one-dimensional arrays according to (5.42) – (5.45).

- $\binom{m}{2}$ two-dimensional arrays according to (5.51) – (5.55)

- $\binom{m}{3}$ three-dimensional arrays according to (5.57)

  $\vdots$

- m $(m-1)$-dimensional arrays calculated recursively according to the algorithm

for $(m-1)$-dimensional data.

Analogously to (5.57) for m-dimensional smoothed data one can write

$$S\left(i_0,i_1,...i_m\right) = \frac{\sum_{j=0}^{m} S\left(i_0,i_1,...,i_j-1,...,i_m\right) P_j^+\left(i_0,i_1,...,i_j-1,\cdots,i_m\right)}{\sum_{j=0}^{m} P_j^-\left(i_0,i_1,...i_m\right)}. \qquad (5.59)$$

5.1.3. Deconvolution Based Peak Searching Algorithms

The above-described algorithm works satisfactorily for well separated peaks. However, the number of peaks in γ-spectra is usually enormous. Often the peaks are located very close to each other. Therefore, to identify the overlapping peaks one has to employ a method based on the decomposition of spectra. We have developed the high resolution algorithm of peak identification based on the Gold deconvolution. The algorithm is based on the assumption that all peaks have the same σ. The procedure to identify peaks in one-dimensional and multidimensional spectra is as follows:

The SSD based methods intrinsically ignore the background. This is not the case of the deconvolution based peak searching methods. Therefore, as the first step using a method of background elimination described in section 3, we remove the background from the spectrum (in case of multidimensional spectra we remove also the lower-fold coincidences).

We apply a deconvolution algorithm (Gold, Richardson-Lucy, Muller, Maximum a posteriori etc.), where the system matrix is composed of mutually shifted Gaussians of the given σ.

In the deconvolved spectrum we find local maxima, which if exceed the specified threshold value, are identified as peaks.

a. One-Dimensional Spectra

The resolution of the above-presented general peak search algorithm is quite limited. For instance for σ =2 it cannot recognize peaks positioned closer than 4, 5 channels to each other. When increasing σ the resolution of the algorithm based on SSD decreases even more. Therefore, to improve the resolution capabilities we have proposed a new algorithm based on Gold deconvolution. However, unlike SSD algorithm before applying the deconvolution algorithm we have to remove the background using one of the background elimination algorithm presented in section 3.

Let us apply the high resolution algorithm to the synthetic spectrum with several peaks located very close to each other. The result for σ =2 and threshold=4% is shown in Fig. 5.35. The method discovers all 19 peaks in the spectrum. It finds also small peak at the position 200 and decomposes the cluster of peaks around the channels 551-568. In the bottom part of the Fig. 5.35, we present the deconvolved spectrum.

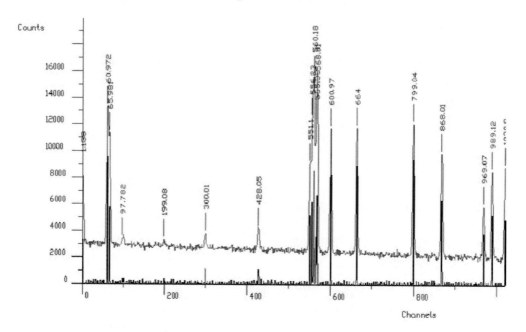

Figure 5.35. Example of synthetic spectrum with doublet and multiplet and its deconvolved spectrum (in bottom part of picture)

The detail of peaks in cluster is shown in the following Fig. 5.36. Again, in the upper part of the figure one can see original data and in the bottom part the deconvolved data. The method finds also the peaks about existence of which it is impossible to guess from the original data.

b. Two-Dimensional Spectra

From the point of view of resolution, naturally also for two-dimensional spectra the method based on SSD has its limitation. In Fig. 5.37 and Fig. 5.38, we show the spectrum of peaks positioned quite close to each other and its SSD spectrum (σ =2, threshold=5%), respectively. It does not recognize the doublet in the foreground of the spectrum (position 46, 48).

Analogously to one-dimensional case to treat the resolution problem, one has to employ the high resolution method based on the Gold (or any other high-resolution) deconvolution algorithm. The peak searching procedure works as follows. First, the background is removed (if desired), then the response matrix is generated according to given σ and deconvolution is carried out. The result of the search using the high resolution algorithm for $\sigma =2$ and threshold=5% applied to the previous example is shown in Fig. 5.39 and its appropriate deconvolved spectrum in Fig. 5.40. It discovers correctly all peaks in the doublet including the small peak at the position 46, 48.

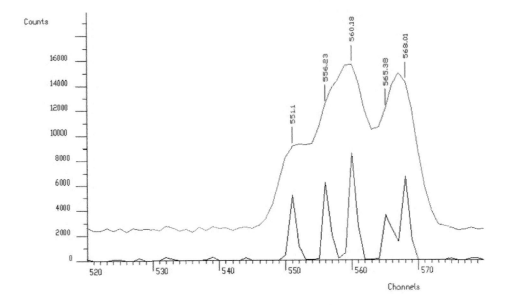

Figure 5.36. Detail of multiplet from Fig. 5.35.

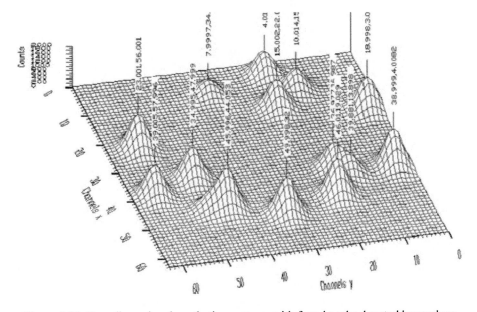

Figure 5.37. Two-dimensional synthetic spectrum with found peaks denoted by markers.

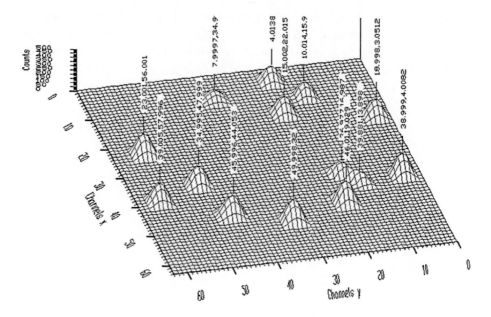

Figure 5.38. SSD spectrum of the data from Fig. 5.37.

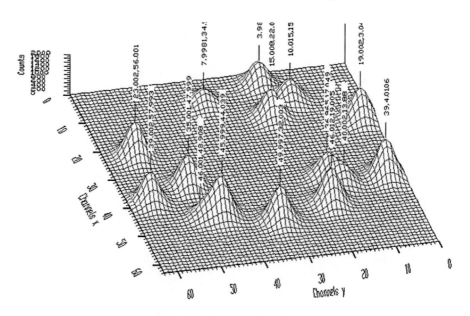

Figure 5.39. Result of the search using high resolution method based on Gold deconvolution.

c. Sigma Range Peak Search

Deconvolution based peak searching methods are selective to the peaks with given σ. There exists a tolerance if the given parameter σ does not match completely the real σ of peaks present in the spectrum. The measure of the tolerance depends on the σ. However, in general, these methods cannot be considered for efficient and reliable in the identification of peaks with σ changing in a large range.

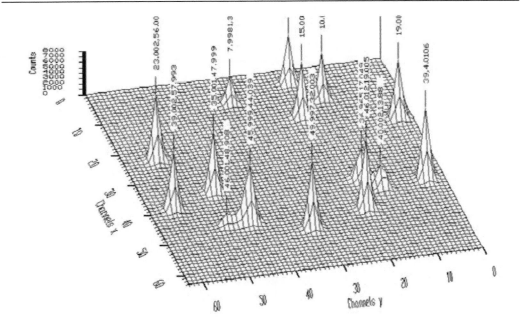

Figure 5.40. Deconvolved spectrum of the data from Fig. 5.39.

In case we are able to express the dependence (relation) of σ on the energy, this fact can be utilized to set up the system matrix. Then one can make use of the above-given procedures employing a method of deconvolution. However, for the data where this dependence is missing and cannot be expressed by a formula (table, rule etc.), we were compelled to develop sophisticated algorithm of successive decompositions with simultaneous estimation of σ.

Analogously to the algorithm presented in the previous section, we shall assume that we have eliminated the background from the spectrum. Further let us assume we are given a range of σ (σ_1, σ_2). For every σ the algorithm comprises two deconvolutions. The principle of the method is as follows:

a.   for $\sigma_i = \sigma_1$ up to σ_2

b.   we set up the matrix of response functions (Gaussians) according to Fig. 5.41. All peaks have the same $\sigma = \sigma_i$. The columns of the matrix are mutually shifted by one position. We carry out the Gold deconvolution of the investigated spectrum.

c.   In the deconvolved spectrum, we find local maxima higher than given threshold value and include them into the list of 1-st level candidate peaks.

d.   Next, we set up the matrix of the response functions for the 2-nd level deconvolution. There exist three groups of positions:

 –   positions where the 1-st level candidate (new) peaks were localized. Here we generate response (Gaussian) with σ_i.

–   positions where the 2-nd level candidate peaks were localized in the previous steps $\sigma_1 \leq \sigma_k < \sigma_i$ (see the next step). For each such a position, we generate the peak with the recorded σ_k.

–   for remaining free positions where no candidate peaks were registered, we have empirically found that the most suitable functions are the block functions with the width $\pm 3\sigma_i$ from the appropriate channel.

The situation for one 2-nd level candidate peak in position j_2 with σ_k and for one 1-st level candidate peak in the position j_1 (σ_i) is depicted in Fig. 5.42. We carry out the 2-nd level Gold deconvolution.

e. Further, in the deconvolved spectrum we find the local maxima greater than given threshold value.
    We scan the list of the 2-nd level candidate peaks:

–   if in a position from this list there is not local maximum in the deconvolved spectrum, we erase the candidate peak from the list.
    We scan the list of the 1-st level candidate peaks

–   if in a position from this list there is the local maximum in the deconvolved spectrum, we transfer it to the list of the 2-nd level candidate peaks.

f. Finally peaks that remained in the list of the 2-nd level candidate peaks are identified as found peaks with recorded positions and σ.

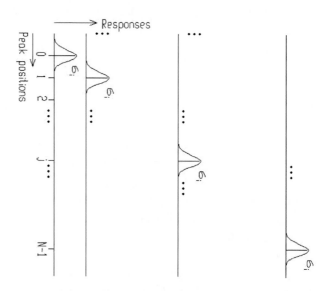

Figure 5.41. Response matrix consisting of Gaussian functions with the same σ.

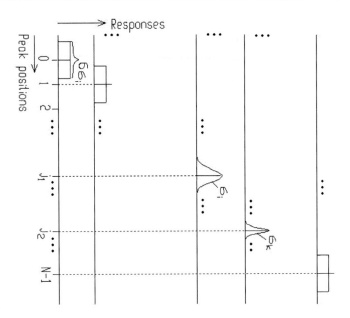

Figure 5.42. Example of response matrix consisting of block functions and Gaussians with different σ.

The method is rather complex as we have to repeat two deconvolutions for the whole range of σ.

In what follows, we illustrate in detail practical aspects and steps during the peak identification. The original noisy spectrum to be processed is shown in Fig. 5.43. The σ of peaks included in the spectrum varies in the range 3 to 43. It contains 10 peaks with some of them positioned very close to each other.

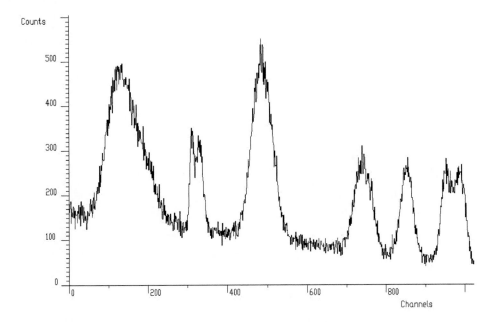

Figure 5.43. Noisy spectrum containing 10 peaks of rather different widths.

As the sigma range search (SRS) algorithm is based on the deconvolution, in the first step we need to remove background. The spectrum without background is shown in 5.44. At every position, we have to expect any peak from the given σ range (3, 43). To confine the possible combinations of positions and σ we generated inverted positive SSD of the spectrum for every σ from the range. We get matrix shown in Fig. 5.45. Further, we consider only the combinations with non-zero values in the matrix.

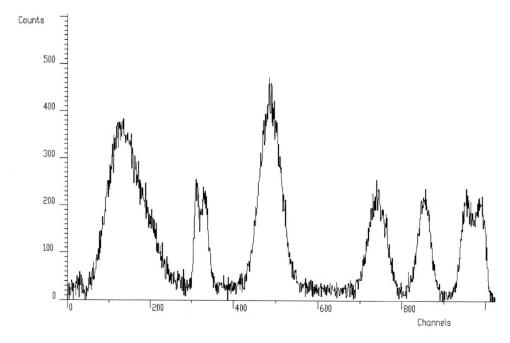

Figure 5.44. Spectrum from Fig. 5.43 after background elimination.

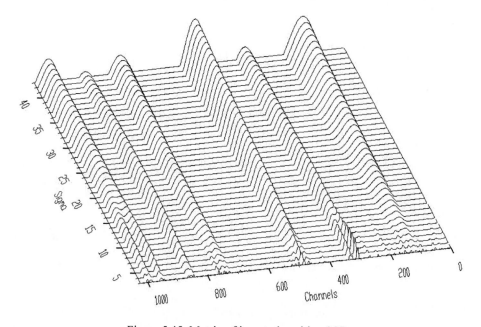

Figure 5.45. Matrix of inverted positive SSD.

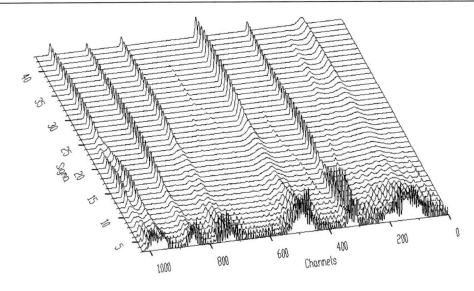

Figure 5.46. Matrix of deconvolved spectra for responses changing σ from 3 to 43.

The SRS algorithm is based on two successive deconvolutions. In the first level deconvolution, we look for peak candidates. We changed σ from 3 up to 43. For every σ we generated response matrix and subsequently we deconvolved spectrum from Fig. 5.44. Again, we arranged the results in the form of matrix given in Fig. 5.46.

Successively, according to the above-proposed algorithm from these data, we pick up the candidates for peaks, construct the appropriate response matrices and deconvolve again the spectrum from Fig. 5.44. The evolution of the result for increasing σ is shown in Fig. 5.47.

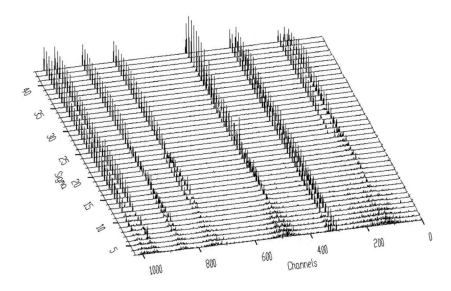

Figure 5.47. Matrix composed of spectra after second level deconvolutions.

From the last row of data, which are in fact spikes, we can identify (applying threshold parameter) the positions of peaks. The found peaks (denoted by markers with channel numbers) and the original spectrum are shown in Fig. 5.48.

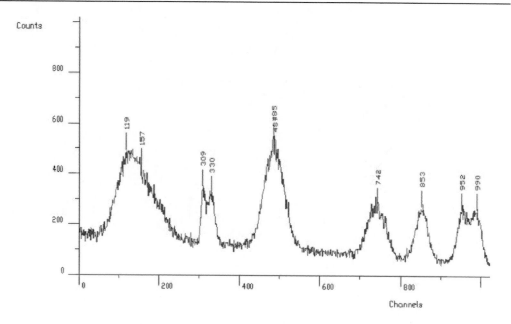

Figure 5.48. Original spectrum with found peaks denoted by markers.

In Tab. 5.1, we present the result of generated peaks and estimated parameters. In addition to the peak positions, the algorithm estimates even the σ of peaks. It is able to recognize also very closely positioned peaks. However, the estimate of the parameters is in some cases rather inaccurate mainly in poorly separated peaks of the spectrum. The problem is very complex and sometimes it is very difficult to decide whether a lobe represents two, eventually more, close positioned narrow peaks or one wide peak.

Table 5.1. Results of the estimation of peaks in spectrum shown in Fig. 5.48.

Peak #	Generated peaks (position/σ)	Estimated peaks (position/σ)
1	118/26	119/23
2	162/41	157/26
3	310/4	309/4
4	330/8	330/7
5	482/22	485/23
6	491/24	487/27
7	740/21	742/22
8	852/15	853/16
9	954/12	952/12
10	989/13	990/12

The algorithm gives satisfactory results, though due to its complexity it is quite time consuming.

5.2. Identification of Isotope Lines in Two-Dimensional Spectra of Nuclear Multifragmentation [85]

In the nuclear reactions well above Coulomb barrier, multiple charged particles are emitted in a wide angular range. In order to detect all charged particles emitted during the reaction, multi-detector arrays consisting of many charged particle telescopes emerged, with optimized choice of angular coverage (ideally $4-\pi$), granularity and detection thresholds. An important step in the off-line analysis of data from the multidetector arrays with large angular coverage is to identify as many fragment species as possible. The method of isotope identification is based on well-known particle telescope technique in which the isotopes are resolved in two-dimensional $\Delta E - E$ spectra of energy losses in two detectors (typically a thin one followed by a thick one). The method needs to be applied to all detectors, which makes the analysis a highly repetitive task and methods allowing automation are preferable.

The main task of the analysis is to parametrize the relation of energy losses to observed electronic signals in two detectors. Charge identification was achieved in the work [86] by obtaining analytical fit of calculated energy losses for specific fragment species and by mapping of resulting matrix on a set of sampled experimental lines using the minimization procedure, with particle charge being the minimization parameter. In other works [26] and [87], the subset of lines with a priori known mass and charge was sampled and the parameters relating the energy loss to electronic signals are obtained by minimization procedure. Such method enables to carry out identification and energy calibrations simultaneously using a minimization procedure applied to two-dimensional spectra. For instance, in the work [26] the lines for three known isotopes (typically the most characteristic isotopes such as ^{1}H, ^{4}He, ^{9}Be) are assigned in the experimental spectra and calibration is carried out by a minimization procedure where these lines are fitted to corresponding calculated energy losses for a given $\Delta E - E$ telescope. The calibrations coefficients are thus obtained as optimum values of minimization parameters and identification can be achieved after superimposing the calculated energy loss lines onto experimental spectra. In this procedure, the initial assignment of the selected isotope lines required human intervention, since the lines had to be drawn by hand. To make such procedure fully automatic, one needs an algorithm which would recognize and tabulate some of the isotope lines in the two-dimensional spectra (manifested as non-linear ridges). In the works [88] and [89], algorithms were proposed to recognize individual spectral lines in the two-dimensional spectra, using the smoothing and differentiation methods [88] or neural network approach [89]. Separation of experimental lines with different atomic number was achieved. An alternative algorithm for recognition of isotope lines is proposed in the present work.

5.2.1. Proposal of the Algorithm

During the process of nuclear multifragmentation, multiple charged particles are produced. An important step in the analysis of data of this kind is to identify correctly the fragment species. As opposed to the identification of self-standing peaks (one-, or multidimensional, e.g. in γ-ray spectra) here we have to find also the correspondence among the identified local maxima belonging to the same ridge.

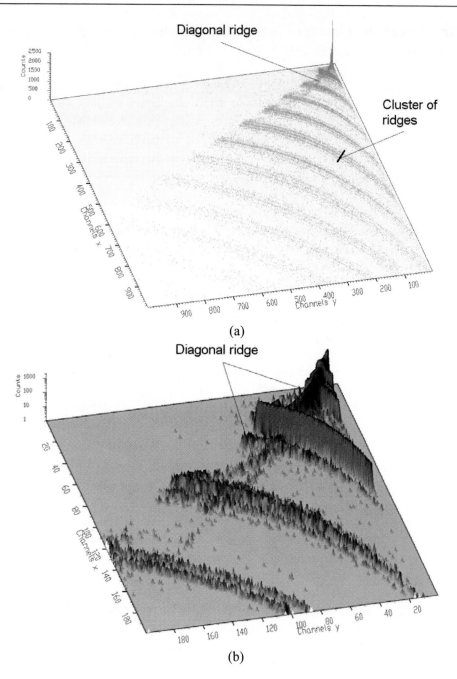

(a)

(b)

Figure 5.49. An example of two-dimensional spectrum of nuclear multifragmentation (a), and its detail (b). The variables x and y represent electronic signals originating from a pair of silicon detectors and recorded by the corresponding ADCs.

Let us now analyze the problems connected with the process of determination of ridges in the spectra of nuclear multifragmentation. To make things clear we shall accompany our considerations with an illustrative example. In Fig. 5.49(a) one can see two-dimensional energy loss spectrum from the telescope, consisting of two silicon detectors, of the size 1000x1000 channels. The data originate from a pair of silicon detectors with respective thickness 150 and 500 microns and the electronic gains were adjusted to equalize amplitudes in both electronic channels. The variables x and y represent electronic signals recorded by the corresponding ADCs. The shape of the spectral lines in the spectrum is a usual one for such type of detectors. The empty region on the top left is caused by the punch through of the second silicon detector, which defines a maximum energy. Particles with higher energies are not shown in the plot, except few events where anticoincidence signal was not detected (represented by weak line where energy losses in both silicon detectors decrease simultaneously). The isotopic resolution in the spectrum is achieved up to the Z=10.

To illustrate the complexity and statistical fluctuations in the data in Fig. 5.49(b) we show detail at the beginning of the coordinate system. We have to determine ridges of corresponding points from very sparsely distributed two-dimensional experimental data. Besides of that the spectra of nuclear multifragmentation are extremely noisy.

At the first glance at the data in Fig. 5.49(a), one can observe that the data vary over many orders of magnitude going from thousands of counts at the beginning of the spectrum to tens or ones in its rest. To compress the dynamic range in channel counts and to suppress the effect of the noise in the data sets a series of mathematical operators were studied in [38]. Taking the square root, then using the natural log operator twice (LLS) was claimed to yield the best results. LLS operator is applied to every channel $y(i)$

$$v(i) = \log\left[\log\left(\sqrt{y(i)+1}+1\right)+1\right]. \tag{5.60}$$

The application of LLS operator is an important starting point in the identification of ridges in the spectra of nuclear multifragmentation. In Fig. 5.50, we present the spectrum from Fig. 5.49 after application of LLS operator.

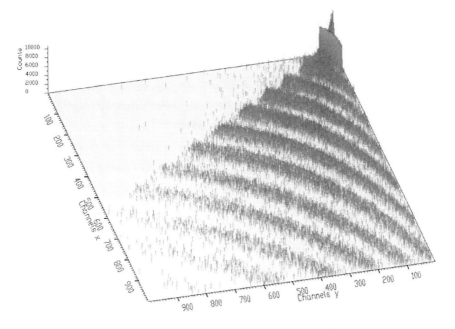

Figure 5.50. Spectrum from Fig. 5.49 after application of LLS operator.

The procedure of the identification of nonlinear ridges can be split down into several points. The first step of the nonlinear ridges identification is their quasi linearization. Though it is possible to do it in various ways (e.g. to transform data to polar coordinates) we proposed rather simple approach based on the slicing of original data from a given point according to Fig. 5.51. Let us start slicing at the point A and go along the diagonal to the beginning of the original coordinate system. The distance from the point A represents the parameter S. In this way we get the slice number 0. When changing successively the end points of the slices to *1, 2,..., N-1* on both axes x_1 and x_2 we get slices R and L, respectively. The new S coordinate is the distance from the point A, while the new R and L coordinates represent the number of the slice or in other words the length of the segment intercepted by the slice on the axis x_1 and x_2, respectively.

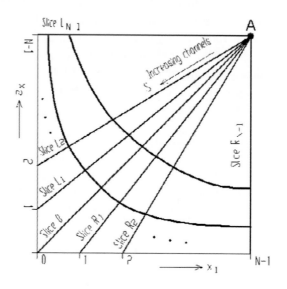

Figure 5.51. Principle of slicing of two-dimensional spectrum.

Due to the outlined coordinate transformation, we obtain data in two halfplanes arranged predominantly in quasi-linear directions. The curvatures on the left and right sides of the lines are different. Therefore, the angles of the arrangements in both halfplanes are also different. Spectrum from Fig. 5.50 sliced according to the algorithm of slicing outlined in Fig. 5.51 is shown in Fig. 5.52. One can observe that the trends in both halfplanes can be well approximated by straight lines one for each halfplane.

On the other hand, the linearized data in Fig. 5.52 are also distributed sparsely. From the above-mentioned considerations, one can conclude that the algorithm of ridges identification should

a.   suppress statistical fluctuations,
b.   glue together points in the linear directions,
c.   separate linearized ridges from each other (in vertical direction),
d.   ignore other artificial objects (e.g. diagonal ridge – see Fig. 5.49).

To suppress statistical fluctuations in the linearized data we employed second derivative filtration technique. Using this technique together with suppression of noise, we can carry out peak searching in one-dimensional slices. There exist a lot of algorithms of this, at the first glance very simple, but in its essence very complicated and complex problem [6], [80], [83], [84]. In [81], [90] so called correlation technique emerged. It is based on the second derivative of the Gaussian as the convolution function, called also the correlator

$$c(j) = \frac{j^2 - \sigma^2}{\sigma^4} \exp\left(-\frac{j^2}{2\sigma^2}\right) = \frac{d^2}{dx^2} \exp\left(-\frac{x^2}{2\sigma^2}\right)_{x=j}, \qquad (5.61)$$

where σ determines the width of searched peaks. The convolution of input data X with correlator C yields

$$y(i) = \sum_{k=0}^{i} x(k)c(i-k), \quad i = 0,1,\ldots,N-1. \qquad (5.62)$$

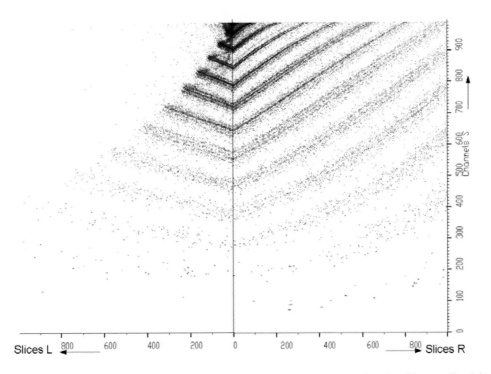

Figure 5.52. Spectrum from Fig. 5.50 sliced and linearized according to the algorithm outlined in Fig. 5.51.

The local minimums of $y(j)$ identify positions of peaks. As we need to carry out the peak smoothing and searching in several directions and moreover we also need to confine the peak regions of potential peak candidates to finite intervals we propose a technique based on inverted positive second derivative (IPSD) of the Gaussian

$$p(j) = \left\langle \begin{array}{ll} -y(j) & \text{if} \quad y(j) < 0 \\ 0 & \text{otherwise.} \end{array} \right.$$

(5.63)

The example of the second derivative and inverted positive second derivative of the Gaussian for $\sigma = 20$ is presented in Fig. 5.53(a) and (b), respectively.

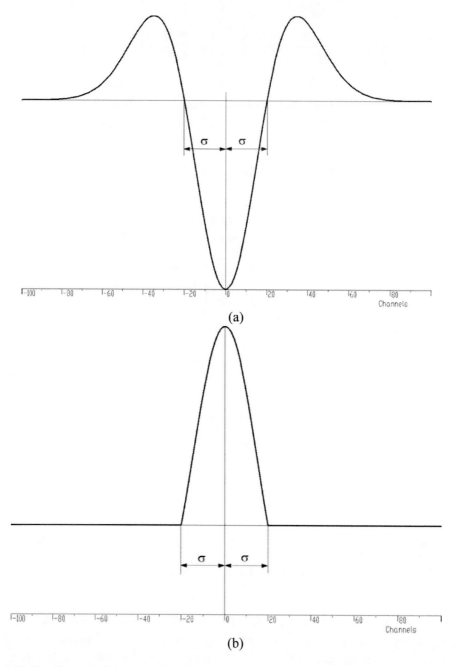

Figure 5.53. The example of the second derivative (a) and inverted positive second derivative (b) of the Gaussian for $\sigma = 20$.

Now let us apply the IPSD algorithm of filtering ($\sigma_1 = 20$) to the columns of linearized data shown in Fig. 5.52. We get data smoothed in vertical directions given in Fig. 5.54.

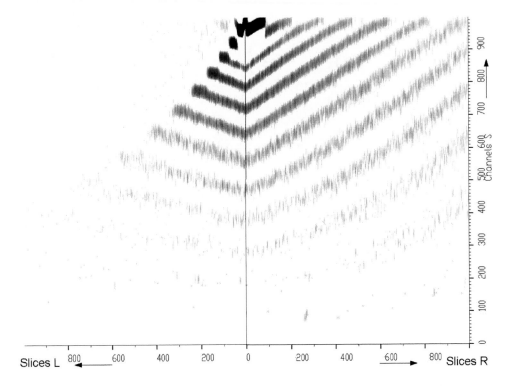

Figure 5.54. Data from Fig. 5.52 smoothed vertically with IPSD filter with $\sigma_1 = 20$.

To glue the slices together we repeat the same procedure in horizontal direction with $\sigma_2 = 30$. We obtain data presented in Fig. 5.55.

Now we need to determine directions of ridges in both halfplanes. To determine seed points of the direction lines we can find maximums in the slice 0 (Fig. 5.56).

We find sequences of maximums in neighboring slices in both directions. By fitting the sequences with lines and taking the estimates with the smallest chi-squares, we get direction angles of dominant ridges in both directions (Fig. 5.57).

In the directions determined by these angles we carry out the smoothing of original (non-smoothed) sliced data employing convolution technique with Gaussian filter with given σ_3 . To span gaps in sparsely distributed data and to determine intrinsically the correspondence among experimental points in the spectrum the parameter σ_3 should be set to an appropriate value. Analogously to (5.62) the convolution of input data x with filter g is

$$y(i) = \sum_{k=0}^{i} x(k)g(i-k), \quad i = 0,1,\ldots,N-1 . \tag{5.64}$$

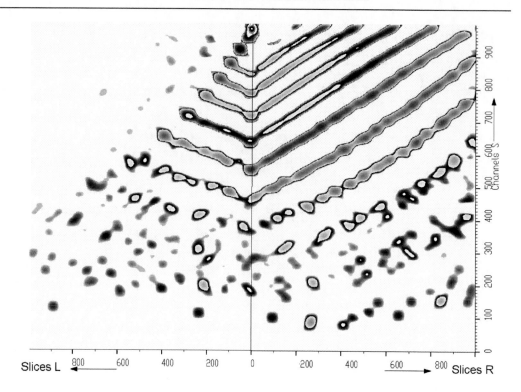

Figure 5.55. Data from Fig. 5.54 smoothed in horizontal direction ($\sigma_2 = 30$).

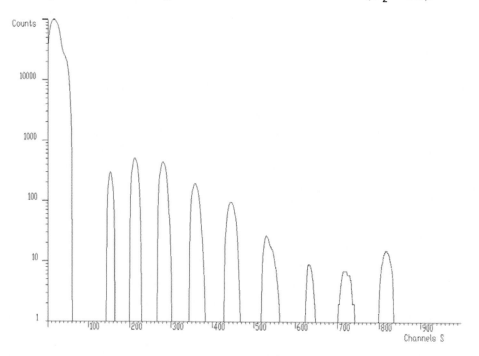

Figure 5.56. Slice 0 from the data from Fig. 5.55.

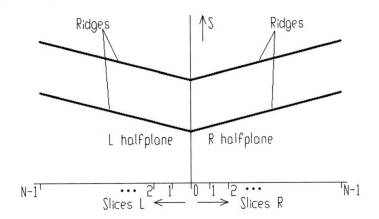

Figure 5.57. Linearized ridges in sliced and twice smoothed spectrum.

Further, let us return back to our example data in Fig 5.52. Let us find the directions in both halfplanes and smooth the data in these directions using (5.64). The result of this operation for $\sigma_3 = 30$ is given in Fig. 5.58

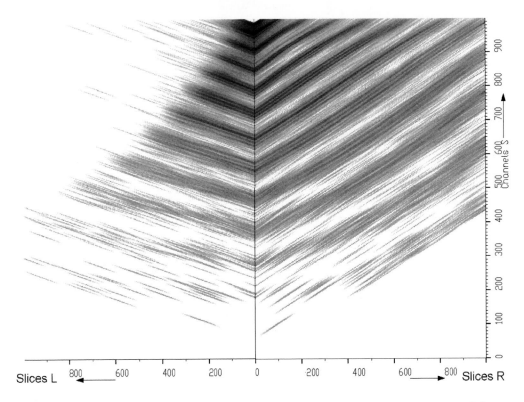

Figure. 5.58. Data from Fig. 5.52 after Gaussian smoothing in the direction of ridges ($\sigma_3 = 30$).

The points in corresponding ridges are connected. However, we need to separate peaks and smooth data in the vertical direction. Moreover, we need to confine data to regions of peaks and to eliminate data outside these regions. Therefore, we employ again the IPSD algorithm given by (5.61-5.63) with $\sigma_4 = 20$. We get smoothed data in both directions with connected points in corresponding ridges (Fig. 5.59).

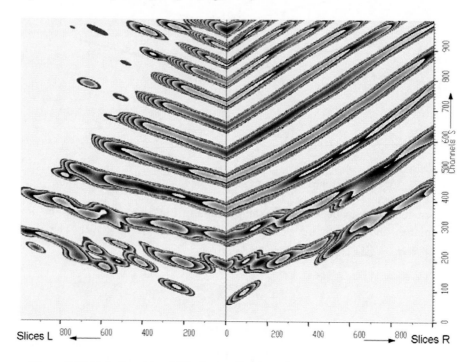

Figure 5.59. Data from Fig. 5.58 after application of IPSD filter with $\sigma_4 = 20$.

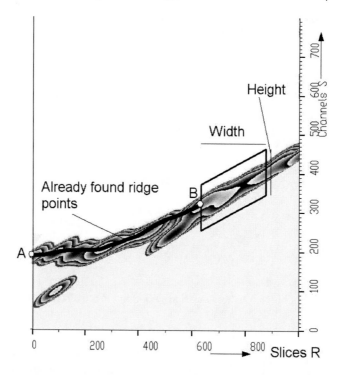

Figure 5.60. Principle of searching for neighboring maximums in vertical slices.

Analogously to Fig. 5.56 we can take slice 0. The local maximums in this data can serve as seed points to find corresponding points in ridges in both halfplanes. We look for the local maximums (in vertical direction) in the neighboring slices. The principle and scanning window for the right halfplane and for the first ridge is given in Fig. 5.60. We assume that starting from the point A we have already found the points belonging to the ridge until point B. Now we have to find the next point within the outlined skew window. The height of the window is $2 \cdot \sigma_4$, its width is σ_3 and its slope is given by the direction angle for the right halfplane. If we find a maximum in this window, we take it to be the ridge point and move the scanning window to right and repeat the scanning for the next point. Otherwise, we take the point B to be the last point of the ridge and we proceed to the next ridge. In this way we can determine all ridges.

Finally, in the last step we transform the ridges back to the original space of the two-dimensional spectrum. The result of our example data is given in Fig. 5.61.

Let us summarize and express the above-described algorithm of the identification of nonlinear ridges concisely in several points:

a. To compress the dynamic range of the channel counts apply the LLS operator to the spectrum data according to (5.60).

b. The next step of the nonlinear ridges identification is their quasi linearization. We propose rather simple approach based on the slicing of original data from a given point according to Fig. 5.51.

c. We obtain data in two halfplanes arranged predominantly in quasi-linear directions. However, the angles of the arrangements in both halfplanes are different. To suppress statistical fluctuations in the data we employed IPSD technique defined in (5.61-5.63). We carry out smoothing using IPSD of Gaussian with given σ_1 separately for every slice (vertically in Fig. 5.51) in both halfplanes.

d. To span gaps and decrease statistical fluctuations in the data in the horizontal direction we employ again IPSD technique (parameter σ_2).

e. We find local maximums in the slice 0 greater than a threshold value (given in percentage of the maximum value in the slice 0) and use them as seed points for the determination of dominant directions in both halfplanes.

f. In the next step, we find dominant directions of lines in both halfplanes by fitting corresponding local maxima in the smoothed inverted positive SSD data. We obtain direction angles for both halfplanes (see Fig. 5.57).

g. In the directions determined by these angles we carry out the smoothing of original (non-smoothed) sliced data employing convolution technique (5.64) with Gaussian filter with a given σ_3. The parameter σ_3 influences smoothness of the estimated ridges.

h. Further, the data smoothed in the dominant directions are submitted to another smoothing in the vertical direction S with IPSD of Gaussian with σ_4.

i. To identify seeds of ridges in the column number 0 we find local maxima.

j. In the last but one step we have to determine neighboring points (maximums) belonging to one ridge. We look for neighboring maximums in both direction angles.

Let us assume we have found maximum in the right halfplane in the point (x_1, y_1) we look for the maximum in the column $(x = x_1 + 1, y \in < y_1 - \sigma_4, y_1 + \sigma_4 >)$. Nevertheless, even among maximums of twice smoothed data, there can be gaps. To span these possible gaps we search for the maximums within the quadrangle $(x = x_1 + i, y \in < y_1 - \sigma_4 + k_r \cdot i, y_1 + \sigma_4 + k_r \cdot i >), \quad i \in < 1, \sigma_3 >$. The width of the quadrangle is given by appropriate coefficients σ for smoothing in the direction of ridges and the last vertical smoothing. The situation for the left halfplane is analogous. The principle of looking for next corresponding local maximum is illustrated in Fig. 5.60.

k. Finally, we transform back the ridges points to original two-dimensional spectrum.

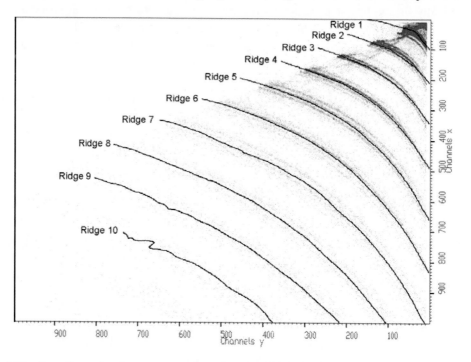

Figure 5.61. Two-dimensional spectrum of nuclear multifragmentation with determined ridges (parameters $\sigma_1 = 20, \sigma_2 = 30, \sigma_3 = 30, \sigma_4 = 20$).

5.2.2. Discussion and Results

Crucial point of the above-described algorithm is linearization of data in the transformed domain. To compare the proposed method with the transformation to polar coordinates, which is an alternative to the suggested algorithm outlined in Fig. 5.51, we introduce the example with the same data from Fig. 5.50. In Fig. 5.62(a) we present data from Fig. 5.50 transformed to polar coordinates. Here the transition from the right halfplane to the left one is much smoother than in the previous method. On the other hand, the data cannot be linearized so smoothly, i.e., one can observe a greater non-linearity in appropriate ridges. This results in non-acceptable estimate of the ridges given in Fig. 5.62(b).

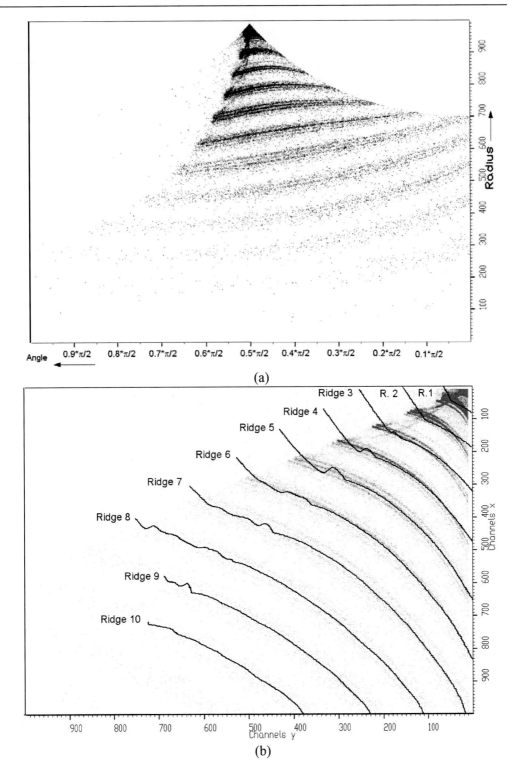

Figure 5.62. Transformation of spectrum from Fig. 5.50 to polar coordinates (a), and identified ridges transformed back to the original spectrum (b).

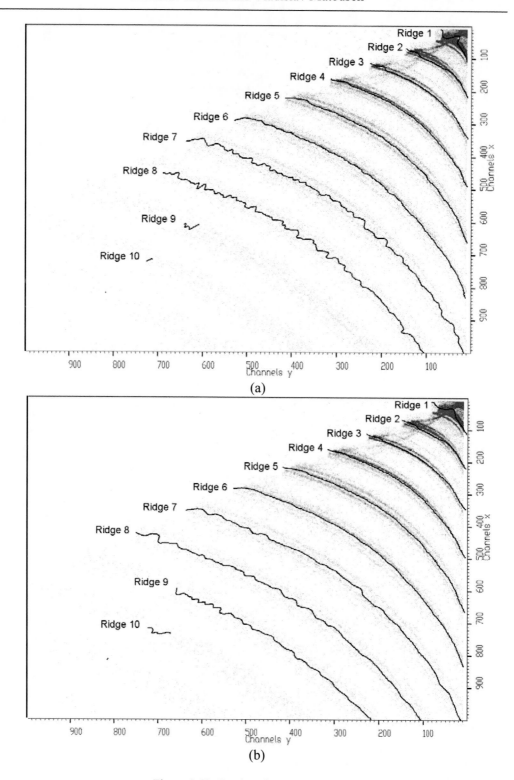

(a)

(b)

Figure 5.63. Continued on next page.

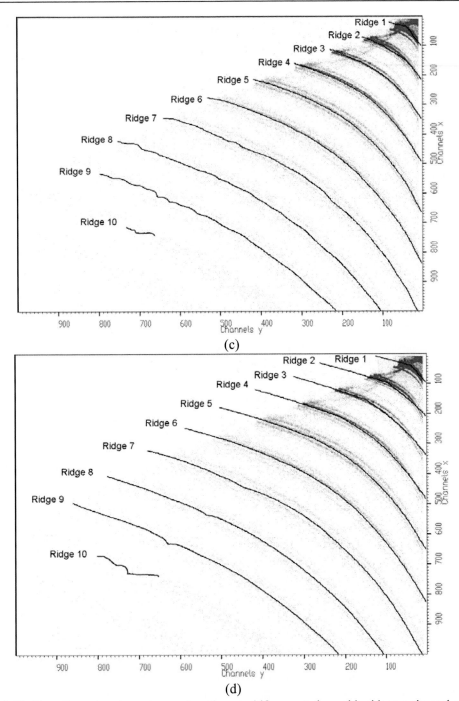

Figure 5.63. Two-dimensional spectrum of nuclear multifragmentation with ridges estimated with parameters $\sigma_3 = 5$ (a), $\sigma_3 = 10$ (b), $\sigma_3 = 20$ (c) and $\sigma_3 = 50$ (d) ($\sigma_1 = 20, \sigma_2 = 30, \sigma_4 = 20$).

Now let us study the influence of the parameters ($\sigma_1, \sigma_2, \sigma_3, \sigma_4$ and threshold) to the estimation of the ridges. Though by tuning these parameters some improvements for polar model of linearization can be achieved in the following examples we shall consider only slicing model of linearization. The first two parameters σ_1, σ_2 are used for the filtration of linearized data in vertical and horizontal direction, respectively, by employing IPSD algorithm. Filtered data are subsequently used for the determination of dominant directions in these data. The algorithm is rather independent and robust to the changes of these parameters. We have studied the influence of σ_1, σ_2 by changing them in the range $\langle 5, 50 \rangle$. The estimates practically coincide with the result shown in Fig. 5.61.

Further, let us analyze the influence of the parameter σ_3. Through the use of this parameter one can control smoothness of the estimated ridges. The examples for $\sigma_3 = 5, 10, 20, 50$ are illustrated in Fig. 5.63. With increasing σ_3 the estimated curves are getting smoother but on the other hand, one can observe undesirable slight distortion of their shapes on the right hand side for the ridges 2, 3, and 4.

When comparing the results achieved in Figs. 5.61, 5.62b and 5.63 we see that the algorithm is able to discover main ridges but it is unable to decompose and identify individual ridges of the clusters (see Fig. 5.49). By changing the parameter σ_4 in the last vertical filtration one can influence the width of identified ridges, or in other words to decompose them to subridges. In Fig. 5.64 we illustrate the results after application of the algorithm for $\sigma_4 = 10$ and $\sigma_4 = 3$. In Fig. 5.64(a) it discovers two subridges in the ridge 7 and identifies seeds of the ridges $R0$ and $R11$. When decreasing σ_4 to 3 we can decompose the main ridges to even more subridges. However, due to the loss of correlation among the points belonging to appropriate ridges, the lines are becoming shorter. The shape of lines fits better the original points in the spectrum than in Fig. 5.61.

The other way to increase resolution in the estimation of ridges is to employ the operation of Gold deconvolution [66-68]. Analogously to the filtration in the columns of the data presented in Fig. 5.59, we slice the data in vertical direction. We get data similar to those presented in Fig. 5.56. Then for $\sigma = \sigma_d$ according to (5.61) we generate the second derivative of the Gaussian ($c(j)$ and we take

$$r(j) = \left\langle \begin{array}{ll} -c(j) & \text{if} \quad c(j) < 0 \\ 0 & \text{otherwise} \end{array} \right. \tag{5.65}$$

to be the response function (data similar to Fig. 5.53(b)). After application of the deconvolution algorithm for $\sigma_d = 5$ we obtain data shown in Fig. 5.65(a). One can observe decomposition of main ridges to their components. After their identification and backward transformation to the space of original spectrum, we present the result in Fig. 5.65(b). Apparently, the deconvolution operation decomposes the clusters of ridges.

The last free parameter in the estimation of ridges is threshold value, which influences the sensitivity of the algorithm. So far, we have processed all the data with threshold value equal to 4%. Furthermore, one can change the starting point of the slicing according to Fig. 5.66. By now, we have used the point A as the starting point of slicing.

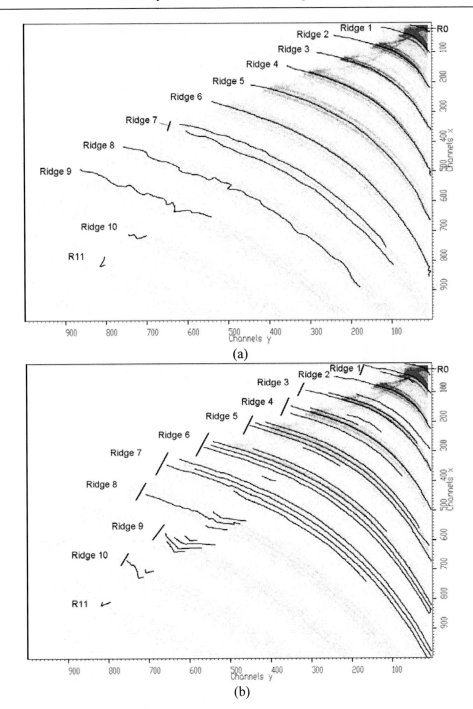

(a)

(b)

Figure 5.64. Two-dimensional spectrum of nuclear multifragmentation with ridges estimated with parameters $\sigma_4 = 10$ (a) and $\sigma_4 = 3$ (b) ($\sigma_1 = 20, \sigma_2 = 30, \sigma_3 = 30$).

In Fig. 5.67, we moved the starting point of slicing to the point A'. In the example we increased the length of the square to $l' = 2N$ and we changed the threshold value to 1%.

One can observe that the identified ridges are smoother and the algorithm discovers also the ridge 11.

Until now we have analyzed two-dimensional spectrum representing the telescope, consisting of two silicon detectors, given in Fig. 5.49. In Figs. 5.68 and 5.69 we present two other results of the identification of ridges in the telescope, consisting of one silicon detector followed by thick CsI scintillator crystal. Again, good fidelity of the estimates of ridges can be observed.

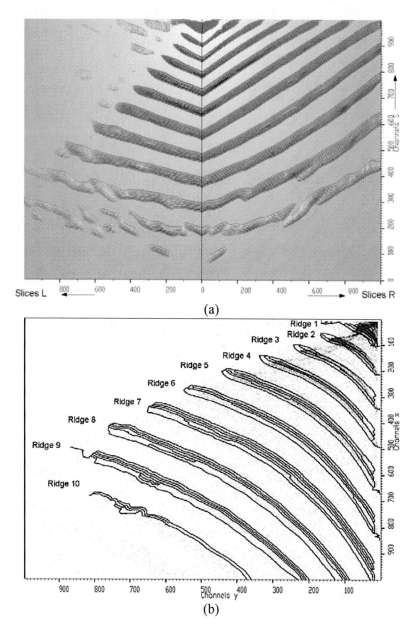

Figure 5.65. Data from Fig. 5.59 ($\sigma_1 = 20, \sigma_2 = 30, \sigma_3 = 30, \sigma_4 = 20$) after application of deconvolution operation to vertical slices ($\sigma_d = 5$) (a) and estimated decomposed slices transformed back to the two-dimensional spectrum of nuclear multifragmentation (b).

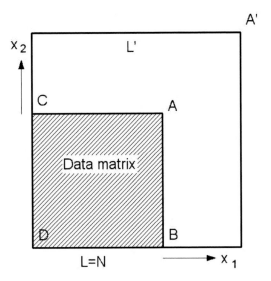

Figure 5.66. Outline of choosing of starting point of slicing in two-dimensional spectrum.

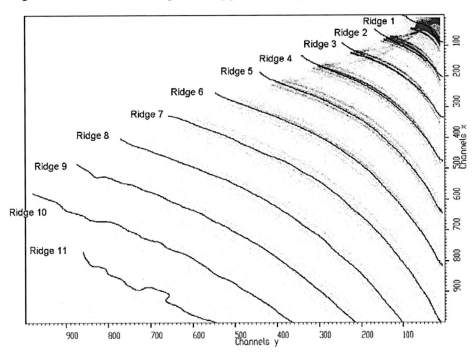

Figure 5.67. Two-dimensional spectrum of nuclear multifragmentation with ridges estimated (threshold=1%) with slicing started at the point A' ($\sigma_1 = 20$, $\sigma_2 = 30$, $\sigma_3 = 30$, $\sigma_4 = 20$).

Sometimes, e.g. for the two-dimensional spectrum, obtained using CsI scintillator crystal via pulse shape discrimination technique, due to the course of the scattered data from the experiment it is necessary to change the starting point of the slicing. In the last example (fig. 5.70), according to visual adjustment of the data, we have moved it to the point C (see Fig. 5.60). Again good agreement of the estimated ridges with experimental measurements can be observed.

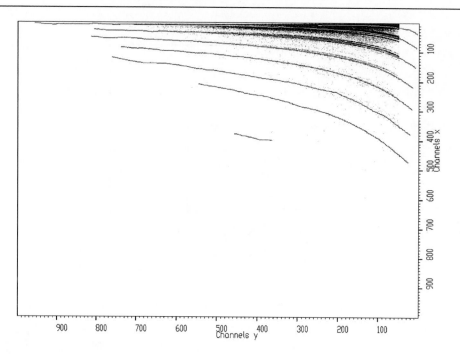

Figure 5.68. Example of two-dimensional spectrum of nuclear multifragmentation with estimated ridges ($\sigma_1 = 20, \sigma_2 = 30, \sigma_3 = 30, \sigma_4 = 5$).

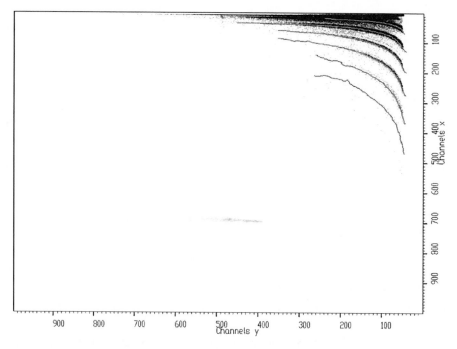

Figure 5.69. Another example of two-dimensional spectrum of nuclear multifragmentation with estimated ridges ($\sigma_1 = 10, \sigma_2 = 10, \sigma_3 = 10, \sigma_4 = 5$).

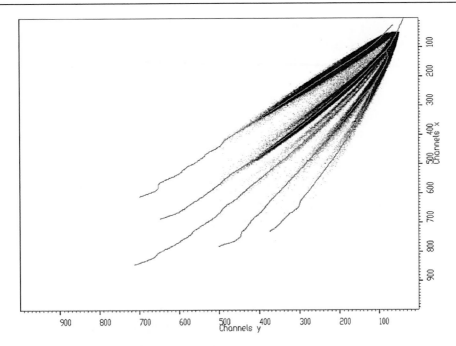

Figure 5.70. Two-dimensional spectrum of nuclear multifragmentation with ridges estimated with slicing started at the point C ($\sigma_1 = 20, \sigma_2 = 30, \sigma_3 = 30, \sigma_4 = 10$).

5.3. Identification of Rings in Spectra from RICH Detectors [91]

A serious task in many experimental setups is the identification of detected charged particles. With the advancement of position sensitive low energy photon detectors with high quantum efficiency, a ring imaging Cherenkov detector (RICH) has been employed in an increasing number of experiments. RICH detector is based on the direct measurement of the Cherenkov angle by measuring emitted Cherenkov photons, which are optically focused on position sensitive photon detectors. The trajectory of a particle with momentum above the Cherenkov photon emission threshold is represented by a ring in a RICH photon detector. The suitable design of the radiator allows particles with a very wide range of momentum to be identified. Detected rings are far from ideal, since the number of detected Cherenkov photons is usually rather low because of radiator design and the low quantum efficiency of the photon detector. The final granularity of the position sensitive photon detector also contributes to the distortion of detected rings. Ring recognition is therefore a serious problem, especially when a fast decision is required, as is usually the case when information from RICH detector is used as a trigger. In this section, we present a method for identifying Cherenkov rings, based on the Gold deconvolution algorithm.

As opposed to the identification of self-standing peaks (one-, or multidimensional, e.g. in γ-ray spectra) in nuclear multi-fragmentation the correspondence between identified local maxima belonging to the same ridge must be found. The algorithm, allowing to automatically recognize isotope lines (manifested as non-linear ridges in the two-dimensional spectra) in the two-dimensional energy loss spectra of charged particles was developed in [85].

Another specific problem in the analysis of objects bearing physical information is the identification of rings, and in general pattern recognition, in the data from Ring Imaging Cherenkov (RICH) detectors. RICH detectors utilize the fact that in some materials charged particles may pass at velocities higher than the speed of light in these materials, leading to emission of Cherenkov radiation (photons) at a constant angle with respect to the direction of flight of the charged particle. After projecting such photons onto a plane covered with photo-sensitive detectors one can observe characteristic ring images, which need to be identified in order to extract the physical properties of the particle.

An algorithm for the recognition of Cherenkov rings based on a triangulation method has been presented in [92]. In [93] and [94] algorithms based on a Metropolis-Hastings Markov chain Monte Carlo sampling to locate rings in experimental data were described. Pattern recognition techniques for charged hadron identification at LHC based on hidden track recognition were described in [95]. A high performance pattern-recognition algorithm using a global likelihood approach [96] has been developed for RICH offline reconstruction [97]. Other kinds of algorithms for the recognition of Cherenkov patterns employ techniques based on the Hough transform method [98] and [99].

An alternative algorithm for pattern recognition applied to identify Cherenkov rings is proposed in the present work. The algorithm is based on the Gold deconvolution method [66], [67] and [68]. The operation of deconvolution can substantially improve resolution in the resulting data, e.g. identification of subridges in the spectra of nuclear multi-fragmentation [85].

5.3.1. Proposal of the Algorithm

The relation between the input value of a linear invariant discrete system and its output can be described by convolution sum

$$y(i) = \sum_{k=0}^{i} x(k)h(i-k) + n(i) = x(i) * h(i) + n(i), \ i = 0,1,...,N-1 \quad (5.66)$$

where $x(i)$ is the input into the system, $h(i)$ is its impulse function (response), $y(i)$ is the output from the system, $n(i)$ is additive noise and '*' denotes convolution operation. In matrix form this system is

$$\boldsymbol{y} = H\boldsymbol{x} + \boldsymbol{n}, \quad (5.67)$$

where the matrix H has dimension $N \times M$, vectors $\boldsymbol{y}$, $\boldsymbol{n}$ have length N and vector $\boldsymbol{x}$ has length M, while $N \geq M$ (overdetermined system). For an invariant convolution system the columns of the matrix H are represented by the response mutually shifted by one position

$$
H = \begin{bmatrix}
h(0) & 0 & \cdots & 0 \\
h(1) & h(0) & \cdots & 0 \\
\vdots & \vdots & \cdots & \vdots \\
h(L-1) & h(L-2) & \cdots & 0 \\
0 & h(L-1) & \cdots & h(0) \\
0 & 0 & \cdots & h(1) \\
\vdots & \vdots & \cdots & \vdots \\
0 & 0 & \cdots & h(L-1)
\end{bmatrix},
\tag{5.68}
$$

where $N = M + L - 1$. This means that the response takes the same form for all columns. For a general one-dimensional system of linear equations, formula (5.66) takes the form of

$$
y(i) = \sum_{k=0}^{i} x(k) h(i,k) + n(i), \quad i = 0,1,...,N-1.
\tag{5.69}
$$

Analogously for a two-dimensional linear invariant discrete system it holds that

$$
\begin{aligned}
y(i_1,i_2) &= \sum_{k_1=0}^{i_1} \sum_{k_2}^{i_2} x(k_1,k_2) h(i_1 - k_1, i_2 - k_2) + n(i_1,i_2) \\
&= x(i_1,i_2) * h(i_1,i_2) + n(i_1,i_2)
\end{aligned}
\tag{5.70}
$$

where $i_1 = 0,1,...,N_1 - 1$, $i_2 = 0,1,...,N_2 - 1$. Now the matrices $\mathbf{y}$, $\mathbf{n}$ have the size $N_1 \times N_2$ and the matrix $\mathbf{x}$ has the size $M_1 \times M_2$, while $N_1 \geq M_1$, $N_2 \geq M_2$ (overdetermined system). For invariant convolution system the columns of the block matrix H are composed of the submatrices mutually shifted by one position

$$
H = \begin{bmatrix}
H_1(0) & 0 & \cdots & 0 \\
H_1(1) & H_1(0) & \cdots & 0 \\
\vdots & \vdots & \cdots & \vdots \\
H_1(L_2-1) & H_1(L_2-2) & \cdots & 0 \\
0 & H_1(L_2-1) & \cdots & H_1(0) \\
0 & 0 & \cdots & H_1(1) \\
\vdots & \vdots & \cdots & \vdots \\
0 & 0 & \cdots & H_1(L_2-1)
\end{bmatrix},
\tag{5.71}
$$

and

$$H_1(i_2) = \begin{bmatrix} h(0,i_2) & 0 & \cdots & 0 \\ h(1,i_2) & h(0,i_2) & \cdots & 0 \\ \vdots & \vdots & \cdots & \vdots \\ h(L_1-1,i_2) & h(L_1-2,i_2) & \cdots & 0 \\ 0 & h(L_1-1,i_2) & \cdots & h(0,i_2) \\ 0 & 0 & \cdots & h(1,i_2) \\ \vdots & \vdots & \cdots & \vdots \\ 0 & 0 & \cdots & h(L_1-1,i_2) \end{bmatrix}, \qquad (5.72)$$

where $N_1 = M_1 + L_1 - 1$, $N_2 = M_2 + L_2 - 1$. Again, for a general two-dimensional system of linear equations, (5.70) can be written as

$$y(i_1,i_2) = \sum_{k_1=0}^{i_1} \sum_{k_2}^{i_2} x(k_1,k_2) h(i_1,k_1,i_2,k_2) + n(i_1,i_2). \qquad (5.73)$$

One of the most efficient deconvolution methods is the Gold algorithm [66], [67] and [68]. Its solution is always positive when the input data are positive, which makes the algorithm suitable for use with naturally positive definite data, i.e., spectroscopic data. For one-dimensional data it takes the form

$$x^{(n+1)}(i) = \frac{y'(i)}{\sum\limits_{m=0}^{M-1} A_{im} x^{(n)}(m)} x^{(n)}(i), \; i \in \langle 0, N-1 \rangle, \qquad (5.74)$$

where $A = H^T H$, $y' = H^T y$, n represents the number of iterations and $x^{(0)} = [1,1,...,1]^T$. For the elements of the matrix A and the vector y' one can write

$$A(i_1,i_2) = \sum_{j=\max(i_1,i_2)}^{M-1+\min(i_1,i_2)} h(j-i_1)h(j-i_2), \quad i_1,i_2 \in \langle 0, M-1 \rangle \qquad (5.75)$$

$$y'(i) = \sum_{j=0}^{L-1} h(j) y(i+j), \; i \in \langle 0, M-1 \rangle. \qquad (5.76)$$

The algorithm converges to the least squares estimate in the constrained subspace of positive solutions.

Analogously to (5.74) for two-dimensional data, the Gold deconvolution algorithm is

$$x^{(n+1)}\left(i_1, i_2\right) = \frac{y'\left(i_1, i_2\right)}{\sum_{m_2=0}^{M_2-1}\sum_{m_1=0}^{M_1-1} A(i_1 N_1 + i_2, m_1 N_1 + m_2) x^{(n)}\left(m_1, m_2\right)} x^{(n)}\left(i_1, i_2\right), \quad (5.77)$$

where

$$A(m_1, m_2) = \sum_{j_1=\max(i_1,k_1)}^{M_1-1+\min(i_1,k_1)} \sum_{j_2=\max(i_2,k_2)}^{M_2-1+\min(i_2,k_2)} h(j_1 - i_1, j_2 - i_2) h(j_1 - k_1, j_2 - k_2), \quad (5.78)$$

and

$$m_1 = i_1 + i_2 M_1, \quad m_2 = k_1 + k_2 M_1, \quad i_1, k_1 \in \langle 0, M_1 - 1 \rangle, \quad i_2, k_2 \in \langle 0, M_2 - 1 \rangle. \ (5.79)$$

Then

$$i_1 = m_1 (\mathrm{mod}\ M_1), \quad i_2 = \left[\frac{m_1}{M_1}\right], \quad k_1 = m_2(\mathrm{mod}\ M_1), \quad k_2 = \left[\frac{m_2}{M_1}\right], \quad (5.80)$$

and $[\ ]$ denotes integer part of division. Analogously to (5.76) it holds that

$$y'(i_1, i_2) = \sum_{j_1=0}^{L_1-1}\sum_{j_2=0}^{L_2-1} h(j_1, j_2) y(i_1 + j_1, i_2 + j_2), i_1 \in \langle 0, M_1 - 1 \rangle, i_2 \in \langle 0, M_2 - 1 \rangle. (5.81)$$

From the computational point of view, Gold deconvolution is rather time-consuming operation. This problem becomes relevant with large two-dimensional spectra. Therefore the implementation of this method requires optimization. The algorithm for one-, and two-dimensional deconvolution resides in the calculation of the vector and matrix $y'(i)$, $A(m)$, respectively, and matrices $y'(i_1, i_2)$, $A(m_1, m_2)$, respectively, before starting iterations, and in successive corrections of the resulting array $\mathbf{x}$. If either $y'(i_1, i_2) = 0$ or $x^{(n)}(i_1, i_2) = 0$ no correction is necessary at this point. If it is not the case, correction is done by multiplying matrix $\mathbf{A}$ with the particular solution $\mathbf{x}^{(n)}$. However, in practice the length of the response is much smaller than the lengths of both input and output arrays. For example if the length of the response is L in (5.75) a $2L-1$-diagonal matrix, A, is obtained. Outside this region zeros are obtained and can be omitted from multiplication in the denominator of (5.74). These facts allow for a substantial reduction in the number of required operations. Optimization of multidimensional Gold deconvolution is analyzed thoroughly in [100]. The optimized Gold deconvolution algorithm allows the completion of the calculation in a reasonable amount of time even for large two-dimensional spectra from RICH detectors with the sizes of hundreds or thousands channels in each dimension.

Moreover the response matrix is almost empty. It contains non-zero values just around the perimeter of the ring. Further reduction of the number of multiplications in the denominator of (5.74) can be achieved by omitting multiplications with zeros.

Iterative positive definite Gold deconvolution converges to a stable state. It is useless to increase the number of iterations; the result obtained practically does not change as it does not lead to sufficient improvement in the resolution and suppression of undesired residues in the deconvolved data. Instead of it, we can stop the iterations, apply a boosting operation and repeat this procedure. In other words, when the solution reaches a stable state it is necessary to change the particular solution $x^{(L)}$ in a particular way and repeat the deconvolution [68]. To change the relationship among elements of the particular solution a non-linear boosting function must be applied to it. The power function proved to work satisfactorily. The algorithm for boosted two-dimensional Gold deconvolution is as follows:

1.   Set the initial solution $x^{(0)}(i_1, i_2) = 1, \quad i_1 \in \langle 0, M_1 - 1 \rangle, i_2 \in \langle 0, M_2 - 1 \rangle$.

2.   Set required number of repetitions R and iterations I

3.   Set the number of repetitions $r = 1$.

4.   According to Eqs. 5.77-5.81 for $n = 0,1,\ldots,I-1$ find solution $x^{(I)}$.

5.   If $r = R$ stop the calculation, else
     a.   apply boosting operation, i.e., set

$$x^{(0)}(i_1, i_2) = \left[x^{(I)}(i_1, i_2) \right]^p, \text{ where } i_1 \in \langle 0, M_1 - 1 \rangle, i_2 \in \langle 0, M_2 - 1 \rangle \quad (5.82)$$

     and p is boosting coefficient >0,
     b. $r = r + 1$,
     c.   continue to 4.

5.3.2. Study and Properties of the Algorithm

Let us investigate the behavior of the algorithm in the presence of noisy input data. Let us take an input matrix composed of 1000 randomly distributed hits arranged in 50 randomly positioned circles (Fig. 5.71).

When applying the Gold deconvolution algorithm (5.77) to these data the result given in Fig. 5.72(a) was obtained. From these data, one can easily identify the centers of 50 circles. Following reverse convolution of the deconvolved data with the response, the smoothed circles shown in Fig. 5.72(b) were obtained. One can observe good accordance between the smoothed circles and the original data in Fig 5.71.

However, when looking at the deconvolved data in more detail (see Fig. 5.73(a)) one can observe some very small peaks due to the present noise as well as that the obtained peaks, which determine the positions of circles, have the widths of several channels. When the boosted deconvolution algorithm ($p = 5, I = 4, R = 5$) was employed the small peaks completely disappear and the peaks determining the centers of circles approach δ-functions – Fig. 5.73(b). In some points, the centers are represented by two neighboring channels meaning that the center is located between them.

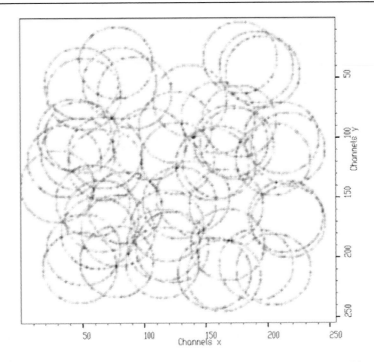

Figure 5.71. Synthetic spectrum composed of 1000 randomly distributed hits arranged in 50 randomly positioned circles.

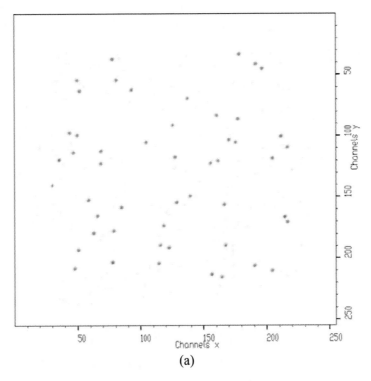

(a)

Figure 5.72. Continued on next page.

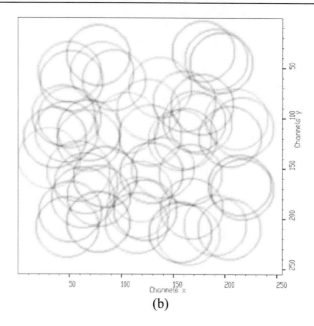

(b)

Figure 5.72. Spectrum from Fig. 5.71 after Gold deconvolution (number of iterations = 20) (a), and after convolution of the result with response (b).

Further, let us study the proposed method of identifying rings depending on the level of noise distributed uniformly in the spectrum. Figs. 5.74 (a) and (b) show a synthetic spectrum composed of 100000 randomly distributed hits arranged in 10 randomly positioned circles with additive noise with a level of 100% of maximum value (in circles without noise) in the three-dimensional and orthogonal view, respectively. Data after boosted deconvolution ($p = 20, l = 5, R = 4$) and convolved data with the response are given in Fig.5.74 (c) and (d). After each boosting operation, a threshold limit 10^{-5} of the maximum data value was applied and all channels with contents smaller than this limit were set to zero.

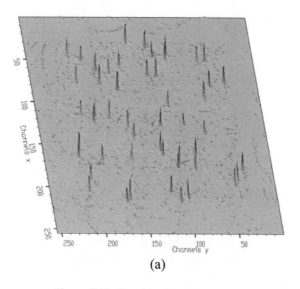

(a)

Figure 5.73. Continued on next page.

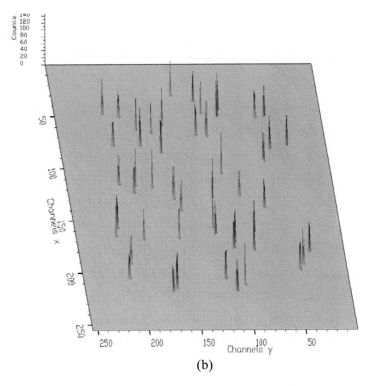

(b)

Figure 5.73. Spectrum from Fig. 5.71 after Gold deconvolution (number of iterations = 20) (a), and after boosted Gold deconvolution ($p = 5, l = 4, R = 5$) (b).

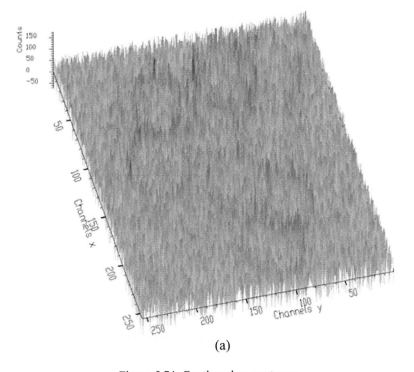

(a)

Figure 5.74. Continued on next page.

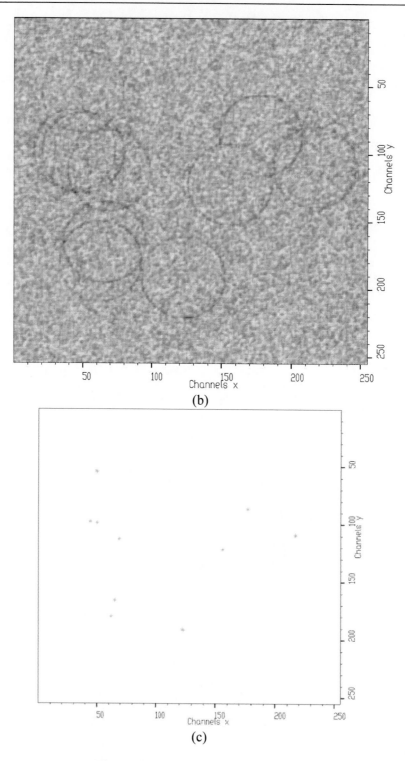

(b)

(c)

Figure 5.74. Continued on next page.

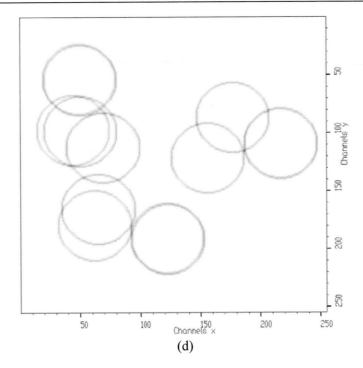

(d)

Figure 5.74. Synthetic spectrum composed of 100000 randomly distributed hits arranged in 10 randomly positioned circles with an additive noise with level of 100% of maximum value, three-dimensional view (a), orthogonal view (b), data after boosted deconvolution ($p = 20, l = 5, R = 4$) (c), and convolved data (d).

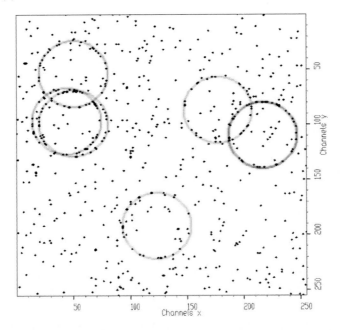

Figure 5.75. Synthetic spectrum composed of 200 randomly distributed hits arranged in 6 randomly positioned circles with 500 additional randomly distributed noise hits and identified circles.

The proposed method correctly identified all 10 circles, even in very noisy data.

In practice, as seen in the next section, besides of hits arranged in circles, the experimental data contain randomly distributed hits. Fig. 5.75 presents a synthetic spectrum composed of 200 randomly distributed hits arranged in 6 randomly positioned circles and 500 added randomly distributed noise hits. The proposed method was robust to these fluctuations and correctly identified the circles in the data.

In addition to randomly distributed hits, some hits in the experimental data are clustered in quasi-Gaussians, i.e. RICH spectra from HADES spectrometer [101]. Next, it was wondered if the proposed method was able to cope with the objects of this type and recognize rings in the presence of both undesired data components. Fig. 5.76 shows the three Gaussians ($\sigma_x = 2, \sigma_y = 2$) added to the data from Fig. 5.75.

Data after boosted deconvolution ($p = 20, l = 5, R = 4$) are given in Fig. 5.77. After each boosting operation, a threshold limit 10^{-10} of the maximum data value was applied and all channels with contents smaller than this limit were set to zero. The algorithm correctly identified the rings in the input data and ignored both randomly distributed hits and clusters of hits.

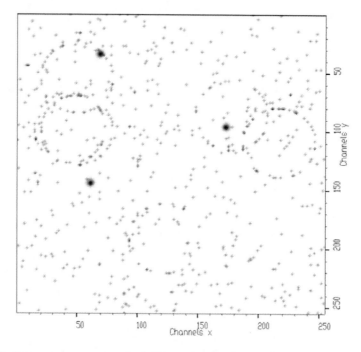

Figure 5.76. Synthetic spectrum composed of 200 randomly distributed hits arranged in 6 randomly positioned circles with 500 additional randomly distributed noise hits and 3 peaks.

The algorithm presented was also tested using more realistic data simulated using the Geant4 simulation package [102]. Among the standard set of examples, an extended example demonstrating the use of optical processes in the simulation was chosen (/examples/extended/optical/LXe) [102], featuring a multi-purpose detector setup implementing a large wall of small photo-multipliers (PMTs) opposite a Cherenkov slab to show the characteristic Cherenkov cone. Simulated individual Cherenkov rings from this example were used to sample two-dimensional spectra with a varying number of Cherenkov rings in order to benchmark the ring recognition algorithms. Examples of Cherenkov rings and rings after convolution with the response are presented in Fig. 5.78(a) and (b), respectively.

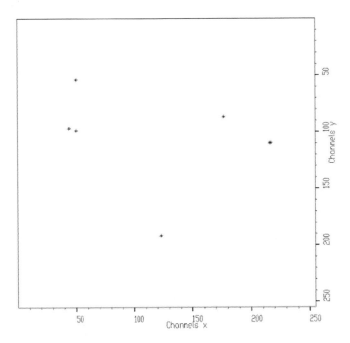

Figure 5.77. Deconvolved spectrum from Fig. 5.76 after complete boosted deconvolution ($p = 20, l = 5, R = 4$).

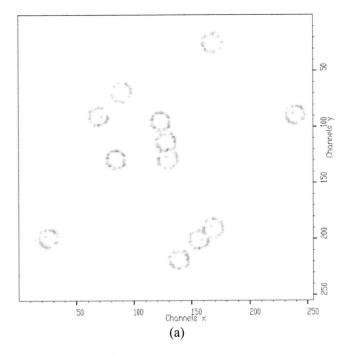

(a)

Figure 5.78. Continued on next page.

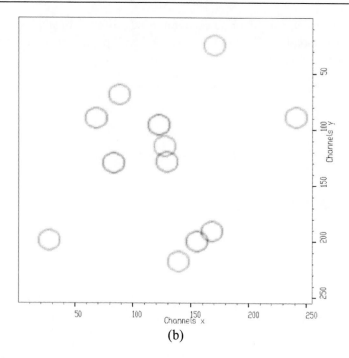

(b)

Figure 5.78. An example of simulated spectrum from GEANT (a), and rings after convolution with the response (b).

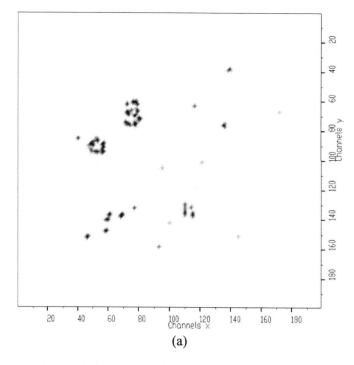

(a)

Figure 5.79. Continued on next page.

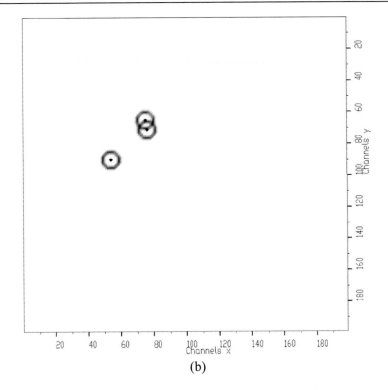

(b)

Figure 5.79. An example of experimental RICH spectrum (a), three identified centers after deconvolution and three rings after convolution with response (b).

The presented algorithm was also used to identify rings in experimental spectra measured with HADES RICH detector [101]. The example containing several rings in the RICH spectrum is shown in Fig. 5.79(a). The spectrum after deconvolution, showing identified centers of the rings and its convolution is shown in Fig. 5.79(b). It can be seen that the algorithm correctly identified the rings and suppressed the noise in the experimental spectrum.

So far, we assumed only circle shape of the objects to be identified. The proposed method is much more universal. It allows identification of the objects of any shape. Let us take, e.g. Lissajous curves [103]

$$x = a\sin(\omega t)$$
$$y = a\sin(\omega t - \pi / 3) \tag{5.83}$$

and

$$x = a\sin(\omega t)$$
$$y = a\sin(2\omega t - \pi / 4). \tag{5.84}$$

Let us generate a synthetic spectrum of 500 randomly distributed events arranged in 10 randomly generated Lissajous curves according to (5.83) and (5.84) and 500 randomly distributed noise events shown in Figs. 5.80 (a) and 5.81 (a), respectively. In Figs. 5.80 (b) and 5.81 (b), we present correctly identified curves using the proposed pattern recognition method and the parameters $p = 20, l = 5, R = 4$.

The algorithm can also be applied to the identification of curves of any, even irregular shape. Such an example of a synthetic spectrum of 500 randomly distributed events arranged in 10 randomly generated curves is shown in Fig. 5.82 (a). Again, in Fig. 5.82 (b), we present correctly identified curves using the proposed pattern recognition method and the parameters $p = 20, l = 5, R = 4$.

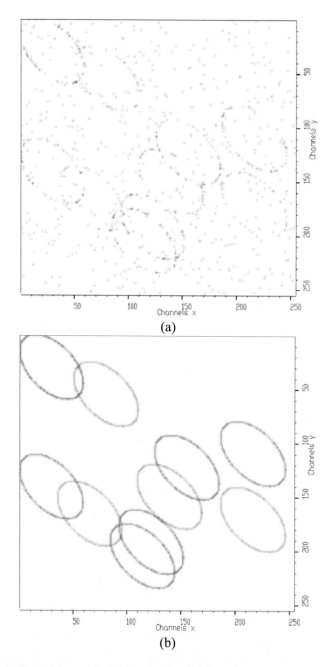

(a)

(b)

Figure 5.80. A synthetic spectrum of 500 randomly distributed events arranged in 10 randomly generated Lissajous curves according to (5.83) and 500 randomly distributed noise events (a) and convolved data after complete boosted deconvolution ($p = 20, l = 5, R = 4$) (b).

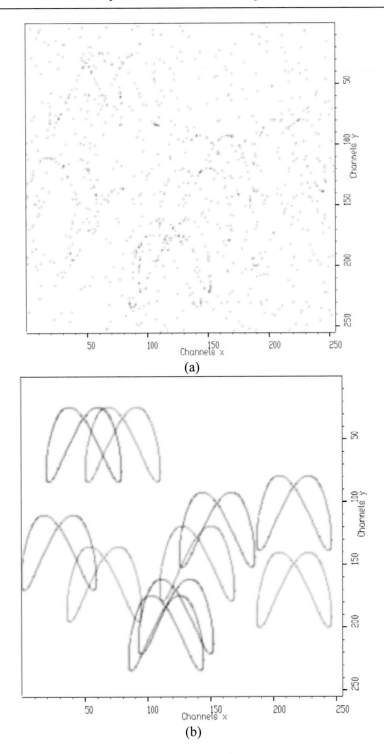

Figure 5.81. A synthetic spectrum of 500 randomly distributed events arranged in 10 randomly generated Lissajous curves according to (5.84) and 500 randomly distributed noise events (a) and convolved data after complete boosted deconvolution ($p = 20, l = 5, R = 4$) (b).

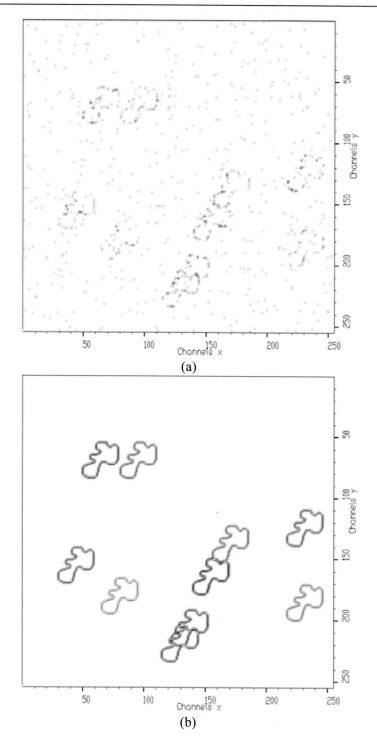

(a)

(b)

Figure 5.82. A synthetic spectrum of 500 randomly distributed events arranged in 10 randomly generated curves of irregular shape and 500 randomly distributed noise events (a) and convolved data after complete boosted deconvolution ($p = 20, l = 5, R = 4$) (b).

6. Fitting

6.1. Introduction

Once we have determined the positions of peaks either in one-, or in multidimensional spectra, we can proceed to the next stage of the analysis - fitting the parameters of the peak shape function. The positions of identified peaks are fed as initial estimates into the fitting procedure.

Fitting procedures are currently used in a large set of computational problems. A lot of algorithms have been developed (Gauss-Newton, Levenberg-Marquart conjugate gradients, etc.) and more or less successfully implemented into programs for analysis of complex spectra [17-19]. Good survey and comparison of different χ^2 statistics and maximum likelihood estimation methods applied to Gauss as well as to Poisson distributed data are given in [20]. Both χ^2 and maximum likelihood based methods turn into well-known successive iterative solution of the set of linear equations. They are based on matrix inversion that can impose appreciable convergence difficulties mainly for large number of fitted parameters.

Peaks can be fitted separately, each peak (or multiplets) in a region or together all peaks in a spectrum. To fit separately each peak one needs to determine the fitted region. However, it can happen that the regions of neighboring peaks are overlapping (mainly in multidimensional spectra). Then the results of fitting are very poor. On the other hand, when fitting together all peaks found in a spectrum, one needs to have a method that is stable (converges) and fast enough to carry out fitting in reasonable time. The gradient methods based on the inversion of large matrices are not applicable because of two reasons

- calculation of inverse matrix is extremely time consuming;
- due to accumulation of truncation and rounding-off errors, the result of the gradient methods based on the inversion of very large matrices can become worthless.

According to numerical analysis, it should be emphasized that the direct inversion of large matrices should be avoided wherever possible [21]. Slavic's algorithm without matrix inversion [8] allows to fit large blocks of data and large number of parameters (several hundreds or thousands) in reasonable time.

6.2. Algorithm without Matrix Inversion [104], [105]

The quantity to be minimized in the fitting procedure for one-dimensional spectrum is defined as:

$$\chi^2 = \frac{1}{N-M} \sum_{i=1}^{N} \frac{[y_i - f(i, a)]^2}{y_i}, \qquad (6.1)$$

where i is the channel in the fitted spectrum, N is the number of channels in the fitting subregion, M is the number of free parameters, y_i is the content of the i-th channel, $\mathbf{a}$ is a vector of the parameters being fitted and $f(i,\mathbf{a})$ is a fitting or peak shape function. Analogously, χ^2 for two-fold coincidence spectra is defined as follows

$$\chi^2 = \frac{1}{N_1 N_2 - M} \sum_{i_1=1}^{N_1} \sum_{i_2=1}^{N_2} \frac{[y_{i_1,i_2} - f(i_1,i_2,\mathbf{a})]^2}{y_{i_1,i_2}}. \tag{6.2}$$

The fitted parameters a_j, $j \in \langle 1,M \rangle$ in the fitting function f in (6.1), (6.2) can be obtained by solving the following equations

$$\frac{\partial \chi^2}{\partial a_j} = 0;\ j \in \langle 1,M \rangle. \tag{6.3}$$

Let us define positive definite derivate squared form quantity

$$c_k = \sum_{i=1}^{N} \left[\frac{\partial f(i,\mathbf{a}^{(t)})}{\partial a_k} \right]^2 \frac{1}{y_i} \tag{6.4}$$

and the derivate-difference mixed form

$$d_k = \sum_{i=1}^{N} \left[\frac{\partial f(i,\mathbf{a}^{(t)})}{\partial a_k} \right] \frac{e_i}{\Delta a_k} \frac{1}{y_i}, \tag{6.5}$$

where $e_i = y_i - f(i,\mathbf{a})$. It holds that $c_k \leq d_k$, i.e.,

$$\sum_{i=1}^{N} \left[\frac{\partial f(i,\mathbf{a}^{(t)})}{\partial a_k} \right]^2 \frac{1}{y_i} \leq \frac{1}{\Delta a_k} \sum_{i=1}^{N} \frac{\partial f(i,\mathbf{a}^{(t)})}{\partial a_k} \frac{e_i}{y_i}. \tag{6.6}$$

From that one can directly write

$$\Delta a_k^{(t+1)} = \alpha^{(t)} \frac{\displaystyle\sum_{i=1}^{N} \frac{e_i^{(t)}}{y_i} \frac{\partial f(i,\mathbf{a}^{(t)})}{\partial a_k}}{\displaystyle\sum_{i=1}^{N} \left[\frac{\partial f(i,\mathbf{a}^{(t)})}{\partial a_k} \right]^2 \frac{1}{y_i}}, \tag{6.7}$$

where the error in the channel i is $e_i^{(t)} = y_i - f(i, a^{(t)})$; $k = 1, 2, ..., M$ and t is the iteration step. The constant $\alpha^{(t)}$ in (6.7) is an arbitrary constant ≤ 1 influencing the speed and convergence of the iterative process. We proposed two algorithms for choosing $\alpha^{(t)}$.

Algorithm A

$\alpha^{(t)} = 1$ if the process is convergent or $\alpha^{(t)} = 0.5\, \alpha^{(t-1)}$ if it is divergent [8].

Algorithm B

1. We set the length I of the interval of the constant $\alpha^{(t)}$ equal to 1;
2. We divide the interval I to appropriate number of subintervals $i \in \langle 0, n-1 \rangle$ (in our case $n = 10$) and for $i \in \langle 0, n-1 \rangle$ we find $\alpha^{(t)}(i_{opt})$ for which χ^2 is minimal.
3. If there is no local minimum within the chosen interval ($i_{opt} = 0$) we set $I = 1/n$ and continue in the point 2.

Obviously, the fitting procedure consists of several steps:

1. choice of suitable optimization fitting algorithm;
2. choice of suitable shape functions of peaks;
3. correct initialization of fitted parameters, i.e., finding peaks and correct estimate of their positions using automatic peak search algorithm. The parameters that are common for all peaks (widths of peaks, correlation coefficients etc.) can be initialized in manual way.

Let us assume that M peaks enter the fitting procedure. Then for one dimensional γ-ray spectrum we employ the following peak shape function

$$f(i, a) = \sum_{j=1}^{M} A(j) \exp\left[-(i - p(j))^2 / 2\sigma^2\right] + b_0 + b_1 i + b_2 i^2, \qquad (6.8)$$

where $A(j)$ is the amplitude of the j-th peak and $p(j)$ is its position. σ represents peak width that is assumed to be the same for all peaks. Finally, the background is represented by the second order polynomial. When the fit is finished the area of the j-th peak can be calculated as

$$V_1(j) = A(j)\, \sigma\, \sqrt{2\pi}. \qquad (6.9)$$

Analogously for two-dimensional peaks, we have chosen the peak shape function of the following form

$f(i_1, i_2, \mathbf{a})$

$$= \sum_{j=1}^{M} \left\{ A_{xy}(j) \exp \left\{ -\frac{1}{2(1-\rho^2)} \left[\frac{(i_1 - p_x(j))^2}{\sigma_x^2} - \frac{2\rho(i_1 - p_x(j))(i_2 - p_x(j)}{\sigma_x \sigma_y} + \frac{(i_2 - p_y(j))^2}{\sigma_y^2} \right] \right\} + \right.$$

$$\left. + A_x(j) \exp \left[-\frac{(i_1 - p_{x_1}(j))^2}{2\sigma_x^2} \right] + A_y(j) \exp \left[-\frac{(i_2 - p_{y_1}(j))^2}{2\sigma_y^2} \right] \right\} + b_0 + b_1 i_1 + b_2 i_2. \quad (6.10)$$

Each peak is represented by one two-dimensional Gaussian with the amplitude $A_{xy}(j)$ and position $p_x(j)$, $p_y(j)$ and two one-dimensional ridges in both directions with the amplitudes $A_x(j)$ and $A_y(j)$. According to our experience the positions of ridges in real coincidence γ-ray spectra slightly differ from the position of two-dimensional peak. This is the reason why we have left the positions of ridges $p_{x_1}(j)$, $p_{y_1}(j)$ independent of $p_x(j)$, $p_y(j)$. The peak widths σ_x, σ_y and correlation coefficient ρ are assumed to be the same for all Gaussians. The background is represented by coefficients b_0, b_1, b_2. The volume of j-th peak can be calculated using

$$V_2(j) = A_{xy}(j) \, \sigma_x \sigma_y \, \sqrt{2^2 \pi^2} \sqrt{(1-\rho^2)}. \quad (6.11)$$

It should be emphasized that very important step in the fitting procedure is the correct finding of the positions of all peaks in the spectrum. The values of peak positions can be used as an initial estimate of the parameters p, p_x, p_y, p_{x_1}, p_{y_1}, p_{x_2}, p_{y_2} in (6.8), (6.10). For the sake of simplicity, let us assume that we have one-dimensional spectrum $S(i)$, $i \in \langle 1, N \rangle$. Then the procedure to analyze and fit γ-ray spectra can be expressed as:

a. finding peak positions $p(j)$, $j \in \langle 1, M \rangle$, in the original spectrum S using the algorithm described in [80]. The amplitudes read-out in the found peak positions can be used as an initial estimate of Gaussian amplitudes, i.e.,

$$A(j) = S\big(p(j)\big).$$

Since the peak amplitudes are in the peak shape function represented as linear parameters such estimate is sufficient. Then we can set common parameters (σ, background). Subsequently the fitting procedure in the original spectrum, i.e., $y_i = S(i)$, $i \in \langle 1, N \rangle$, using one of the above described algorithm can be started.

b. However sometimes the shape of the background is very complicated and for large fitting intervals cannot be described sufficiently by the second order polynomial. In this case before the fitting procedure starts one can determine background $B(i)$ employing the algorithm [43] and subtract it from the original spectrum, i.e.,

$$G(i) = S(i) - B(i), \; i \in \langle 1, N \rangle.$$

Further the pure peaks without background can be found and fitted in the spectrum $G(i)$ in the way analogous to the point a.

c. In γ-ray spectra it frequently happens that the number of peaks is too big and peaks are located very close to each other – they are overlapping. In these situations after the background elimination one can apply the method of deconvolution [67] to the spectrum of pure peaks $G(i)$. It substantially improves the resolution. The deconvolution decomposes multiplets to single peaks and allows the discovery of even hidden peaks in the original spectrum. Thus after the deconvolution we obtain spectrum $D(i)$. In this spectrum one can find peaks again using algorithm presented in [80]. Since the peaks after deconvolution are very narrow, it is possible to find peaks simply by determination of local maxima. However, to avoid an identification of too small peaks we can set a threshold value and consider only the peaks higher than this value. Further one can proceed in an analogous way according to point a. or b., i.e., to apply the fitting procedure either to the original spectrum or to the spectrum of pure peaks after background elimination.

6.3. Discussion and Results

The fitting procedure has been applied to the determination of relative yields of correlated fragment pairs in ternary fission of ^{252}Cf. One example of the results of the fit of the one-dimensional spectrum obtained by gating on ^{138}Xe $2^+ \rightarrow 0^+$ (588.8 keV) transition in γ-γ matrix of coincidences in ^{4}He ternary fission of ^{252}Cf is shown in Fig. 6.1 and Fig. 6.2, respectively. The high accuracy and resolution of the fitting procedure can be observed. The transitions in fragments Mo, their doublets and multiplets are well resolved.

We have extended the fitting procedures to higher dimensions. Let us illustrate results of two-dimensional fitting. In Fig. 6.3, we present original two-dimensional $\gamma - \gamma$ coincidence spectrum (256 x 256 channels) with relatively small number of peaks (15).

General peak search algorithm, presented in previous section and in [80], identifies correctly the peaks in the spectrum. The estimates of their positions are fed into the fitting procedure. After the fit (χ^2=0.8761) we obtain the result presented in Fig. 6.4. One can observe good correspondence between both data.

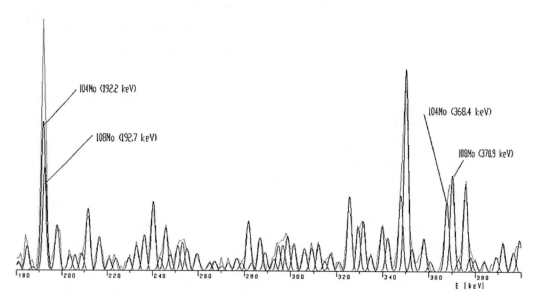

Figure 6.1. One-dimensional slice from γ- γ-coincidence matrix and its fit decomposed to individual peaks.

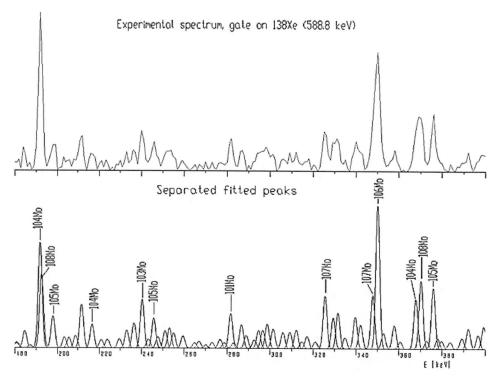

Figure 6.2. Experimental spectrum and separated fitted peaks (below).

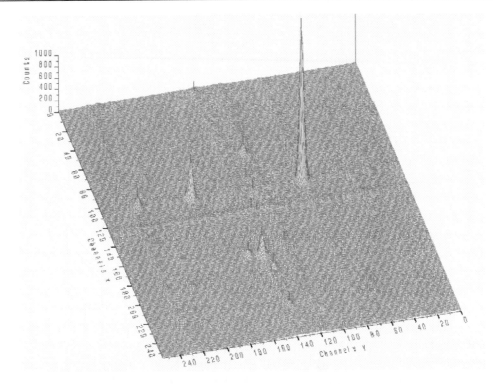

Figure 6.3. Two-dimensional γ- γ-coincidence spectrum with identified peaks (using above-described peak searching procedure) denoted by markers.

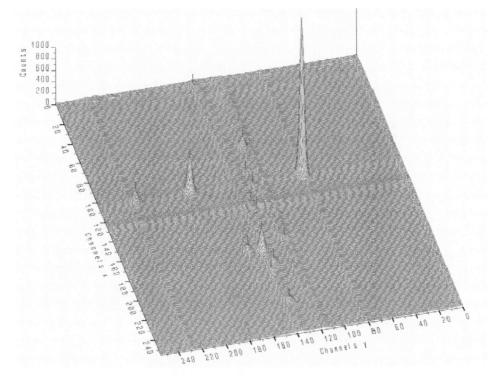

Figure 6.4. Two-dimensional fitted spectrum of the data from Fig. 6.3.

Now, we shall analyze a more complicated two-dimensional spectrum (256 x 256 channels) with a large number of overlapped peaks shown in Fig. 6.5. Using this example let us illustrate the entire procedure of the analysis, which consists of the following steps

- background elimination
- deconvolution – improving the resolution
- identification of peaks
- fit.

Because of the large number of overlapped peaks, to find their positions in the spectrum one needs to employ a more sophisticated peak searching technique. We improve the resolution in the spectrum through the use of the Gold deconvolution method. However, before doing that it is necessary to determine and eliminate continuous background as well as coincidences of the background with peaks in both dimensions (one-fold coincidences), which are represented by the ridges in the spectrum. We employ the algorithm described in [43]. After elimination of the background, we get the spectrum of pure two-fold coincidences is shown in Fig. 6.6.

Then we can deconvolve these two-dimensional peaks using the optimized Gold deconvolution algorithm [67]. We get the spectrum of narrow decomposed peaks shown in Fig. 6.7. The positions of the peaks coincide with the positions of peaks in the original spectrum.

Now, it is a simple task to find local maxims higher than a given threshold value and to feed their positions as initial estimates into the fitting procedure. The perpendicular view of the deconvolved spectrum with markers denoting the found peaks is given in Fig. 6.8. 1108 peaks have been identified in this spectrum.

Figure 6.5. An example of two-dimensional γ- γ-coincidence spectrum with large number of peaks (some of them are overlapping).

Figure 6.6. Spectrum from Fig. 6.5 after background elimination.

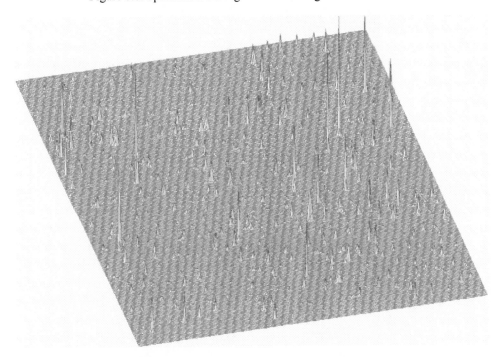

Figure 6.7. Deconvolved spectrum of the data from Fig. 6.6.

According to the peak shape model proposed in [104] (where each peak is represented by 7 parameters) together with σ_x, σ_y and background, it represents 1108 x 7+3=7759 estimated parameters. Thanks to fast algorithms without matrix inversion derived in [104] it can be carried out in a reasonable time. The fitted spectrum after 1000 iterations is shown in Fig. 6.9. It coincides well with the original spectrum given in Fig. 6.5.

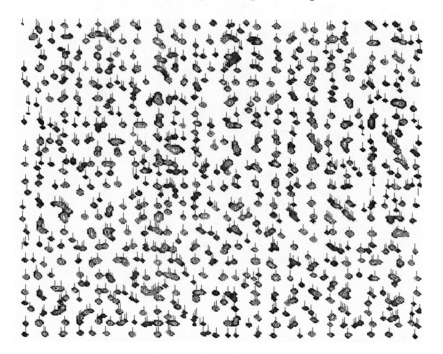

Figure 6.8. Perpendicular view of deconvolved spectrum with identified peaks denoted by markers.

Figure 6.9. Two-dimensional fitted spectrum of the data from Fig.6.5.

Finally, in Fig. 6.10 we present an example of the original three-dimensional spectrum. The cube (64 x 64 x 64 channels) was cut out of the data cube of γ- γ- γ-coincidences from the GAMMASPHERE [1].

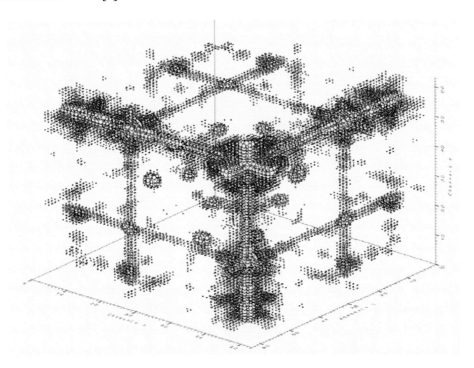

Figure 6.10. Three-dimensional γ- γ- γ-coincidence spectrum.

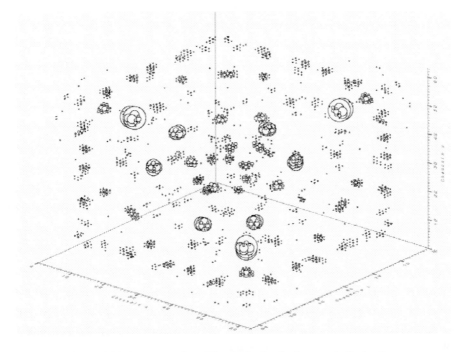

Figure 6.11. Spectrum from Fig. 6.10 after background elimination.

After applying the algorithm of the background elimination for three-dimensional spectra [80], we get pure three-fold coincidence peaks in Fig. 6.11. The estimates of the positions of three-fold coincidences (207 peaks) are fed into the fitting procedure and together with the initial estimate of common parameters they represent 3316 parameters After the fit (200 iterations, $\chi^2 = 1.182$) and background elimination we obtain the result presented in Fig. 6.12. One can observe good correspondence with the data given in Fig. 6.11.

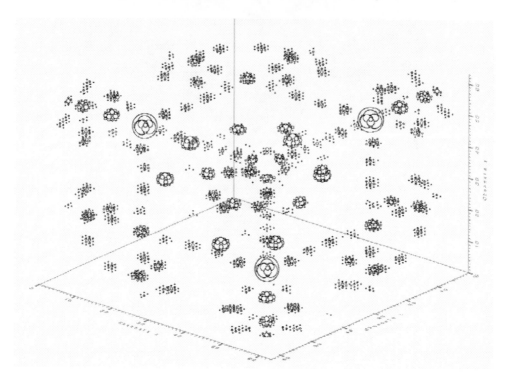

Figure 6.12. Three-dimensional fitted spectrum (after background elimination) of the data from Fig. 6.10.

7.Visualization of Multidimensional Experimental Spectra

7.1. Introduction

Visualization is one of the most powerful and direct ways how the huge amount of information contained in multidimensional histograms can be conveyed in a form comprehensible to a human eye. Visualization is concerned with representing data in a graphical form that improves understanding. The major objective is to provide insight into the underlying meaning of the data. To achieve this goal visualization relies heavily on the human's ability to analyze visual stimuli to convey the information inherent to the data. It is the process of converting scientific data into visual information. Data visualization uses computer graphic effects to reveal the patterns, trends, relationships out of datasets. It is all about understanding ratios and relationships among numbers.

Many techniques have been proposed for visualization of higher dimensional data and applied in various scientific fields. Isosurfaces, slicing, and volume rendering are the three main techniques for visualizing 3D scalar fields on a 2D display. A survey presented in [106], [107] describes the maturation of these techniques since the mid 1980s. Progress in visualizing 3-D structures and generally in scientific visualization of observed and simulated data has been described in [108]. It concentrates on molecular visualization, volume rendering, contour surfaces and flow visualization. In [109] the authors describe and classify the glyph-based techniques with application to medical visualization. The same technique applied for multivariate data is described in [110]. Another approach based on isosurfacing has been presented in [111]. It is based on creating surfaces from the set of points with identical scalar values. The authors suggest generalization of the algorithm for higher-dimensional data. In [112] an interface for isocontouring, which provides the user with a collection of data characteristics such as surface area, volume, and gradient integral, is proposed. In [113] time-varying volumetric data are considered as 4D data and proposed algorithm makes it possible to extract specific space-time features. Hierarchical representation of the same kind of data with multiple levels of detail is given in [114]. Survey of various above mentioned algorithms for visualization of multidimensional, multivariate and time-varying data are presented in [115]-[118].

In [119], [120] a concept of a contour tree as an abstraction of a scalar field that encodes the nesting relationships of isosurfaces has been proposed and studied. Structural enhancement of topology in scalar fields and topology controlled volume rendering based on contour tree segmentation were described in [121] and [122], respectively. Generalized approach to data visualization critical for the correlative analysis of distinct, complex, multidimensional datasets in the space and earth sciences is described in [123]. Other techniques based on low-dimensional projections (up to 3D) are proposed in [124], [125]. Large hyper-volume visualization of physical data and its interactive exploration is highly challenging. In [126] the authors describe a hyperslicing-based interactive visualization technique designed to explore large hyper-volumetric 4D scalar fields. In [127] the authors designed a higher dimensional generalization of the splatting algorithm for 3D volume rendering that provides a global view of scalar fields. They emphasized importance of user interface, which should scale linearly with the dimension of displayed space. Appropriateness of methods and algorithms mentioned above strongly depends on the application and type of data.

In today's nuclear and high-energy experiments the number of detectors being included in the measurements is going up to one hundred or more (e.g. Gammasphere spectrometer [1] is composed of 110 detectors). If during a predefined time interval two or more detectors register particles or γ-ray quanta we say about coincidence event. In the analysis of coincidence measurements we take the data from coincidence events (values from ADCs), we use them as coordinates of the address in multidimensional array (histogram) and increment the addressed memory cell (channel) by one. Consequently, all, independent variables (addresses in histogram) and dependent variable (counts), are positive integers. Theoretically, the number of dimensions can go up to the number of detectors. Moreover, the resolution of employed ADCs goes today up to 14 bits, i.e., 16384. Apparently, the volumes of the data we have to analyze are enormous. It should be stressed that increasing the dimensionality of the measurements can discover new interesting physical phenomena. Consequently, with increasing dimensionality of histograms (nuclear spectra) the requirements in developing of multidimensional scalar visualization techniques become striking. The emphasis in such a tool is mainly on its interactivity. In practice, one can hardly employ any of commercially available products. Therefore, the experimenters have to resort to specialized real time analysis and visualization systems tailored for nuclear physics community.

Historically, from its earliest beginnings to scientific experiments in nuclear physics there is an endeavor to establish effective algorithms and tools to visualize measured, processed and analyzed data. CERN, the European Laboratory for Particle Physics, which is one of the largest scientific research laboratories in the world, plays a leading role in development of packages and software libraries to solve problems related to High Energy Physics (HEP). One of the first systems for graphics developed at CERN was a system GD3 [128]. It was followed by a large software package where graphics plays a key role (PAW—Physics Analysis Workstation) [129]. At present, a large object oriented data analysis framework ROOT, which includes also graphical interface, is being developed in CERN [130]. However, graphical capabilities of the ROOT system are limited to 3D volumetric data and particle scattering display mode.

The aim of the paper is to propose and implement visualization algorithms, which should

- be scalable to higher dimensions (theoretically without limitations),
- allow users interactive handling of the acquired or analyzed data,
- be able to visualize enormous volumes of the histograms on-line,
- allow users to localize multidimensional peaks in the presence of noise and background (originating for example from cosmic rays) and to determine their positions and intensities.

7.2. Direct Visualization Techniques of Scalar Fields

A scalar variable is a single quantity, in the case of nuclear spectra - counts, which can be represented as a function of independent variables - particle energies. Most scalar visualization techniques use a consistent approach across one-, two-, or three-dimensional fields. More recent techniques of the visualization of three-dimensional fields attempt to show the full three-dimensional variations of a scalar variable within a volume field. The goal is to propose a technique that allows one to localize and scan interesting parts (peaks) in multidimensional spectra. Moreover it should permit to find correlations in the data, mainly among neighboring points, and thus to discover prevailing trends around multidimensional peaks.

3D-histograms

One of the techniques to display 3D volumetric arrays (histograms) is creation of an isosurface in the 3D space using marching cube algorithm [131]-[133]. The surface is composed of triangles. Let us divide each cube to five tetrahedrons according to Fig. 7.1(a). Then we can say about marching tetrahedra. Further, let us arrange divisions of neighboring cubes in a way suggested in Fig. 7.1 (b). Then the surface will tend to create closed shapes. In Fig. 7.1 (c) we can see an example of the shape if only one node has the value greater than isosurface value (drawn in red color).

Let us arrange the values (counts) at the vertices of the tetrahedron according to their magnitudes V_1, V_2, V_3, V_4 and let b is chosen isosurface value. Then if $b \geq V_1 \geq V_2 \geq V_3 \geq V_4$ or $V_1 \geq V_2 \geq V_3 \geq V_4 \geq b$ nothing is drawn. If $V_1 \geq b \geq V_2 \geq V_3 \geq V_4$ the triangle outlined in Fig. 7.2(a) is drawn where $l_1 \approx \dfrac{V_1 - b}{V_1 - V_2} \cdot l$. If $V_1 \geq V_2 \geq b \geq V_3 \geq V_4$ we get a quadruple which can be divided to two triangles as illustrated in Fig. 7.2 (b). The situation for $V_1 \geq V_2 \geq V_3 \geq b \geq V_4$ is analogous to that given in Fig. 7.2 (a)

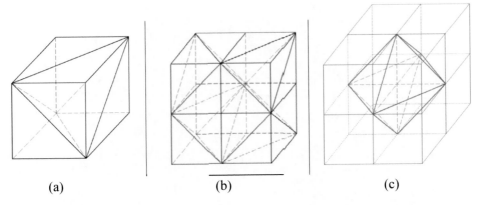

(a)   (b)   (c)

Figure 7.1. Division of a cube to five tetrahedrons (a), division of neighboring cubes (b), and an example of 8 neighboring cubes with only one node greater than the isosurface value (c).

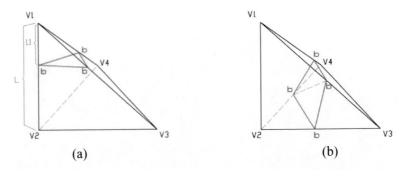

(a)   (b)

Figure 7.2. Red triangle is drawn (a) red quadruple is divided to two triangles, which are drawn (b).

An example of the 3D γ-γ-γ-ray nuclear spectrum displayed in isosurface mode is shown in Fig. 7.3(a) Another technique based on volume rendering ([106]-[108]) combined with volume slicing for the same kind of data is illustrated in Fig. 7.3(b).

As mentioned in the Introduction in nuclear spectra all, independent and dependent variables are positive integers. Isosurface method was originally designed for real variables. However, by setting the boundary b to a real value and interpolating the counts along the edges of the tetrahedra (see Fig. 7.2) we can get continuous transitions between counts in neighboring channels. In this way, the isosurface method becomes applicable also for discrete data in histograms.

Another very frequently employed technique is based on representing the counts using glyphs [109]. As an example, we can mention stick icons implemented in [133], [134]. In the case of the 3D histograms, the counts (scalars) can be represented by various attributes. In the example given in Fig. 7.4(a) the channels (histogram entries) are represented by small spheres (with equal diameter) with the color proportional to the counts contained. In Fig. 7.4(b) the diameter of a sphere is proportional to the counts in the corresponding channel. Sometimes it is called particle scattering display mode [131]. In both cases we can set a range of counts of displayed channels. Channels with the counts outside of this range are not displayed.

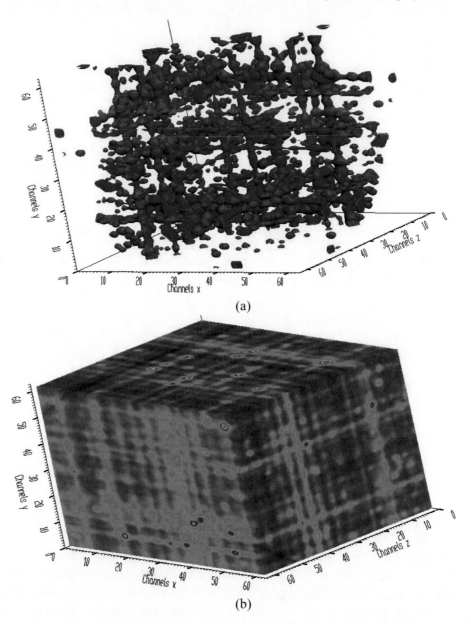

Figure 7.3 Coincidence $\gamma - \gamma - \gamma$ ray nuclear spectrum displayed in isosurface mode (a), volume rendering display mode using volume slicing (b).

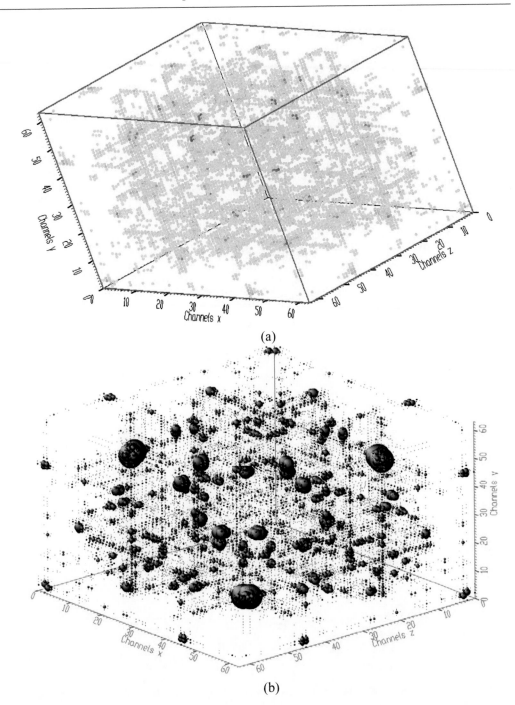

(a)

(b)

Figure 7.4. 3D spectrum with channels shown as spheres with colors proportional to counts (a), and diameters proportional to counts (b).

When processing and analyzing the nuclear spectra a user needs to see details in the spectrum in an interactive way. To satisfy the requirement one can make two or one-dimensional slices in the 3D data and to analyze these distributions. Examples of the display of two- and one-dimensional slices are given in Fig. 7.5(a) and (b), respectively.

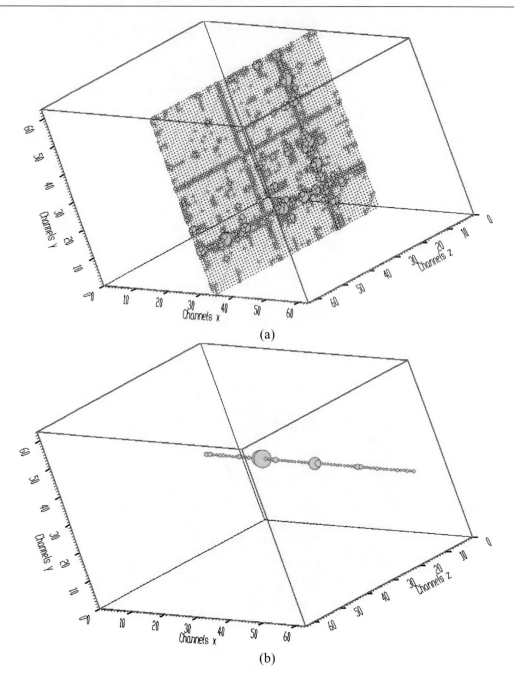

(a)

(b)

Figure 7.5. Two-dimensional (a), and one-dimensional (b) slices in three-dimensional spectrum.

However, to find interesting locations in the spectrum one can move interactively with the slices.

4D-histograms

In 3D spectrum the glyph represents one channel. Now it represents a slice in the fourth parameter, i.e., $c_{i,j,k}(v) = f(x_i, y_j, z_k, v)$, where v, c denote the fourth independent variable and the counts, respectively. We depict each slice as a closed polygon with the center positioned analogously with three-dimensional data [132] at the location

$$x^{\cdot} = t_{11} \cdot i + t_{12} \cdot j + t_{13} \cdot k + t_{14}$$
$$y^{\cdot} = t_{21} \cdot i + t_{22} \cdot j + t_{23} \cdot k + t_{24},$$
(7.1)

where transform coefficients reflect scaling, shifts and rotations of 3D space. For the positions of points on screen belonging to the polygon i, j, k we define

$$x^{\cdot\cdot} = x^{\cdot} - r_{max} \cdot \frac{c_{i,j,k}(v) - c_{min}}{c_{max} - c_{min}} \cdot \cos\left(\frac{2\pi(v - v_{min})}{v_{max} - v_{min} + 1} + \phi_0\right)$$

$$y^{\cdot\cdot} = y^{\cdot} + r_{max} \cdot \frac{c_{i,j,k}(v) - c_{min}}{c_{max} - c_{min}} \cdot \sin\left(\frac{2\pi(v - v_{min})}{v_{max} - v_{min} + 1} + \phi_0\right),$$
(7.2)

where the fourth variable $v \in\, < v_{min}, v_{max} >$, r_{max} (constant value) is maximum distance of the polygon point from its center, ϕ_0 is starting angle of the display of the first point of polygon and c_{min}, c_{max} determine the range of the displayed counts. The principle of the algorithm for one slice in 4D histogram data is depicted in Fig. 7.6(a) The slice is represented by a kiviat diagram. In Fig. 7.6(b) we present an example of display of 4D spectrum with four peaks.

In what follows we present another glyph-based algorithm for visualization of 4D scalar data. In the slice in the fourth dimension one can change the color (level of shading) while keeping the radius of circle constant. According to the resolution in the fourth independent variable, the circle is divided to the pies. The size of the sphere is proportional to the sum of counts in the fourth dimension

$$\sum_{v=v_{min}}^{v_{max}} f(x_i, y_j, z_k, v).$$
(7.3)

An example of 4D spectrum displayed in this mode is given in Fig. 7.7(a)

Finally analogously to isosurface mode (see 3D histograms), for 4D histograms we proposed isovolume display mode. Now the surface is composed from the channels with the same volume (7.3) in the fourth dimension. An example for this kind of display is presented in Fig. 7.7(b).

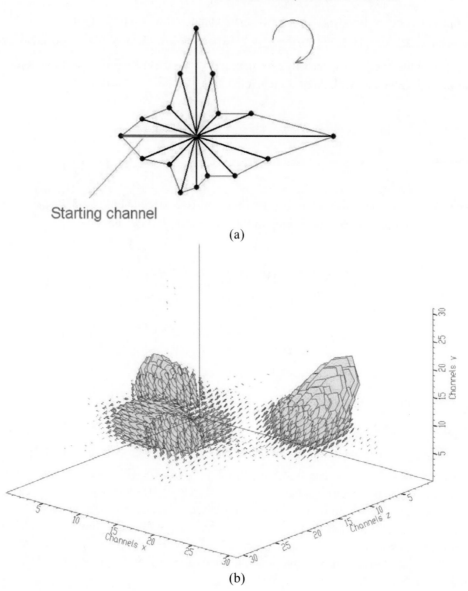

Figure 7.6. Principle of the algorithm for one slice (kiviat diagram) in 4D data (a), and the example of 4D spectrum displayed in this mode (b).

Analogously to 3D data, one can move and analyze interactively slices in the 4D histograms as well.

In this section we presented several algorithms for visualization of 3 and 4D histograms. Unfortunately, none of these algorithms can be extended to higher dimensions. In the next sections we will develop visualization algorithms scalable also to higher dimensions.

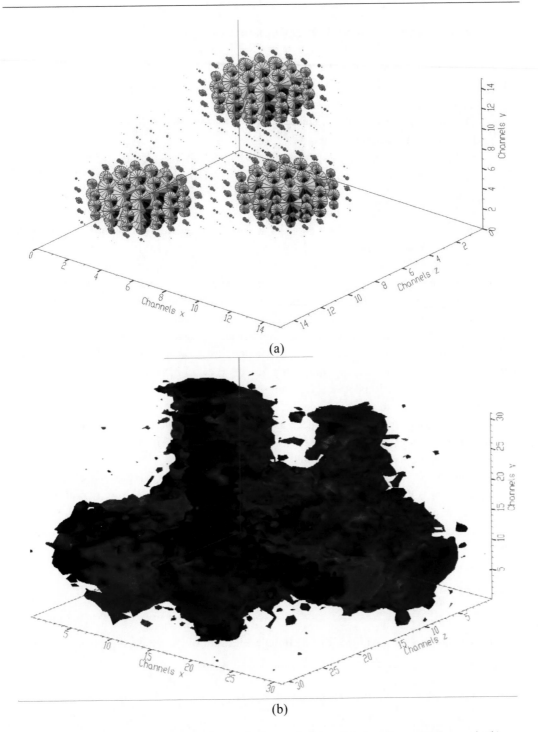

(a)

(b)

Figure 7.7. 4D histogram displayed in pie display mode (a), and in isovolume display mode (b).

7.3. Technique of Successive Projections of Embedded Subspaces

The above-mentioned algorithms can be used even for higher dimensions by employing technique of embedded subspaces. The goal is to propose a technique that allows one

- to localize interesting parts (peaks) in multidimensional spectra,
- to scan multidimensional spectra,
- to find correlations in the data, mainly among neighboring points,
- to discover prevailing trends around multidimensional peaks.

A similar method for the visualization of scalar functions of many variables was described in [136]. The basic idea consists in the representation of a multi-dimensional function as a matrix of orthogonal two-dimensional slices. In [135] the author describes a system named XmdvTool (originally coding from [137]) which integrates several of the most common methods for projecting multivariate data onto a two-dimensional screen and introduces dimension stacking technique. A view enhancement mechanism called an N-dimensional brush allows users to gain insights into spatial relationships over N dimensions by highlighting data, which falls within a user-specified subspace.

The proposed technique makes benefit of specific character and features of nuclear spectra. It utilizes the fact that the interesting objects (peaks) have shape of quasi Gaussians. In enormous multidimensional space the events are distributed very sparsely, which allows to preserve main features of the data even after reducing the dimensionality by employing a projection functional. Successive decreasing the dimensionality makes it possible to determine the positions of appropriate multidimensional peaks. Similar approaches based on low-dimensional projections are employed in [124], [125].

Without loss of generality, we shall assume the reduction of the space up to two-dimensional one. Other alternatives are also possible, but the display of two-dimensional array using perpendicular view allows utilizing screen area the most efficiently. Let us start with three-dimensional spectrum. Let us apply a projection functional reducing dimensionality by one to two-dimensional array, e.g.

$$f^{(1)}(x,y) = F[f(x,y,z)].\tag{7.4}$$

One can employ various types of functionals, e.g. sum of channels contents in a slice

$$f^{(1)}(x,y) = \sum_{z=z_{min}}^{z_{max}} f(x,y,z),\tag{7.5}$$

or maximum in a slice

$$f^{(1)}(x,y) = \max\{f(x,y,z)\},\ z \in <z_{min}, z_{max}>.\tag{7.6}$$

Interpretation of a glyph mentioned in previous section can vary in a very large range. Let us display each channel i, j in the form of glyph with the size proportional to $f^{(1)}(i, j)$. Because of the most efficient way of utilizing the screen, in place of the glyph we choose a rectangle. The rectangle represents a "window" into the subspace. Inside of the rectangle, we can display the slice $f(i, j, z)$, $z \in < z_{min}, z_{max} >$.

Apparently, the technique of embedded subspaces lends itself to generalization for p-dimensional nuclear spectra employing several level merging and projections. Without loss of generality, we shall assume that p is even. Analogously to the above-given relations one can write

$$f^{(1)}(x_1, x_2, \ldots, x_{p-2}) = \sum_{x_{p-1}=x_{(p-1)min}}^{x_{(p-1)max}} \sum_{x_p=x_{pmin}}^{x_{pmax}} f(x_1, x_2, \ldots, x_p)$$

$$f^{(2)}(x_1, x_2, \ldots, x_{p-4}) = \sum_{x_{p-3}=x_{(p-3)min}}^{x_{(p-3)max}} \sum_{x_{p-2}=x_{(p-2)min}}^{x_{(p-2)max}} f^{(1)}(x_1, x_2, \ldots, x_{p-2})$$

$$\vdots$$

$$f^{(j)}(x_1, x_2, \ldots, x_{p-2j}) = \sum_{x_{p-j-1}=x_{(p-j-1)min}}^{x_{(p-j-1)max}} \sum_{x_{p-j}=x_{(p-j)min}}^{x_{(p-j)max}} f^{(j-1)}(x_1, x_2, \ldots, x_{p-2j-2})$$

$$\vdots$$

$$f^{(p/2-1)}(x_1, x_2) = \sum_{x_3=x_{3min}}^{x_{3max}} \sum_{x_4=x_{4min}}^{x_{4max}} f^{(p/2-2)}(x_1, x_2, x_3, x_4),$$

(7.7)

where j is the level of merging.

3D-histograms

Using this technique, we can divide 3D space to outer 2D space and inner 1D subspaces (slices in the third variable). An example of the projection of 3D spectrum to the 2D outer subspace is given in Fig. 7.8(a) The example of a detail from this spectrum is illustrated in Fig. 7.8(b).

The channels in the slices in the third dimension are summed according to (7.5). Then the sizes of rectangles in both Figs. 7.8a and b are proportional to the contents inside of the slice. In Fig. 7.8(a) the display inside of rectangles is senseless because of poor resolution. However, in Fig. 7.8(b) one can see simultaneously the distribution of the two-dimensional projection (green squares) together with the one-dimensional slices inside of the squares (blue one-dimensional spectra). One can observe correlations among neighboring points inside of the rectangles as well as correlation of corresponding points in the rectangles in both **x** and **y** directions.

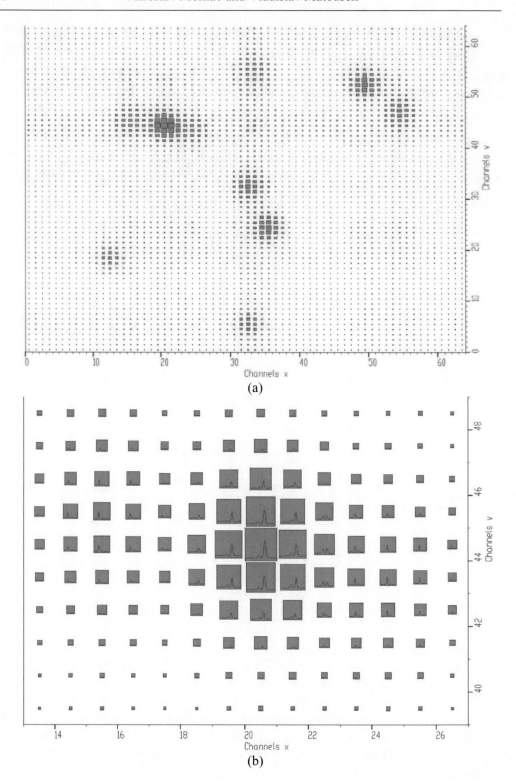

Figure 7.8. 3D spectrum projected to 2D outer subspace xy (a), and detail of the projection with displayed 1D slices in the third dimension (b).

4D-histograms

Now let us divide 4D space to 2D outer space and 2D inner subspaces. To every point of the 2D outer space there belongs one 2D inner subspace. An example of the projection of 4D histogram to the 2D outer subspace is shown in Fig. 7.9(a) A detail of the spectrum with the displayed inner subspaces is given in Fig. 7.9(b).

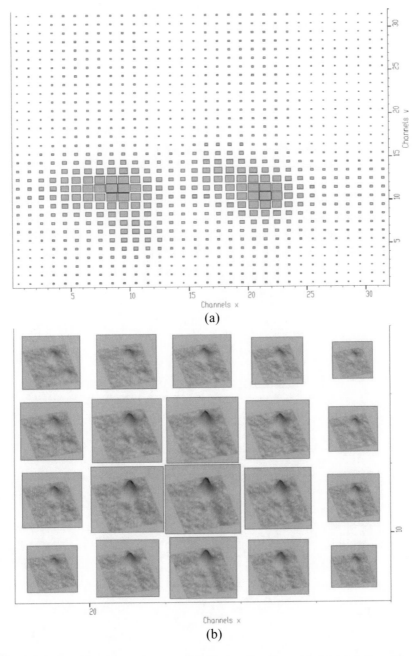

(a)

(b)

Figure 7.9. 4D spectrum projected to 2D outer subspace xy (a), and detail of the projection with displayed 2D slices in the third and fourth dimensions (b).

5D-histograms

5D space can be divided in two ways, i.e., either 2D outer space plus 3D inner subspaces or 2D outer space plus 2D inner subspaces of the first level plus 1D subspaces of the second level. In Fig. 7.10(a) we present 5D histogram projected to 2D outer subspace. A chosen region of interest (ROI) is shown in Fig. 7.10(b) together with the displayed 3D subspaces in isosurface mode.

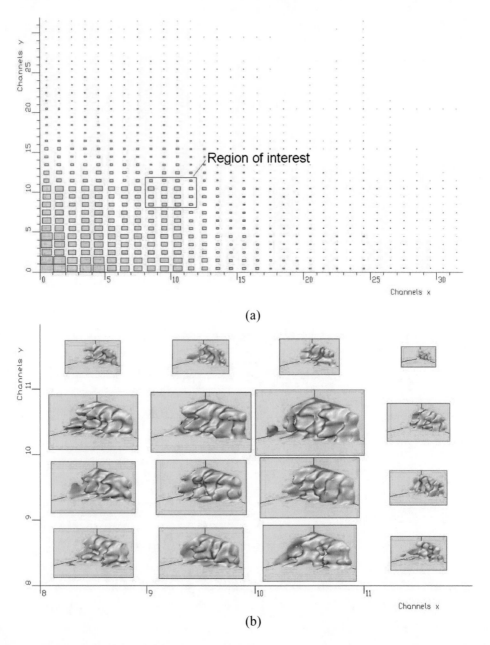

(a)

(b)

Figure 7.10. Outer subspace of 5-fold γ-ray spectrum (a), zoomed region of interest with 3D inner subspaces displayed in isosurface display mode (b).

The alternative way is to proceed in the projections and to divide inner 3D subspaces to 2D plus 1D ones. While yellow rectangles (Fig. 7.10) represent the outer subspace, the light blue ones represent the first level inner 2D subspaces (Fig. 7.11(a)). If we enable the display of the second level subspaces, we can see all three levels simultaneously. One may focus attention to the channel $x = 10, y = 10$ where the volume of data is the biggest (the largest yellow rectangle, see Fig. 7.11(b)).

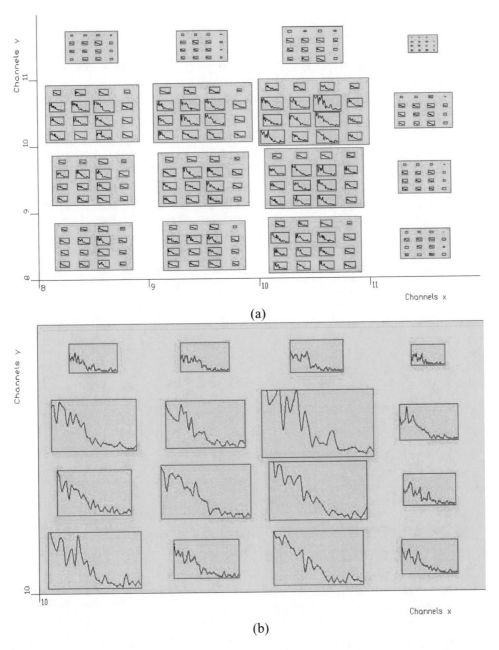

(a)

(b)

Figure 7.11. Zoomed ROI with shown outer, the first level and the second level inner subspaces (a), zoomed one channel of the outer subspace (b).

This technique is recursive and can be extended to any dimension. We can focus to any region of the spectrum (even to one channel) and analyze the distribution in the lower level subspaces. It allows user to choose any combination of axes of outer space or higher level subspaces. Nevertheless, when choosing the display limits some interesting parts can be hidden and missed. It does not provide us with the compact integrated view about the distribution of counts in multidimensional space.

7.4. Hypervolume Visualization

Hypervolume visualization is designed to provide simply and fully explanatory images that give comprehensive insight into the global structure of scalar fields of any dimension [127]. In what follows we shall consider only particle scattering display mode (see section 7.2). The projection of m-dimensional space to 2D space can be defined as mapping

$$\Pi\left(\mathfrak{R}^m \to \mathfrak{R}^2\right),\tag{7.8}$$

which is a linear transformation given in the matrix form as

$$
\begin{bmatrix} x^{\cdot} \\ y^{\cdot} \end{bmatrix} =
\begin{bmatrix} t_{11} & t_{12} & \cdots & t_{1m} & t_{1,m+1} \\ t_{21} & t_{22} & \cdots & t_{2m} & t_{2,m+1} \end{bmatrix}
\begin{bmatrix} i_1 \\ i_2 \\ \vdots \\ i_m \\ 1 \end{bmatrix},\tag{7.9}
$$

where

$$i_1 = \frac{x_1 - x_{1,\min}}{k_1};\ldots i_m = \frac{x_m - x_{m,\min}}{k_m}\tag{7.10}$$

and

$$k_1 = \frac{x_{1,\max} - x_{1,\min}}{n_1};\ldots k_m = \frac{x_{m,\max} - x_{m,\min}}{n_m}.\tag{7.11}$$

$n_1, \ldots n_m$ are numbers of nodes of regular grid and $x_1, \ldots x_m$ are the channels of mD spectrum, where $x_1 \in \left\langle x_{1,\min}, x_{1,\max} \right\rangle; \ldots x_m \in \left\langle x_{m,\min}, x_{m,\max} \right\rangle$. By changing the numbers of nodes, we can change the density and thus the speed of the visualization. This feature is important mainly in on-line applications, e.g. in visualization of the acquired data during acquisition. Transform coefficients $t_{1,i}, t_{2,i}, \quad i = 1, 2, m+1$ reflect shifts, in both original and transformed space, scaling in all dimensions as well as rotations in any direction.

Geometrical Imagination of Multidimensional Space – Evolution from 3D to *m*D

Let us start from the 3D space (cube). Number of visible planes is $\binom{3}{2} = 3$. Let us add

the fourth v-axis to 3D space (Fig. 7.12(a)). We obtain the projection of the coordinate system of the 4D space to the 2D plane and the cube xyz. In the 4D space we can define three other cubes xyv, xzv, yzv (Fig. 7.12(b-d)).

Let us move the cubes from Fig. 7.12 in their free dimensions. We get mesh of points (Fig. 7.13(a)), in which we can hardly get oriented. Further, we have redrawn internal edges (in 3D sense) as dotted lines (Fig. 7.13(b)) and subsequently removed (Fig. 7.13(c)). Finally in Fig. 7.13(d) we have colored visible planes of 4D space. Number of visible planes is $\binom{4}{2} = 6$.

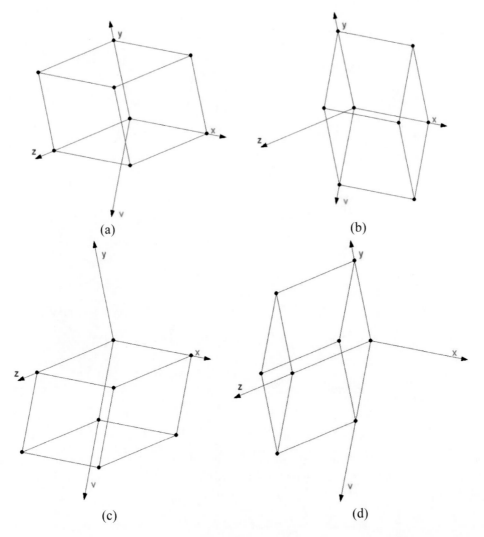

(a)

(b)

(c)

(d)

Figure 7.12. Addition of v-axis to 3D space, cube xyz (a), cube xyv (b), cube xzv (c), cube yzv (d).

We can proceed in analogous way by adding the fifth w-axis to the 4D space (Fig. 7.14). Number of visible planes is $\binom{5}{2} = 10$. Number of invisible planes is the same. The surface of the envelope (in 3D sense) is composed of 20 plates.

Obviously, this algorithm can be extended directly to higher dimensions. Number of visible planes for mD space is $\binom{m}{2}$. The surface of the envelope (in 3D sense) is composed of $m(m-1)$ plates.

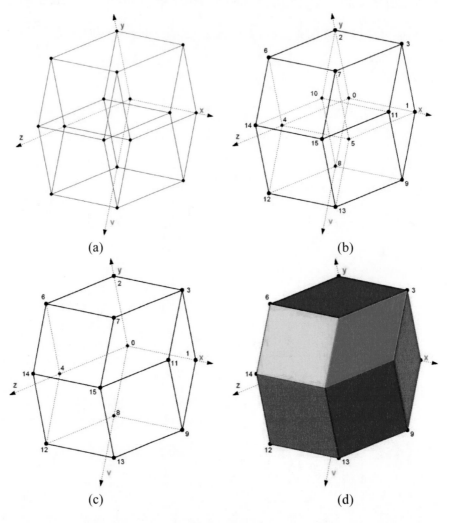

(a) (b)

(c) (d)

Figure 7.13. Creation of 4D space by moving the cubes from Fig. 7.12 in the fourth dimension (a). Invisible edges are drawn as dotted lines (b), 4D space after removal of internal vertices and edges (c), and after coloring of visible planes (d).

Occlusion and Invisibility

In the previous section in 4D and 5D data we mentioned that some vertices and edges are "invisible". We automatically assumed the conception of the real 3D space. From human point of view, this is the highest dimension we can imagine. Therefore when projecting data from 3, 4, 5, …, m-dimensional array we need to transform them not only to 2D plane (screen) but also to 3D space. In 3D space we can decide which vertices are visible and which are not. Let us define 3D space with xy plane coincident with 2D space (screen). Hence, the z-axis gets automatically to be orthogonal to the screen. Then z-coordinates of the vertices give measures of distances from the observer. From them we can also determine which vertices are inside of the 3D space and which are on its surface. In Fig. 7.15 we illustrate the situation for 5D space.

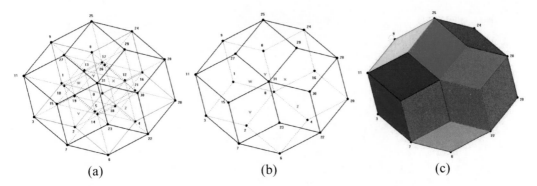

Figure 7.14. 5D space with visible edges drawn as thick lines and invisible edges drawn as dotted lines (a), 5D space with removed internal vertices and edges (b), and after coloring of visible planes (c).

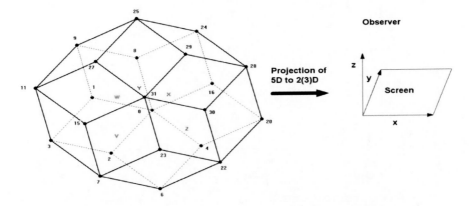

Figure 7.15. Principle of the projection of 5D data to 2D (screen) and 3D spaces.

Examples

In Fig. 7.16(a) we present an example of synthetic 4D Gaussian together with a so-called marker. On the coordinate backplanes, one can see the position of the marker (red crosses). Moving with the marker the user can read out values in interesting points (channels). In Fig. 7.16(b) we give an example of five 4D Gaussians with added ridges, which arise as a consequence of the fact that some detectors fire in the region of a peak and the other ones in the region of background, and noise.

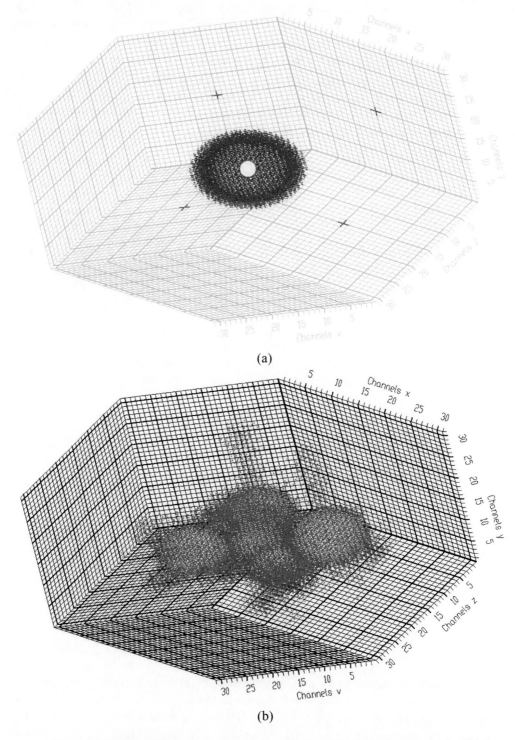

(a)

(b)

Figure 7.16. 4D Gaussian with marker (a), five 4D Gaussians with ridges and noise (b).

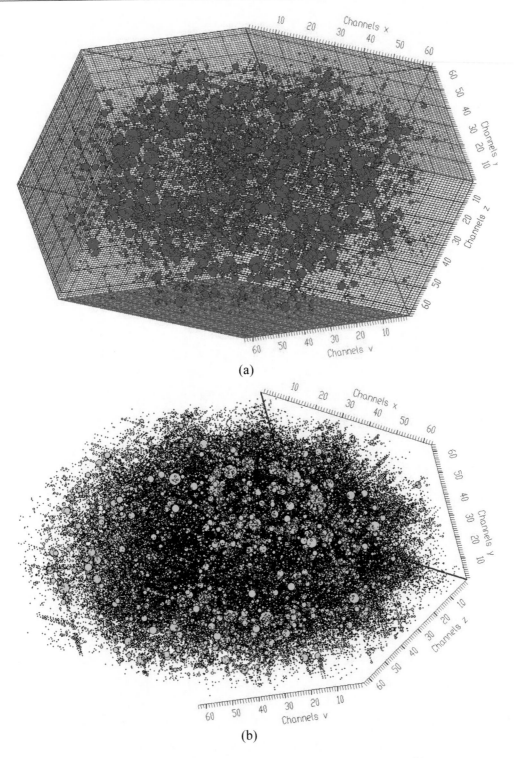

Figure 7.17. Experimental 4D spectrum with (a), and without rasters (b).

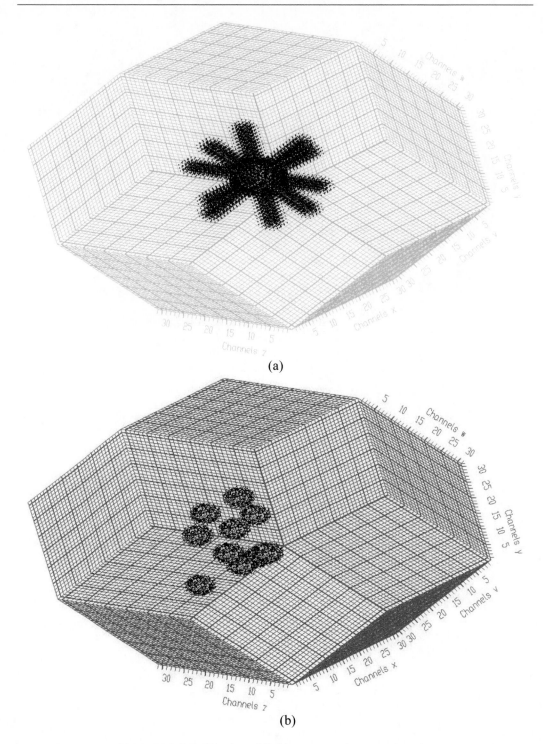

(a)

(b)

Figure 7.18. 5D Gaussian with ridges (a), ten 5D Gaussians (b).

In Figs. 7.17(a), (b) we present experimental 4D spectrum using different display modes and parameters. Again, one can observe ridges in the data in all four directions, which are sometimes called lower order coincidences.

Again, to find interesting locations and to identify peaks in spectra one can utilize various motions during the visualization.

To illustrate 5D visualization in Fig. 7.18(a) and (b) we introduce the display of 5D synthetic Gaussian with ridges and ten 5D Gaussians, respectively.

8. Conclusions

The chapter is devoted to the actual problems connected with the processing of experimental data in the field of nuclear physics. From the subject point of view, the work is divided into several parts, starting from event data sorting up to sophisticated processing and analysis of multidimensional spectra. In the work, on one hand, we propose the extensions and generalizations of the conventional methods for multidimensional data and on the other one, we define new original approaches, methods and algorithms. The proposal of new numerical algorithms covers all theoretical part, implementation aspects and subsequent application for processing of nuclear spectra.

In the work, we have divided the topics of processing of experimental multidimensional nuclear spectra into several items:

- – event data sorting
- – background estimation
- – deconvolution, unfolding
- – identification of peaks and other spectroscopic information carrier objects
- – fitting
- – visualization of multidimensional spectra.

In section 2, we have presented the conventional as well as new developed algorithms of experimental data sorting. An attention must be paid when unfolded events are further decomposed into lower fold events. By using the unsuitable methods, the statistics can become artificially high so that peak positions and statistics may be wrong. The sorting methods presented can:

- – decrease the required storage space,
- – improve the resolution of the analyzed γ-ray spectra.

After thorough analysis, in section 3, we have generalized and extended the existing basic SNIP algorithm for additional parameters and possibilities that make it possible to improve substantially the quality of the background estimation. We have modified the algorithm so that it was possible to select higher orders of clipping filter, to choose the direction (strategy) of changing of clipping window, to give possibility to estimate the background with simultaneous filtration (smoothing). We have included these modifications and derived the algorithms for two-, three-, up to n-dimensional spectra.

For two-dimensional spectra, we have developed new sophisticated algorithms for estimation of the background containing ridges of specific shapes, i.e., different widths of peaks (clipping window) in both directions (e.g. $\gamma - X$-ray spectra), skew ridges and nonlinear ridges. We introduce series of examples illustrating the use of the algorithms of background elimination for all one-, two-, three-, up to four-dimensional spectra.

In the field of processing of spectroscopic information, the deconvolution methods represent an efficient tool to improve the resolution in the data. It is of great importance mainly in the tasks connected with decomposition of overlapped peaks (multiplets). In section 4, we have discussed and analyzed a series of deconvolution methods. At the beginning, we have presented a survey of existing deconvolution methods and we emphasize the relations among them. We focused on possible suitability and applications of the algorithms for positive definite data, i.e., spectroscopic data.

In the next section, we illustrated the behavior of the deconvolution methods using a testing synthetic spectrum. First of all, we have concentrated on the positivity of the solution and on the resolution capabilities of the algorithms. As a conclusion in this section, we have restricted our investigations only to positive definite methods. Though they improve substantially the resolution in the spectra they are not efficient enough to decompose closely positioned peaks.

Therefore, in the subsequent section we proposed boosted deconvolution algorithms. We have studied the decomposition capabilities of boosted Gold, Richardson-Lucy and MAP algorithms, respectively. We have illustrated their properties using the testing synthetic spectrum. We have analyzed and compared achieved results, first and foremost the positions and areas of peaks. We can conclude that all three boosted algorithm are able to decompose the overlapped peaks practically to δ functions while concentrating the peak areas to one channel. Richardson-Lucy and MAP algorithms give slightly better result than one-fold Gold algorithm.

Further, we have investigated the influence of the noise (Gaussian distribution with $\sigma=1$) on the deconvolved data, i.e. robustness of the algorithms. We increased the amplitude of the noise vector added it to the testing spectrum and deconvolved the data using the chosen positive definite algorithms. The algorithms preserved the positions of dominant peaks, but for higher levels of noise, the Richardson-Lucy and MAP algorithms generated non-realistic fake peaks.

We have extended the algorithms to two-, and three-dimensional data. However, the chosen iterative deconvolution algorithms are quite time-consuming. Due to this fact, we studied only one-fold Gold algorithm, which is very simple and its implementation can be easily extended to higher dimensions. We demonstrate the efficiency of boosted deconvolution algorithm on both synthetic and experimental spectra. The result achieved proves in favor of the deconvolution methods employed.

The quality of the analysis of nuclear spectra consists in the correct identification of the existence of peaks and subsequently in the good estimation of their positions. After the analysis of the known peak searching methods in [80], we extended the conventional SSD algorithm to two-, three-, up to n-dimensional spectra. Further, we proposed high-resolution peak searching algorithm based on Gold deconvolution for one-, and two-dimensional spectra. All these algorithms are derived only for a given σ parameter. Though, to some extent, they are robust to the change of this parameter nevertheless, for wide range of σ they fail to work correctly. Therefore, we have developed sophisticated algorithm of σ-range peak searching.

The integral part of experimental data processing is elimination of the influence of noise contained in experimental data to the results of the analysis. When processing experimental nuclear spectra one can meet the problem of the noise suppression at all levels, i.e., starting from background estimation, through deconvolution, peak searching up to fitting and final analysis. Therefore, one has to pay special attention to the problems connected with this topic. We have focused our attention to specific problems born by data filtration in nuclear spectra. In section 5.1.2, we have derived the algorithms of the smoothing based on Markov chains for two and three-dimensional nuclear spectra. The algorithm was generalized also for n-dimensional spectra. The algorithms proposed were tested for both synthetic and experimental γ-ray one-, two-, and three-dimensional spectra. They were incorporated in the methods of peak identification.

An algorithm, allowing to automatically recognize isotope lines (manifested as non-linear ridges in the two-dimensional spectra of nuclear multifragmentation) in the two-dimensional energy loss spectra of charged particles, is proposed in section 5.2. Crucial point of the algorithm is linearization of data in the transformed domain followed by smoothing in both directions and possible deconvolution. The method after proper tuning can further minimize the human intervention, reducing it to supervision of the procedure. Due to its simplicity, the method lends itself for the application during on-line acquisition where one often needs a preliminary fast evaluation of the yield.

A deconvolution based pattern recognition algorithm, allowing automatic ring recognition in two-dimensional spectra from RICH detectors is proposed in the present work. A boosted Gold deconvolution algorithm has been proposed, which makes it possible to concentrate the contents of one ring into one point located in its centre. The proposed method is universal. It allows identification of objects of any shape, e.g. the Lissajous curves or in general curves of irregular shape.

The basic problem of the fitting of γ-ray spectra resides in the existence of a large number of peaks in the spectrum and thus large number of fitted parameters. When fitting two-, and multidimensional spectra in general, additional problems are connected with huge data volumes and consequently with practical realization of such calculations (time, rounding-off and truncation errors etc.). In the work, we have analyzed the existing conventional methods and Algorithm Without Matrix Inversion (AWMI). We proposed several modifications of this algorithm that allow increasing its efficiency. We introduce examples of the fitting of very large number of parameters in one-, two-, and three-dimensional spectra.

Very important problem during the acquisition and interactive analysis of the nuclear spectra is their operative on-line visualization. Basic visualization algorithms of two-, three-dimensional nuclear spectra using surface and volume rendering techniques were presented in [132]. We have proposed several original ways of visualization of two-, three-, and four-dimensional spectra. The illustrating examples of the results of processing are spread throughout the whole work Furthermore, we have derived new technique of visualization of multidimensional spectra based on projections of inserted subspaces. This allows one, in interactive way, to localize interesting parts in the data of this kind.

We have suggested a new algorithm of hypervolume visualization that is based on particle scattering display mode. It is extendible to any dimension and makes it possible to provide user with the compact global view of multidimensional data. To see details one can slice data in any direction and dimension.

The conventional as well as new developed algorithms of data acquisition, sorting, compression processing and visualization were implemented in DaqProVis system [137],

[138] and [139].The presented processing algorithms were employed to analyze data in the framework of the collaboration with the GAMMASPHERE spectrometer [140], [141], [142], [143], [144], [145], and [146]. The processing algorithms (background estimation, peak searching, deconvolution, fitting, and filtration) were also implemented in ROOT system [130] in the form of TSpectrum class, developed in collaboration with CERN, which is widely and extensively used in the community of nuclear and high energy experimental data processing.

References

[1]   In *GAMMASPHERE a National Gamma-Ray Facility, a proposal*; Delaplanque, M.A., Diamond, R.M.; Eds.; LBL Report 5202, 1988.

[2]   Beck, F.A. *Prog. Part. Nucl. Phys.* 1992, 28, 443-461.

[3]   Rossi-Alvarez, C. *Nucl. Phys. News* 1993, 3, 10-13.

[4]   Theisen, C.; Kintz, N.; Stezowski, O.; Vivien, J.P. *Nucl. Instrum. Meth.* 1999, A 432, 249-264.

[5]   Ryan C.G., Clayton E., Griffin W.L., Sie S.H. and Cousens D.R. *Nucl. Instrum. Meth.* 1988, B 34, 396-402.

[6]   Mariscotti, M.A. *Nucl. Instrum. Meth.* 1967, 50, 309-320.

[7]   Slavic, I.A. *Nucl. Instrum. Meth.* 1973, 112, 253-260.

[8]   Slavic, I.A. *Nucl. Instrum. Meth.* 1976, 134, 285-289.

[9]   Slavic, I.A. *Nucl. Instrum. Meth.* 1970, 84, 261-268.

[10]   Coote, G.E. *Nucl. Instrum. Meth.* 1997, B 130, 118-122.

[11]   Sanchez-Avila, C. *Nonlinear Anal-Theor* 1997, 30, 4909-4914.

[12]   van Cittert, P.H. *Z. Phys.* 1933, 298-308.

[13]   Weese, J. et al. *Nucl. Instrum. Meth.* 1996, *A 378*, 275-283.

[14]   Groetch, C.W. *The theory of Tikhonov regularization for Fredholm equations of the first kind,* Pitman, Boston, 1984.

[15]   Tikhonov, A.N.; Arsenin, V.Y. *Solutions of ill-posed problems*, Wiley, New York, 1977.

[16]   Nashed, M.Z. *IEEE Trans. Antenn. Propagat.* 1981, 29, 220-231.

[17]   Gmuca, Š.; Ribanský, I. *Jaderná energie* 1983, 29, 56-59.

[18]   Awaya, T. *Nucl. Instrum. Meth.* 1979, 165, 317-323.

[19]   Phillips, G.W.; Marlow, K.W. *Nucl. Instrum. Meth.* 1976, 137, 525-536.

[20]   Hauschild, T.; Jentschel, M. *Nucl. Instrum. Meth.* 2001, A 457, 384-401.

[21]   Conte, D.S. *Elementary numerical analysis,* McGraw-Hill, New York, 1965.

[22]   Cole J. D. et al. *Using New Fission Data with the Multi-Detector Analysis System for Spent Nuclear Fuel,* INEEL/CON-98-01126 preprint Idaho National Laboratory, 1998.

[23]   Beausang, C.W., et al. *Nucl. Instrum. Meth.* 1995, A364, 560-566.

[24]   Wilson, J.N., et al. *Nucl. Instrum. Meth.* 1997, A399, 147-151.

[25]   Crowell, B., et al. *Nucl. Instrum. Meth.* 1995, A355, 575-581.

[26]   Veselský M. et al. *Progress in Research, 2000-2001,* Cyclotron Institute, Texas A&M University College Station, 2001.

[27]   Groszwendt, B. *Nucl. Instrum. Meth.* 1971, 93, 461-472.

[28]   Kawarasaki, Y. *Nucl. Instrum. Meth.* 1976, 133, 335-340.

[29] Kennet, T.J.; Prestwich, W.V.; Tervo, R.J. *Nucl. Instrum. Meth.* 1981, 190, 313-323.

[30] Westmeier, W. *Nucl. Nucl. Instrum. Meth.* 1981, 180, 205-210.

[31] Clayton E. *PIXAN - The Lucas Heights PIXE Analysis Package*, Australian Atomic Energy Commission, report AAEC-M113, 1986.

[32] Burgess, D.D.; Tervo, R.J. *Nucl. Instrum. Meth.* 1983, 214, 431-434.

[33] Ryder, P.L. In *Proc. Workshop on Energy Dispersive X-ray Spectrometry;* Heinrich, K.F.J.; Newbury, D.E.; Myklebust, R.L.; Fiori, C.E.; NBS Special publ. 604; 1981, 177.

[34] Kajfosz, J.; Kwiatek, W.M. *Nucl. Instrum. Meth.* 1987, B 22, 78-81.

[35] Statham, P.J. *X-ray Spectrom.* 1976, 5, 16-28.

[36] Ralston, H.R.; Wilcox, G.E. In *Modern Trends in Activation Analysis*; Devoe, J.R.; NBS Special Publication 312; 1969, Vol. II, 1238-1243.

[37] Schamber, F.H. In *X-ray fluorescence analysis of Environmental Samples*; Dzubay T.G.; Ann Arbor Science Publishers, Ann Arbor, MI, 1977, 241-257.

[38] Hampton, C.V.; Lian, B.; McHarris, W.C. *Nucl. Instrum. Meth.* 1994, A 353, 280-284.

[39] Bergström, M.; Ekström, P. *Nucl. Instrum. Meth.* 1991, A 301, 132-137.

[40] Radford, D.C. *Nucl. Instrum. Meth.* 1995, A 361, 297-305.

[41] Palameta, G., Waddington, J.C. *Nucl. Instrum. Meth.* 1985, A 234, 476-478.

[42] Radford, D.C. *Nucl. Instrum. Meth.* 1995, A 361, 306-316.

[43] Morháč, M.; Kliman, J.; Matoušek, V.; Veselský, M.; Turzo, I. *Nucl. Instrum. Meth.* 1997, A 401, 113-132.

[44] Morháč, M.; Matoušek, V. *Appl. Spectrosc.* 2008, 62, 91-106.

[45] Nussbaumer, H.J. *Fast Fourier transform and convolution algorithms*, Springer Verlag, Berlin, 1981.

[46] Quittner, P. *Gamma-ray Spectroscopy*, Halsted Press, New York, 1972.

[47] Veselský, M., et al. *Phys. Rev.* 2000, C 62, 646131.

[48] Garcia-Talavera, M.; Ulicny, B. *Nucl. Instrum. Meth.* 2003, A 512, 585–594.

[49] Levitt, D. G. *BMC Clin Pharmacol.* 2003, 3, 1-29.

[50] Los Arcos, J.M. *Nucl. Instrum. Meth.* 1996, A 369, 634-636.

[51] Starck, J.L.; Pantin, E.; Murtagh, F. *PASP* 2002, 114, 1051–1069.

[52] Zou, M.; Unbehauen, R. *Meas. Sci. Technol.* 1995, 6, 482-487.

[53] Morháč, M. *Appl. Math. Comput.* 2008, 202, 1-23.

[54] Morháč, M.; Matoušek, V. *Appl. Math. Comput.* 2005, 164, 155-166.

[55] Morháč, M. *Appl. Math. Comput.* 1994, 61, 135-149.

[56] Press, W.H.; Flannery, B.P.; Teukolsky, S.A.; Vetterling, W.T. *Numerical recipes*, Cambridge University Press, New York, 1986.

[57] Firsov, D.; Lui, S.H. *Appl. Math. Comput.* 2006, 183, 285-291.

[58] Morozov, V.A. *Regularization methods for ill-posed problems*, CRC Press, Boca Raton, 1993.

[59] Tikhonov, A.N.; Goncharsky, A.V. *Ill-posed problems in the natural sciences*, Mir Publishers, Moscow, 1987.

[60] Tikhonov, A. N.; Goncharsky, A. V.; Stepanov, V. V.; Yagola, A. G. *Numerical Methods for the Solution of Ill-Posed Problems;* Kluwer Academic Publishers: The Netherlands, 1995.

[61] Takiya, C.; Helene, O.; do Nascimento, E.; Vanin, V. R. *Nucl. Instrum. Meth.* 2004, A 523, 186-192.

[62] Miller, K. *SIAM J. Math. Anal.* 1970, 1, 52-74.

[63] Johnston, R. A.; Connolly, T. J.; Lane R. G. *Opt. Commun.* 2000, 181, 267-278.

[64] Morháč, M.; Matoušek, V. *Digit. Signal Process.* 2009, 19, 372-392.

[65] Janson, P. A. *Deconvolution with Applications in Spectroscopy;* Academic Press: New York, 1984.

[66] Gold, R. *An Iterative Unfolding Method for Response Matrices*; ANL-6984 Report; Argonne National Laboratories: Argonne, 1964; Vol. Ill.

[67] Morháč, M.; Kliman, J.; Matoušek, V.; Veselský, M.; Turzo, I. *Nucl. Instrum. Meth.* 1997, A 401, 385-408.

[68] Morháč, M. *Nucl. Instrum. Meth.* 2006, A 559, 119-123.

[69] Lucy, L. B. *Astron. J.* 1974, 79, 745 754.

[70] Richardson, W. H. *J. Opt. Soc. Am.* 1972, 62, 55-59.

[71] Lin, Y.; Lee, D. D. *IEEE Trans. Signal Process.* 2006, 54, 839-847.

[72] Cadzov, J. A.; Li, X. *Digit. Signal Process.* 1995, 5, 3-20.

[73] Hunt, B. *Int. J. Mod. Phys. C* 1994, 5, 151-178.

[74] Pham, D. T. *Signal Process.* 2007, 87, 2045-2060.

[75] Backus, G. E.; Gilbert, F. *Geophys. J. of the Royal Astr. Soc.* 1968, 16, 169-205.

[76] Ozkan, M. K.; Tekalp, A. M.; Sezan, M. I. *IEEE Trans. Image Process.* 1994, 3, 450-454.

[77] Sanchez-Avila, C.; Figueiras-Vidal, A. R. *J. Comput. Appl. Math.* 1996, 72, 21-39.

[78] Morháč, M.; Matoušek, V.; Kliman, J. *J. Comput. Appl. Math.* 2002, 140, 639-658.

[79] Jandel, M.; Morháč, M.; Kliman, J.; Krupa, L.; Matoušek, V.; Hamilton, J. H.; Ramaya, A. V. *Nucl. Instrum. Meth.* 2004, A 516, 172-183.

[80] Morháč, M.; Kliman, J.; Matoušek, V.; Veselský, M.; Turzo, I. *Nucl. Instrum. Meth.* 2000, A 443, 108-125.

[81] Routti, J.T.; Prussin. S.G. *Nucl. Instumr. Meth.* 1969, 72, 125-142.

[82] Krištiak, J. et al. *J. Phys.* 1993, IV 3, 265-270.

[83] Silagadze, Z.K. *Nucl. Instrum. Meth.* 1996, A 376, 451-454.

[84] Morháč, M. *Nucl. Instrum. Meth.* 2007, A 581, 821-830.

[85] Morháč, M.; Veselský M. *Nucl. Instrum. Meth.* 2008, A 592, 434–450.

[86] Mastinu, P.F.; Milazzo, P.M.; Bruno, M.; D'Agostino, M. *Nucl. Instrum. Meth.* 1996, A 371, 510-513.

[87] Leneindre, N. et al. *Nucl. Instrum. Meth.* 2002, *A 490*, 251-262.

[88] Benkirane, A.; Auger, G.; Bloyet, D.; Chbihi, A.; Plagnol, E. *Nucl. Instrum. Meth.* 1995, A 355, 559-574.

[89] Alderighi, M. et al. *IEEE Trans. Nucl. Science* 2001, 48, 385-390.

[90] Black, W.W. *Nucl. Instrum. Meth.* 1969, 71, 317-327.

[91] Morháč, M.; Hlaváč, S.; Veselský, M.; Matoušek, V. *Nucl. Instrum. Meth.* A 621, 2010, 539-547.

[92] Altherr, T.; Seixas, J. *Nucl. Instrum. Meth.* 1992, A 317, 335-338.

[93] Linka, A.; Picek, J.; Volf, P.; Ososkov, G. *Czech. J. Phys.* 1999, 4952, 161-168.

[94] Lester, C. G. *Nucl. Instrum. Meth.* 2006, A 560, 621-632.

[95] Di Bari, D. *Nucl. Instrum. Meth.* 2008, A 595, 241-244.

[96] Forty, R. for the LHCb Collaboration *Nucl. Instrum. Meth.* 1999, A 433, 257-261.

[97] Buszello, C. P. *Nucl. Instrum. Meth.* 2008, A 595, 245-247.

[98] Elia, et al. *Nucl. Instrum. Meth.* 1999, A 433, 262-267.

[99] Braem, A. et al. *Nucl. Instrum. Meth.* 2003, A 499, 720-724.

[100] Morháč, M.; Matoušek, V.; Kliman, J. *Digit. Signal Process.* 2003, 13, 144-171.

[101] HADES Collaboration, HADES website http://www-hades.gsi.de.

[102] Agostinelli, S. et al. *Nucl. Instr. Meth.* 2003, A 506, 250-303.

[103] Lawrence, J. D. *A Catalog of Special Plane Curves*, New York, Dover, 1972.

[104] Morháč, M.; Kliman. J.; Jandel, M.; Krupa, L.; Matoušek, V. *Appl. Spectrosc.* 2003, 57, 753-760.

[105] Morháč, M.; Kliman, J.; Jandel, M.; Krupa, L.; Matoušek, V.; Hamilton, J. H.; Ramaya, A. V. *Comput. Phys. Commun.* 2005, 172, 19-41.

[106] Brodlie, K.; Wood, J. *Comput. Graph. Forum* 2001, 20, 125-148.

[107] Brodlie, K. *Nucl. Instrum. Meth.* 1995, A354, 104-111.

[108] Max, N. *Visual Comput.* 2005, 21, 979-984.

[109] Ropinski, T.; Preim, B. In *Proc. 19-th Conference on Simulation and Visualization (SimVis 2008)*; 2008, 121-138.

[110] Bürger, R.; Hauser, H. In *Proc. Eurographics 2007, State of the Art Report*; 2007, 117-134.

[111] Bhaniramka, P.; Wenger, R.; Crawfis, R. In *IEEE Visualization '00*; 2000, 267-273.

[112] Bajaj, C. L.; Pascucci, V.; Schikore, D. R. In *IEEE Visualization '97*; 1997, 167-173.

[113] Woodring, J.; Wang, C.; Shen, H. W. In *IEEE Visualization '03*, 2003, 417-424.

[114] Linsen, L.; Pascucci, V.; Duchaineau, M. A.; Hamann, B.; Joy, K. I. In *Proc. 10-th Pacific Conference on Computer Graphics and Applications*; 2002, 346-355.

[115] Woodring, J. L.; Shen, H. W. *IEEE T. Vis. Comput. Gr.* 2006, 12, 909-916.

[116] dos Santos, S. R. *A Framework for the Visualization of Multidimensional and Multivariate Data*, PhD Thesis, University of Leeds, 2004.

[117] Woodring, J. L. *High Dimensional Direct Rendering of Time-Varying Volumetric Data*, MSc Thesis, Ohio State University, 2003.

[118] Jirka, T. *Multidimensional Data Visualization*, Technical Report No. DCSE/TR-2003-03, University of West Bohemia in Pilsen, Czech Republic, 2003.

[119] Carr, H.; Snoeyink, J.; van de Panne, M. In *Proc. 14-th Annual Fall Workshop on Computational Geomety '04*; 2004, 51-52.

[120] Carr, H.; Snoeyink, J.; van de Panne, M. In *IEEE Visualization '04*; 2004, 497-504.

[121] Bajaj, C. L.; Pascucci, V.; Schikore, D. R. In *IEEE Visualization '98*; 1998, 51-58.

[122] Weber, H.; Dillard, S. L.; Carr, H.; Pascucci, V.; Hamann, B. *IEEE T. Vis. Comput. Gr.* 2007, 13, 330-341.

[123] Treinish, L. A.; Goettsche, C. *Comput. Graph.* 1991, 35, 184-204.

[124] Thakur, S.; Hanson, A. J. In *3-rd International Symposium on Visual Computing*; 2007, 804-815.

[125] dos Santos, S. R.; Brodlie, K. In *ACM International Conference Proceeding Series 22*; 2002, 173-182.

[126] Haefele, M.; Zara, F.; Latu, G.; Dischler, J. M. In *The 1-st Workshop on Super Visualization (IWSV08)*; 2008.

[127] Bajaj, C. L.; Pascucci, V.; Rabbiolo, G.; Schikore, D. R. In *IEEE Symposium on Volume Visualization*; 1998, 95-102.

[128] Howie, J.M. *Comput. Graph.* 1979, 4, 161-164.

[129] Vandoni, C. E. *Comput. Graph.* 1989, 13, 243-252.

[130] Brun, R. et al. *An Object-Oriented Data Analysis Framework*, Users Guide 3.02c, CERN, 2002.

[131] R. S. Gallagher, *Computer Visualization. Graphics Techiques for Scientific and Engineering Analysis*, CRC Press, Boca Raton, Ann Arbor, London, Tokyo, 1995.

[132] Morháč, M.; Kliman, J.; Matoušek, V.; Turzo, I. *Acta Phys. Slovaca* 2004, 54, 385-400.

[133] Morháč, M.; Matoušek, V. In *ACAT2007 XI. International Workshop on Advanced Computing and Analysis Techniques in Physics Research*; 2007, 1-34.

[134] Grinstein, G.; Pickett, R. M.; Williams, M. G. In *Graphics Interface'89*; 1989, 254-261.

[135] Ward, O. In *Proceedings IEEE Visualization'94*; 1994, 326 - 333.

[136] van Wijk, J.J.; van Liere, R. In *Proceedings IEEE Visualization'93*; 1993 *IEEE Computer Society Press*, Los Alamitos, CA, 1993, 119-125.

[137] Morháč, M.; Turzo, I.; Krištiak, J. *IEEE Trans. Nucl. Science* 1995, 42, 1-7.

[138] Morháč, M.; Matoušek, V.; Turzo, I. *IEEE Trans. Nucl. Science* 1996, 43, 140-148.

[139] Morháč, M.; Matoušek, V.; Kliman, J.; Turzo, I.; Krupa, L.; Jandel, M. *Nucl. Instrum. Meth.* 2003, A 502, 728-730.

[140] Butler-Moore, K.; Kliman, J.; Morháč, M. et al. *J. Phys. G. Nucl. Partic.* 1999, 25, 2253-2269.

[141] Hamilton, J. H.; Ter-Akopian, G. M.; Kliman, J.; Polhorský, V.; Morháč, M. et al. *J. Phys. G. Nucl. Partic.* 1994, 20, L85-L89.

[142] Jandel, M.; Kliman, J.; Krupa, L.; Morháč, M. et. al.. *J. Phys. G. Nucl. Partic.* 2002, 28, 2893-2905.

[143] Kliman, J.; Polhorský, V.; Morháč, M. et al. *Phys. Atom. Nucl.* 1994, 57, 1108-1111.

[144] Ter-Akopian, G.M.; Kliman, J.; Morháč, M. et al. *Phys. Rev. C* 1997, 55, 1146-1161.

[145] Ter-Akopian, G.M.; Kliman, J.; Morháč, M. et al. *Phys. Rev. Lett.* 1996, 77, 32-35.

[146] Ter-Akopian, G.M.; Kliman, J.; Morháč, M. et al. *Phys. Rev. Lett.* 1994, 73, 1477-1480.

In: Computer Physics
Editors: B.S. Doherty and A.N. Molloy, pp. 237-266

ISBN: 978-1-61324-790-7
© 2012 Nova Science Publishers, Inc.

Chapter 2

Solving the Vibrational Schrödinger Equation without a Potential Energy Surface Using a Combined Neural Network Collocation Approach

Sergei Manzhos[1,*], Tucker Carrington, Jr.[2,†] and Koichi Yamashita[3,‡]

[1]Research Center for Advanced Science and Technology, University of Tokyo, 4-6-1, Komaba, Meguro-ku, Tokyo 153-8904, Japan
[2]Chemistry Department, Queen's University, Kingston, Ontario, K7L 3N6, Canada
[3]Department of Chemical System Engineering, School of Engineering, University of Tokyo, 7-3-1, Hongo, Bunkyo-ku, Tokyo 113-8656

Abstract

We review approaches for solving the vibrational Schrödinger equation focusing on collocation and neural network (NN) methods that do not require computing integrals. A Radial Basis Neural Network (RBNN) - based method we developed is given particular attention. It allows one to compute several vibrational levels with an accuracy of the order of 1 cm^{-1} by expanding the wavefunction in an extremely small and flexible basis. This approach combines non-linear optimization of basis (neuron) parameters with a linear rectangular matrix method. It improves dimensionality scaling and permits computing several levels from a single RBNN. The algorithm avoids the calculation of integrals and of a potential energy function – the vibrational spectrum may be obtained directly from ab initio data. Owing to the fact that the construction of a PES is bypassed, it is possible to directly use this approach with an ab initio code to compute vibrational spectra. When normal model coordinates are used, the method is molecule-independent. It is particularly useful for computing vibrational spectra of molecules adsorbed at surfaces and finally provides surface scientists with a black-box and easily usable tool to compute spectra of adsorbate complexes without neglecting

---

[*] E-mail address: Sergei@tcl.t.u-tokyo.ac.jp
[†] E-mail address: Tucker.Carrington@queensu.ca
[‡] E-mail address: Yamasita@chemsys.t.u-tokyo.ac.jp

anharmonicity and coupling effects. Applications to synthetic problems of different dimensionalities, the water molecule, and to H_2O on Pt(111) are reviewed.

1. Introduction

Multidimensional ordinary differential equations (ODE) are ubiquitous in the modeling of physical and chemical processes [1]. In particular, the behavior of molecular and reactive systems is governed by a Schrödinger equation [2⁻8] for the motion of the nuclei. It is therefore not surprising that much effort has been dedicated to the development of methods for its solution. In particular, to calculate the vibrational spectra of molecules, one often solves the time-independent Schrödinger equation (SE) for the nuclei

$$\hat{H}\psi(x) \equiv \left(\hat{T} + \hat{V}\right)\psi(x) = E\psi(x) \tag{1}$$

where $\hat{H}$ is the Hamiltonian operator which consists of a kinetic energy operator (KEO) and a potential part $\hat{V} = V(x)$, where $V(x)$ is the potential energy surface (PES) [9, 10], $x = (x_1, x_2, ..., x_d)$. Here d is the number of vibrational coordinates. For a given $V(x)$, one wants to find wavefunctions $\psi(x)$ and energies E that satisfy Eq. (1). In the multidimensional case, simple approaches like Runge-Kutta[11] and Numerov [12, 13] methods do not work, and one usually resorts to a basis set expansion of the solution [3, 5, 8, 14]

$$\psi(x) = \sum_{\vec{s}} c_{\vec{s}} \Phi_{\vec{s}}(x) \tag{2}$$

with basis functions $\Phi_s(x)$ of known and fixed shape. Substituting Eq. (2) into Eq. (1), one sees that for a particular wavefunction, the coefficients $c = (c_1, c_2, ..., c_N)^T$ can be found by solving $Hc = Ec$, where $H_{ij} = \left\langle \Phi_i \middle| \hat{H} \middle| \Phi_j \right\rangle$ is the Hamiltonian matrix. The simplest basis functions are products of primitive, orthogonal univariate basis functions:

$$\Phi_{\vec{s}}(x) = \prod_{i=1}^{d} \varphi_{s_i}(x_i)$$
$$\int \varphi_{s1_i} \varphi_{s2_i} dx_i = \delta_{s1_i, s2_i} \tag{3}$$

where $\vec{s} \equiv \{s_1, s_2, ..., s_d\}$ is now a composite index. Using a product basis, it is possible to compute hundreds of vibrational levels to sub-cm^{-1} accuracy for molecules with as many as 5 atoms [3, 7, 8, 15-20]. To employ this method, one must evaluate quadratures and solve a large eigenvalue problem. This quadrature approach is the benchmark against which all other methods are compared.

Two major difficulties impede the application of this method: the number of required basis functions is large, and an analytic representation of the PES is required. First, if the form of the basis functions is fixed, the number of required functions increases drastically with dimension d. Consequently, both the memory and the CPU cost of a calculation are large for molecules with more than four atoms ($d=6$). It is possible to use basis functions that are better than simple product functions [5, 17, 21⁻26], but basis size remains a significant problem. Second, one requires potential values at all the quadrature points, i.e., at millions of points. This means that one needs an analytic PES. It is usually obtained by fitting a functional form to ab initio points (typically $\sim10^4$ points are used) and the sampling of the configuration space [9, 10, 27⁻31] is often very sparse. For larger molecules or molecules adsorbed on a surface, one is forced to use less accurate quantum chemistry methods, often the density functional theory (DFT) [32⁻34]. Both poor quantum chemistry and fitting can introduce significant error, up to ~1000 cm^{-1}. [28, 35⁻38]

The quadrature problem can be alleviated by representing the potential via low-dimensional functions. Examples are the high-dimensional model representations [29, 39⁻42], in particular, the multimode expansion [43⁻46], and representations over sums of products of univariate terms using the potfit algorithm [47⁻49] or neural networks [50]. If product basis functions are used, it is then possible to factor the potential integrals. Representing the potential in this form introduces (an often unknown) error and may itself be costly.

Due to the importance of these problems, there is interest in alternative approaches to computing vibrational spectra [51⁻61]. The purpose of this work is to review methods which alleviate one or both of the aforementioned problems. The two classes of alternative approaches we focus on are collocation methods [61⁻67], which obviate the need to compute integrals, and neural network based methods [51⁻53, 68⁻70], which both do not use quadratures and also allow for the use of extremely small basis sets. The collocation method is briefly reviewed in Section 0.

Artificial neural networks (NN) [71⁻73] have been employed to solve the Schrödinger equation for bound state problems [51⁻53, 57, 58]. One approach is to learn the mapping between potential parameters and the spectrum [57, 58, 74, 75]. The principle disadvantage of such mapping methods is that one must still solve Schrödinger equation, using some other method, for some set of reference potentials. In this work, we review stand-alone NN methods for solving the Schrödinger equation. Such methods represent a vibrational wavefunction with a NN [51⁻53]. Neural network representations of functions have several advantages. NNs are universal approximators [76⁻79], and there are mathematical arguments for favorable dimensionality scaling of NNs [80⁻85]. These advantages stem from the fact that NNs represent multidimensional functions with univariate functions, but not with sums of products of functions. Traditional methods using direct product basis expansions can also build wavefunctions from simple univariate functions, but the arguments are simply the coordinates [8]. The number of basis functions can be reduced drastically by using simple but flexible functions – neurons of a neural network. With a NN, the exponential scaling of the number of functions is avoided. The most costly part of an NN calculation is the multidimensional NN parameter optimization problem. In the past, this has limited the application of NN based methods to 1, 2, and 3-dimensional problems [51⁻53] and to solving for one vibrational level at a time. We give a brief overview of simple neural eigensolvers in Section 0.

We have shown that it is possible to improve NN-based eigensolvers by (i) using one set of hidden neurons to compute simultaneously several wavefunctions and energies and thereby exploit the implicit information sharing between different wavefunctions via the PES and (ii) determining the spectrum and NN output weights (basis set expansion coefficients) using linear algebra instead of standard minimization methods. To realize (ii), we must find pseudo eigenvectors of so-called rectangular matrix pencils: recent progress in this field [86] makes this approach feasible. This reduces the dimensionality of the parameter space subject to non-linear optimization methods. In contrast to previous works, we use radial basis neural networks (RBNN) and not sigmoid neural networks in line with the argument that RBNN are better suited to approximate band-limited functions, e.g. vibrational wave functions [87]. The resulting approach combines elements of both neural network and collocation methods; it avoids integrals and the need to use a large number of basis functions and PES points. This method is the subject of Section **Error! Reference source not found.**, where we review its development and applications based on Refs. [68⁻70].

It is difficult to achieve sub-cm^{-1} accuracy with a NN-based approach, but the NN approach we favor is general and easy to use. Because the basis functions are adaptable, the basis is small, and calculations for systems with d as large as 6 can be done on a desktop computer. The method is most promising when it is (i) difficult or unfeasible to obtain an accurate PES because of the errors of either ab initio or fitting methods and (ii) only several, usually the lowest, vibrational levels are required with an accuracy of the order of 1 cm^{-1}. These two conditions are satisfied when the purpose of the calculation is species identification for interacting polyatomic molecules or extended systems.

Of particular interest is the application to infrared (IR) spectroscopy of molecules on surfaces. IR spectra are sensitive to geometry and can serve to identify adsorbed species and to determine their structure and the nature of the adsorption site [88⁻93]. Vibrational analysis is needed to compute thermal corrections to the enthalpy and entropy of a reaction [94]. The ability to compute vibrational spectra of adsorbed molecules is therefore important for heterogeneous and photo-induced catalysis [33, 89, 92, 95⁻104], sensing [105], and material quality assessment [106]. Vibrational spectra have been used for structure determination of bioactive metal-organic complexes [106, 107]. Here the interest is mostly in the lowest lying vibrational states, and it is desirable to identify frequencies with an accuracy not worse than the experimental error, which is usually of the order of several cm^{-1} [89, 108]. The quality of computed spectra is, however, critical for the interpretation of experiments. Calculations on molecule-surface complexes are most often done by using methods of electronic structure theory and assuming that all vibrations are harmonic and uncoupled. The reasons for this are (i) difficulties of sampling and fitting a global PES: electronic structure calculations of molecules on surfaces [109] are more costly than for molecules in vacuum, and the dimensionality of the configuration space is larger due to frustrated transition and rotational degrees of freedom (DOF); (ii) difficulties of computing the vibrational spectrum on a realistic PES. Most experimental surface science groups do not attempt such calculations [43, 103, 110⁻112].

Ref. [70] lists several examples where errors due to the neglect of anharmonicity and coupling are likely large [33, 89, 97⁻102]. Without an objective estimate of the magnitude of anharmonicity and coupling effects, conclusions about frequency shifts upon adsorption become pure speculation. In fact, one can expect an increased role of anharmonicity and coupling upon adsorption, as the interaction with the substrate tends to weaken intramolecular bonds which should make the PES deviate more from a sum of uncoupled harmonic wells. A comparison between DFT's uncoupled harmonic, experimental frequencies, and frequencies computed on a more realistic PES for a polyatomic molecule-surface system is offered in Ref. [113], where a PES including only the 2nd order coupling was produced using a modified PM3 method [114, 115]. For many modes, a significant improvement of frequencies was reported even when a low level of electronic structure calculations and coupling was used. In Ref. [116], anharmonic frequencies were computed for selected stretches (i.e. 1D) of NH_3 adsorbed on Ni(111) with a variational method using a simple Morse or Chebyshev polynomial expansion of the potential curve. The inclusion of anharmonicity significantly improved the agreement with experimental data. Especially the red shifts due to adsorption were much improved (error decreasing from about 30 cm^{-1} in harmonic calculations to about 10 cm^{-1} in the anharmonic case). It is therefore clear that it is highly desirable to account for anharmonicity and coupling when computing spectra of molecules adsorbed on surfaces, yet there are currently no readily available tools to do so.

The method introduced in Section 0 makes it possible to compute anharmonic frequencies of molecules directly from ab initio points. The points can be obtained with the general scheme outlined in Section 0. Tests on the synthetic potentials of Section 0 predict a favorable scaling of the basis size with the number of degrees of freedom, and applications to gas phase and adsorbed molecules reviewed in Sections 0 and 0 have established the validity of the method for real-life applications. This approach looks especially promising for surface science applications. The accuracy achieved to date of about 1 cm^{-1} is very good considering the experimental uncertainties [89, 108] and ab initio errors [33, 34, 117] typical for such systems. We believe the method could be routinely used in surface science labs.

2. Methods to Solve the Vibrational Schrödinger Equation

A. Standard Basis Set + Quadrature Approach

It is most common to calculate a spectrum by replacing the wavefunction in the Schrödinger equation with a basis representation

$$\psi(x) = \sum_{i=1}^{N} c_i \Phi_i(x), \tag{4}$$

multiplying on the left with a basis function, integrating to compute elements of the Hamiltonian matrix

$$H_{ij} = \left\langle \Phi_i \middle| \hat{H} \middle| \Phi_j \right\rangle \tag{5}$$

and the overlap matrix $S_{ij} \equiv \left\langle \Phi_i \middle| \Phi_j \right\rangle$, and finally solving an eigenvalue problem:

$$Hc = E'Sc \tag{6}$$

It is customary to use orthogonal basis functions, in which case $S = I$. When integrals are computed exactly, eigenvalues E' of (6) are upper limits for exact values E of (1) [118]. This statement is based on the Rayleigh-Ritz variational theorem [119], and if integrals are exact, the method is variational. In this chapter, we denote the method outlined above as "variational", regardless of the exactness of the integrals. As the basis size N is increased to

minimize the expansion error $\varepsilon = \left\| \psi(x) - \sum_{i=1}^{N} c_i \Phi_i(x) \right\|$, the eigenvalues E' converge to E;

moreover, an error ε in a wavefunction corresponds to an error ε^2 in the energy [120]. If methods of direct linear algebra [11, 121–123] are used to solve the eigenvalue problem, it is necessary to compute the matrix elements and then diagonalize the matrix. Both these steps may be costly. A direct eigensolver is useful when N is $\approx 10^4$, i.e. for molecule with fewer than 4 atoms. For four-atom molecules and for larger systems, the basis size is too big for a direct eigensolver to be practical, as the Hamiltonian matrix has to be stored in memory (N^2 elements), and the CPU cost is $\propto N^3$. If an iterative eigensolver [18, 22, 124–131] is used, it is possible to avoid computing Hamiltonian matrix elements [17, 18, 132–134]. The Lanczos algorithm [130, 131], the filter diagonalization method [124, 125, 127, 129], and the Davidson algorithm [135] are often used. When using an iterative method, it is not necessary to store the Hamiltonian matrix.

The dimensionality of the quadratures can be reduced by representing both the basis functions $\Phi_i(x)$ and the potential function $V(x)$ with lower-dimensional functions [43, 136]. When the d-dimensional basis functions Φ of Eq. (4) are products of one-dimensional factors, Eq. (3), it is most common to use 1-D basis polynomial basis functions. The so-called primitive basis functions $\varphi_{s_i}(x_i)$ form a complete and orthonormal set in each x_i. The number of $\Phi_i(x)$ scales exponentially with the number of degrees of freedom. The potential terms can also be represented via low-dimensional terms. An expansion over orders of coupling can be used, such as the popular multimode expansion [43–46] or the more general high dimensional model representation [29, 39–42]. A multimode expansion of the potential is often used with the VSCF method [43, 137–139]. A representation as a sum of products of one-dimensional functions can be obtained from a known PES function using the potfit algorithm [47–49] or directly from ab initio samples using a neural network based algorithm [50]. This form facilitates calculations with the MCTDH method [140–143].

If it is costly or impossible to fit the potential to a sum of lower dimensional functions, then high dimensional quadratures are necessary, and the number of points is huge. It is common to use direct product quadrature grids that have about 10^d points. If d is larger than about six, the grid is unmanageable. Using a non-product basis does help, but the number of quadrature points is still very large [144]. Using the discrete variable representation [18, 145–150] (DVR) does not solve the problem because the number of DVR points is equal to the number of quadrature points. Because the number of points is so large, it is impossible to compute the potential at each of the points. Instead, one fits a PES and evaluates it at the quadrature points. Building a PES is, by itself, a significant task and a source of error [28, 29, 31, 35–38, 50, 151–169].

Two common ways of reducing the cost of a calculation of this kind are pruning and contraction, both of which reduce the size of the matrix. To prune, one discards basis functions from the full direct product basis [3, 170‑179]. Often this is done by keeping only basis functions with the smallest zeroth-order energies. Contraction means using basis functions that are solutions of low-dimensional problems [24, 180, 181]. A significant cost reduction can be achieved. Both contraction and pruning will be more effective for some molecules than for others, and both require some knowledge of the molecule whose spectrum is to be computed.

State of the art vibrational calculations combine a Lanczos eigensolver and contracted basis functions and are able to compute hundreds of levels with a sub-cm^{-1} accuracy for up to $d=12$ [16, 133, 182]. The standard basis set + quadrature approach provides benchmark accuracy against which to compare other approaches.

In connection with the radial basis neural network method introduced below, we note the use of Gaussian functions for the basis. They have the advantage of a simple functional form, which results in analytic overlap and KEO matrix elements. Gaussians are localized and constitute an efficient [183, 184] and problem-independent basis set. Efficient wavefunction representations have been obtained even with Gaussian functions having pre-selected parameters [5, 185‑189]. Poirier et al. have optimized Gaussian parameters via minimization of the trace [190]

$$\mathrm{tr}\left(\hat{H}\hat{\rho}\right) = \sum_{i=1}^{N} H_{ii} \tag{7}$$

where

$$\hat{\rho} \equiv \sum_{i,j=1}^{N} \left[S^{-1}\right]_{ij} |g_i\rangle\langle g_j| \tag{8}$$

and $S_{ij} \equiv \langle g_i | g_j \rangle$, $H_{ij} = \langle g_i | \hat{H} | g_j \rangle$. The Gaussian basis functions $g_i(x) = (2\alpha_i/\pi)^{1/4} e^{-\alpha_i(x-x_i)^2}$ are similar to the ones used with RBNNs in Section **Error! Reference source not found.**, but the parameter optimization is done differently. In particular, due to numerical instabilities, it is necessary to limit the overlap of the basis functions in (8).

B. The Collocation Method

The collocation method is a conceptually simple method for solving ODEs that has the advantage that it requires no integrals. Other similar formulations exist [62, 191‑193], but we focus here on collocation, as it is a simple way of obtaining a matrix eigenvalue equation from the SE and has been successfully applied both to both spectroscopic [64‑66] and quantum scattering problems [67]. A solution $\psi(x)$ to Eq. (1) is expanded in terms of basis functions:

$$\psi(x) = \sum_{i=1}^{N} c_i \varphi_i(x) \tag{9}$$

The coefficients c are obtained by requiring that the SE, Eq. (1), be satisfied at N (collocation) points in configuration space $x^{(j)}$, $j=1,...,N$. This can be achieved, similarly to the method of Section 0, my multiplying on the left by $\varphi_j^\delta(x) = \delta(x - x^{(j)})$ and integrating,

$$\langle \varphi_j^\delta | R \rangle = \langle \varphi_j^\delta(x) | \hat{H} - E | \psi(x) \rangle = 0, \quad i = 1,...,N \tag{10}$$

This yields a square but non-symmetric generalized matrix eigenvalue problem:

$$\sum_{i=1}^{N} \left(\langle \varphi_j^\delta(x) | \hat{H} - E | \varphi_i(x) \rangle \right) c_i = 0, \quad j = 1,...,N \tag{11}$$

In Eq. (11), the Hamiltonian matrix is in a mixed representation with $\{\varphi_j^\delta(x)\}$ on the left and $\{\varphi_i(x)\}$ on the right. $V(x)$ is therefore required at the points $x^{(j)}$, and no integrals are performed. A disadvantage is that the error in E is now proportional to [120] $\varepsilon \cdot \varepsilon_\delta$, where

$$\varepsilon = \left\| \psi(x) - \sum_{i=1}^{N} c_i \varphi_i(x) \right\|, \quad \text{and} \quad \varepsilon_\delta = \left\| \psi(x) - \sum_{i=1}^{N} c_i^\delta \varphi_i^\delta(x) \right\| \quad \text{(cf. Section 0). For the}$$

expansion error ε_δ in the delta basis $\{\varphi_\delta^{(j)}(x)\}$ to be small requires a large N. The number of basis functions N must also be equal to the number of potential points. A dense sampling of the configuration space will often be required, which precludes computing the PES at the collocation points and forces one to fit a PES as in the "variational" method. The need to use a large number of points makes collocation scale badly with dimension. In addition, iterative methods for solving large non-symmetric eigenvalues problems are less robust.

Whether one uses the DVR, does quadrature, or uses collocation, a set of points is introduced. There is however a fundamental distinction. When quadrature or the DVR is used, kinetic matrix elements are evaluated without approximation. This is not true with collocation. Collocation is equivalent to a method in which both the potential and the kinetic matrix elements are done with quadrature. It is sometimes called a pseudospectral method [191⁻195].

The collocation method has been successfully applied to compute low lying as well as highly excited vibrational states and ro-vibrational states of small molecules with only two or three degrees of freedom considered and for these problems was shown to be about as accurate as the variational approach [65, 66, 144]. A variant of the collocation method for scattering problems was also presented and tested on a triatomic system [67]. In electronic structure theory, the collocation method has been used to solve the (three-dimensional) Hartree-Fock [191⁻193] and Kohn-Sham [196] equations. We are not aware of any applications in more than 3-D. In more than 3-D, the size of the matrix becomes large and iterative methods, required for large matrices, are more difficult to use. In the next section, we explain that by taking more points than basis functions and minimizing residuals using a fitting method, it is possible to solve more difficult problems.

C. The Neural Network Approach

Neural network (NN) approaches for solving the Schrödinger equation are based on a parameterized representation of wavefunction(s). The parameterized representation is a neural network, and the non-linear NN parameters are adjusted to minimize residual(s). Because the NN representation is efficient, the number of parameters is small, and wavefunction(s) can be determined with a smaller number of potential points than would be required with a quadrature or collocation method. Neural networks representations are general and have a solid mathematical foundation.

A neural network [71, 73] is a representation of a multivariate function $f(x)$ with univariate functions σ called "neurons":

$$f^{NN}(x) = \sigma_0\left(\sum_{q=1}^{N} c_q \sigma\left(y_q\left(b_q, w_q, x\right)\right) + b\right) \tag{12}$$

σ is also called the activation or transfer function. When

$$y_q\left(b_q, w_q, x\right) = \sum_{p=1}^{d} w_{qp} x_p + b_q, \tag{13}$$

the NN is called a "ridge activation function neural network" [81], and when

$$y_q\left(b_q, w_q, x\right) = b_q\left\|w_q - x\right\|, \tag{14}$$

the NN is called a "radial basis neural network" (RBNN) [183, 184]. The NN of Eq. (12) is the so-called single hidden layer neural network. It is possible to have multiple layers when each y_q also has the form of Eq. (12), i.e.

$$y_q = \sum_{q_2=1}^{N_2} c_{q_2} \sigma_2\left(z_{q_2}\left(b_{2q_2}, w_{2q_2}, x\right)\right) + b_q, \tag{15}$$

and so on. The key NN parameters are input weights w, biases b, and output weights c. To represent a known function, they are adjusted with non-linear methods until the residual is minimized at M samples of the function $f(x)$:

$$\min\left\{\sum_{i=1}^{M}\left(f\left(x^{(i)}\right) - f^{NN}\left(x^{(i)}\right)\right)^2\right\} \tag{16}$$

Because of the layered architecture of an NN, it is possible to minimize the error consecutively in each layer with the backpropagation algorithm [73], which makes it possible to optimize a large number of non-linear parameters (thousands) in a computationally effective way. For each layer, methods like BFGS or Levenberg-Marquardt[11] are used to find the parameters.

The most common NN architecture used in applications in physics and chemistry [27‑29, 35, 37, 38, 51, 52, 72, 152‑154, 197‑210] is a single hidden layer ridge activation function NN with a sigmoid transfer function [79, 211] $\sigma(x) = \left(1 + e^x\right)^{-1}$. Specifically, this architecture was used in NN based methods to solve the SE [51, 52, 57, 58]. It was first proven that a NN with this architecture can fit any smooth function to arbitrary accuracy provided enough samples are available, i.e. neural networks are universal approximators [76‑79]. Later, it was shown that any smooth non-linear function can be used as an activation function [212]. Mathematicians have also argued that the number of neurons scales well with dimensionality [80‑85]. This has to do with the fact that although NN use univariate functions, they do not fit with sums of products of functions and therefore should avoid the exponential scaling of the number of functions from which direct product bases suffer. The disadvantageous scaling of the basis size with dimension in the variational and collocation methods is partly due to the fact that the basis functions are simple and pre-determined. Eq. (12) can be viewed as an expansion over N basis functions $\sigma\left(y_q(x)\right)$, where the location, orientation, and width of each "ridge" $w_q \cdot x + b_q = 0$ are varied via the parameters w and b until the wavefunction represented by the NN nearly satisfies the SE. Using adaptable, i.e. parameterized, albeit simple basis functions should therefore reduce the basis size. Contraction schemes make optimized basis function [8, 21, 22] from linear combinations of primitive basis functions. In contrast, the shape of NN's neural basis functions is optimized directly by varying non-linear parameters.

To determine parameter values that nearly solve the SE, one minimizes the norm of the residual [51‑53]

$$R = \frac{\left\langle \psi(x) \middle| \left(\hat{H} - E\right)^2 \middle| \psi(x) \right\rangle}{\left\langle \psi(x) \middle| \psi(x) \right\rangle} \approx \frac{\sum_{i=1}^{M} \left| \left(\hat{H} - E\right)\psi\left(x^{(i)}\right) \right|^2}{\sum_{i=1}^{M} \left| \psi\left(x^{(i)}\right) \right|^2} \tag{17}$$

or maximizes e^{-R} [51, 52]. If the wavefunction is represented as an NN, the kinetic energy operator (KEO) can be applied to the wavefunction analytically. In previous NN methods, NN parameters and E [51, 52, 213] or only NN parameters [53, 213] (E being computed from the wavefunction) were fitted for a single state. For one-dimensional problems, the best values of $\sqrt{R}$ were typically 2.3-19 % of the corresponding energies [51, 52]. For a two-dimensional Henon-Heiles potential, the first four wavefunctions and levels were reported in Ref. [53] with a relative error of the order of 10^{-5}. Only the ground state for a Hamiltonian representing three coupled anharmonic oscillators was reported[53]. Recently, the ground state of the three-dimensional Kohn-Sham equation was computed with a modest accuracy [213].

In these papers, the number of potential values was small enough that they could be computed directly with ab initio methods obviating the need to fit a potential function. This NN method therefore obviates the need to use a PES. Note that there is no matrix algebra involved. These methods, however, compute one state at a time and have been applied only to low-dimensional systems. It is ironic that a method that is not hampered by the problems associated with large matrices and which might require fewer potential points cannot compete with the variational method in accuracy and the number of levels computed even for triatomic molecules. What limits it to low-dimensional systems? The main culprit is the cost of the non-linear fitting required to minimize the residual (17). The cost of this fit cannot be reduced by using backpropagation, because one is not fitting an NN to a known function. The fact that the sigmoid neurons do not satisfy the boundary conditions is a second problem. To deal with this problem, an amplitude-phase decomposition [51, 52] or pre-multiplication with a function that does satisfy the boundary condition [53] were used. Both these approaches, however, make the application of the KEO cumbersome for a multidimensional problem. In the next section, we introduce a method that mitigates these and other disadvantages.

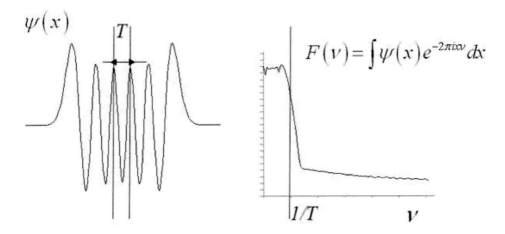

Figure 1. A vibrational wavefunction of a harmonic oscillator and its Fourier spectrum.

3. The Method of Manzhos and Carrington

A. Theory

In Ref. [68], we introduced a method which combined the use of a small basis of adaptable NN-type univariate functions and collocation. Radial basis functions rather than ridge activation functions are used. RBNNs have been shown to be universal approximators [183, 184]. Moreover, it is argued that an expansion over adaptable Gaussian functions of band-limited functions is "free of the curse of dimensionality" [87], i.e. that the scaling of the number of neurons with dimension is favorable. Vibrational wavefunctions usually represent oscillations under a smooth envelope and are band-limited; for example, in Figure 1, a one-dimensional harmonic oscillator wavefunction is shown together with its Fourier spectrum whose amplitude decays quickly for frequencies higher than the inverse period of wavefunction oscillations. Gaussian basis functions satisfy zero boundary conditions at infinity by construction. Gaussian functions had been previously used to represent vibrational wavefunctions [5, 185-190]. The expansion for the kth wavefunction is

$$\psi_k(x) = \sum_{i=1}^{N} c_{ki} g_i(b, w, x) = \sum_{i=1}^{N} c_{ki} \prod_{j=1}^{d} \frac{b_{ij}}{\sqrt{\pi}} e^{-b_{ij}^2 (w_{ij} - x_j)^2} \tag{18}$$

In contrast to previous NN methods, where a separate NN was fitted to each vibrational wavefunction [51-53], we use the same basis $\{g_i(b, w, x)\}$ to determine several levels via the parameter vectors c_k, $k=1,...,L$. This corresponds to the RBNN architecture shown in Figure 2 for the case of two levels. As the c_k are linear parameters, they can be determined by means of linear algebra.

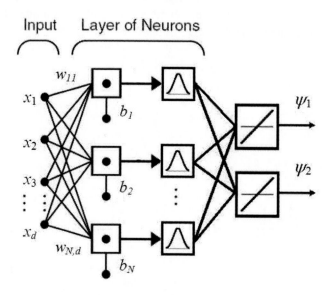

Figure 2. The architecture of a RBNN corresponding to Eq. (18).

We apply $\left(\hat{H} - E_k\right)$, where $\hat{H}$ is the Hamiltonian operator, to Eq. (18). The kinetic energy operator (KEO) is applied exactly by differentiating Eq. (18). If the Schrödinger equation is satisfied exactly at a set of points $x^{(n)}$, $n=1,...,M$, chosen in the region where the wavefunction is expected to be non-negligible, then a matrix equation obtains

$$\left(M - E_k S\right) c_k = 0 \tag{19}$$

where M and S are rectangular matrices (of the dimension $M \times N$) depending on values of the potential and of the Gaussian neurons at different points in the configuration space:

$$S_{ni} = \prod_{j=1}^{d} \frac{b_{ij}}{\sqrt{\pi}} e^{-b_{ij}^2 \left(w_{ij} - x_j^{(n)} \right)^2},$$ (20)

$$M_{ni} = \Gamma_{ni} S_{ni} + V_n S_{ni}$$

where $V_n = V\left(x^{(n)}\right)$ are the values of the potential energy, and the elements of the matrix Γ depend on the form of the KEO and are therefore molecule-dependent. However, if mass-weighted normal mode coordinates Q are used, the KEO can be written in a problem-independent way as

$$\hat{T} = -\frac{1}{2} \sum_{i=1}^{d} \frac{\partial^2}{\partial Q_i^2}$$ (21)

by neglecting the π–π coupling terms [214]. This approximation is expected to result only in small errors, and the KEO of Eq. (21) has been successfully used to solve vibrational problems [137, 215‑218]. Applying Eq. (21) to Eq. (18), one obtains an expression for the matrix Γ of Eq. (20):

$$\Gamma_{ni} = \sum_{j=1}^{d} b_{ij}^2 \left(1 - 2b_{ij}^2 \left(Q_j^{(n)} - w_{ij} \right)^2 \right)$$ (22)

Using this Γ with Eq. (21), the matrix equation (19) is the same for all molecules. A molecule-independent code simplifies setting up calculations.

The parameters and E_k are chosen so that E_k and ψ_k minimize a sum of residuals of all levels:

$$R' = \frac{1}{L} \sum_{k=1}^{L} \frac{\left\| (M - E_k S) c_k \right\|^2}{E_k^2 \left\| S c_k \right\|^2}$$ (23)

A key difference between this method and other collocation methods is that we solve a rectangular matrix equation. This allows for a much smaller number of basis functions than of PES points. The critical distinction between our approach and other neural network methods is that we use linear algebra to find the linear parameters-coefficients c. The non-linear parameters w, b determine the basis and are the same for all vibrational levels. This reduces the dimensionality of the parameter space in which non-linear optimization is done, which is the Achilles' heel of neural eigensolvers.

The non-linear parameters and the linear parameters c are optimized simultaneously. The rectangular matrix problem of Eq. (19) is solved for the linear coefficients c for each set of values of w and b as they are updated by the non-linear fitting algorithm. The energies E_k could be included into the set of non-linear parameters or be computed from the wavefunction as

$$E_k^{\text{int}} = \frac{\langle \psi_k | \hat{H} | \psi_k \rangle}{\langle \psi_k | \psi_k \rangle} \tag{24}$$

We found that the algorithm is most stable when E_k are fitted parameters which are made to track E_k^{int}, i.e. when we minimize

$$R'' = \frac{1}{L} \sum_{k=1}^{L} \frac{\|(M - E_k S)c_k\|^2}{E_k^2 \|Sc_k\|^2} + \lambda \sum_{k=1}^{L} \left(E_k - E_k^{\text{int}} \right)^2 \tag{25}$$

where λ is an empirical constant chosen so that the order of magnitude of both terms in (25) is similar. This is particularly useful when the anharmonicity is strong or the density of states is high, which increases the danger of being trapped in a local minimum with a high R'. Even when E_k^{int} are not used, the fitted energies E_k were found to converge to E_k^{int} and, therefore, similar to the variational method, the error in energy is expected to scale as ε^2 where ε is the error of the expansion (18).

We found that fixing the ratio of different j-components of b_{ij} to the ratio of corresponding coordinate ranges has a negligible effect on the quality of the best attainable fit while being much faster (due to a reduced dimensionality of the non-linear parameter space). All calculations presented below were performed in this way. Another way to reduce further the dimensionality of the parameter space while maintaining high basis set quality is to place several basis functions at the position w but optimize independently their widths b. This idea is similar to the split-valence or multiple-ζ bases commonly used in electronic structure calculations:

$$\psi_k(x) = \sum_{i=1}^{N} \sum_{s=1}^{\zeta} c_{kis} \prod_{j=1}^{d} \frac{b_{isj}}{\sqrt{\pi}} e^{-b_{isj}^2 (w_{ij} - x_j)^2} \tag{26}$$

For each Gaussian function, there are d parameters determining the location and only one width parameter if the ratio of b_{ij} is fixed. Adding a function at the same center introduces therefore only one additional parameter while greatly increasing the flexibility of the basis. This leads to a favorable scaling of the method with the size of the system. A sum of neurons at the same center can be thought of as a composite basis function whose shape is level-dependent, as components with different width b_i have independent coefficients c_{ki}. For example, in Figure 3, we show example shapes which can be obtained by positioning more than one Gaussian at the same center for $d=2$. The new basis is therefore extremely flexible. The flip side is that the calculation becomes less stable, and in practice, we obtained good fits only with $\zeta=2$ or 3. This strategy was used to compute the spectrum of the water molecule [69] reviewed in Section 0.

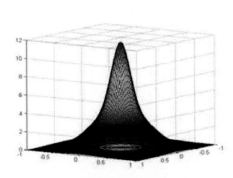 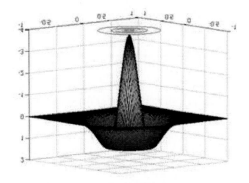

Figure 3. Examples of shapes that can be obtained by positioning two Gaussian functions at the same location (i.e. w) for $d=2$ by varying only b_q and c_q.

To solve the rectangular matrix equation (19), it is necessary to find pseudo eigen-pairs of the so-called "rectangular matrix pencil", and the solution is approximate [86, 219, 220]. Our algorithm is based on that proposed in Ref. [86], according to which an approximate right pseudo eigenvector c and the corresponding pseudo eigenvalue E of the rectangular pencil $M - ES$ can be computed with an iterative procedure that minimizes $\|(M - ES)c\|^2 / (1 + |E|^2)$. The approximate right pseudo eigenvector is the eigenvector of $(M - E_k S)^H (M - E_k S)$ corresponding to the smallest eigenvalue. Here, superscript "H" stands for the Hermitian conjugate. It is only necessary to compute the eigen-decomposition of a (small) matrix of size $N \times N$. E is then updated as the (+)-root of the scalar quadratic equation $c^H (M^H + ES^H)(M - E_k S)c$, and the cycle repeats until convergence. See Ref. [86] for more details. In our case, the energy is determined as a fitting parameter and from the wavefunction, Eq. (25). Only c_k are determined as in Ref. [86] at each iteration over non-linear parameters w, b, and E_k.

The algorithm was programmed in Matlab [221]. The matrix w was initialized by setting each row to the coordinate vector of a randomly chosen data point. The jth component of the initial vector b_{ij} for each i was set to the inverse of a fraction (1/3~1/4) of the range of the jth coordinate. The parameters w, b, E_k were fitted with the Levenberg-Marquardt algorithm [11] from Matlab. The parameter values change wildly during early stages of the fit, and it is necessary to restrain the range of their values. The values of w are limited to the coordinate ranges, and those of b to changes by a factor of 3 from the initial values; E_k are allowed to change in a window $E_{ini} \pm \Delta E$, where the initial guess E_{ini} can be guessed or taken either from the experiment or from a solution on a similar potential (see below about "parameter walking"), and ΔE is a few % of the expected energy

B. Sampling of the Potential Energy Surface

We recommend using a general method to sample the PES. First, a maximum value V_{max} is decided which ensures a reliable determination of the highest vibrational level $E_{v_{max}}$, for example, $V_{max} = 2E_{v_{max}}$. We assume $V_{min}=V(x_{equil})=0$. Next, the function $\max\left[V_{max} - V_{approx}(x)\right]$ (appropriately normalized) is used as a probability density function from which random points are drawn. With NN based methods, there is no need to sample the PES on a grid, and samples can be chosen to emphasize specific regions of space. The sampling is done by generating a random configuration x and accepting the configuration into the database if $\left(V_{max} - V_{approx}(x)\right)/V_{max} > r$, where r is a random number in the range [0, 1]. Rather than starting from configurations obtained with a conventional random number generator, we recommend using a multidimensional pseudo-random Sobol sequence [222]. This provides a more even point distribution. This sampling approach overweights low-energy points and was proven to be efficient in previous work [69, 223].

The function $V_{approx}(x)$ needs only approximately follow $V(x)$. If the SE is solved for a known potential [69], it can be used also for $V_{approx}(x)$. If the potential function is unknown, a simple approximation can be built by fitting a PES function to a small set of points [152]. We prefer using an uncoupled approximation to $V_{approx}(x)$:

$$V_{approx}(x) = \sum_{i=1}^{d} V_i^{slice}(x_i), \text{ where } V_i^{slice}(x_i)$$ are obtained by fitting ab initio slices along x_i

up to V_{max} to simple functional forms, e.g. polynomials [70]. $V_i^{slice}(x_i) = V_{max}$ will then also determine coordinate ranges for random configurations. The uncoupled harmonic PES based on normal mode frequencies can also be used instead of $V_{approx}(Q)$, if the ranges of normal mode coordinates Q required to reach V_{max} are realistic (i.e. if anharmonicity is not too large). Finally, ab initio calculations are performed at the retained points.

C. Tests on Synthetic Potentials

To study the performance and scaling of the algorithm, test calculations on coupled harmonic potentials of different dimensionality were performed in Ref. [68]. In one test, starting from a harmonic potential with a known solution,

$$V(x) = \sum_{i=1}^{d} x_i^2$$

$$\hat{T}(x) = -\sum_{i}^{d} \frac{\partial^2}{\partial x_i^2} \tag{27}$$

$$E = 2\sum_{i=1}^{d} \left(n_i + \frac{1}{2}\right)$$

a coordinate transformation was performed to create a coupled problem to which the algorithm of Section 0 was applied:

$$V(\mathbf{y}) = \sum_{i,j=1}^{d} a_{ij} y_i y_j$$

$$\hat{T}(\mathbf{y}) = -\sum_{i,j=1}^{d} \gamma_{ij} \frac{\partial^2}{\partial y_i \partial y_j}$$

$$(28)$$

where $a_{ij} = 1$ for $i=j$, $a_{ij} = 0$ for $i<j$, and $a_{ij} = 0.2/d$ for $i>j$. The results established a ballpark estimate of the number of PES points and basis functions that one should expect to use for systems of different size. For $d=2$, with 2500 PES points and 70 neurons, the average $\sqrt{R}$ (Eq. (17)) was about 1.5% of the corresponding energies, and the average energy error was under 0.3%, which was comparable to or better which previous neural eigensolvers obtained for one-dimensional problems and solving for one level at a time [51, 52]. There is only one report of a neural network solution for only the ground state of a three-dimensional coupled harmonic oscillator [53]. With our method, we could compute the three lowest vibrational levels of a four- and six- dimensional system. For $d=4$, with 5000 PES points and 70 neurons, the average $\sqrt{R}/E$ was about 1.7%. For $d=6$, from 5000 points and with 50 neurons, we obtained $\langle \sqrt{R}/E \rangle \approx 3.4\%$ and an energy error of about 2.3%. It was therefore established that the method is able to solve higher-dimensional problems than the standard NN method and with a favourable dimensionality scaling. The number of PES points was small enough so that they could be computed with electronic structure methods.

Another test was done for a case where the exact solution and therefore a good guess for E_{ini} was not known. The potential used was

$$V(\mathbf{x}) = \sum_{i=1}^{d} x_i^2 + \alpha \left(\sum_{i=1}^{d} x_i \right)^2$$

$$(29)$$

where $\alpha=1$, and the KEO was the same as in Eq. (27). By solving the SE for a series of $\alpha=0.0, 0.05, \ldots, 1.0$ and using the energies and wavefunctions from the previous solution as initial guesses for the next, it was possible to obtain the solution of (29) by effectively "walking" the parameters into the global minimum by adiabatically changing the potentiall [68]. As a harmonic approximation can be easily obtained for the energies and vibrational wavefunctions on any bound potential, this strategy represents an effective general approach, and we used it for a highly anharmonic molecule-surface system in Section 0.

D. Real Molecule in Vacuum

The first test of the algorithm on a real molecule was presented in Ref. [69], which was also the first application of an NN based method to a molecule. For the calculations of Ref. [69], a previously published PES [224] of H_2O was used ($d=3$) with an exact KEO, which allowed for comparisons with previous variational calculations on the same potential [225]. This established the accuracy of the method for a real-life application. The PES of Ref. [224]

was sampled as described in Section 0 up to $V_{max}=15000$ cm^{-1}. The exact KEO in symmetrized Radau coordinates (R,r,c) was used:

$$-2m_H T(R,r,c) = \frac{\partial^2}{\partial R^2} + \frac{\partial^2}{\partial r^2} + \left(\frac{1}{R_1^2} + \frac{1}{R_2^2}\right)\frac{\partial}{\partial c}(1-c^2)\frac{\partial}{\partial c} \tag{30}$$

where $R \equiv (R_1 + R_2)/\sqrt{2}$, $r \equiv (R_1 - R_2)/\sqrt{2}$, R_1 and R_2 are unsymmetrized Radau coordinates [226] and $c \equiv \cos\theta$. The matrix Γ of Eq. (20) then has the form

$$\Gamma_{ni} = \frac{2b_i^2}{m_H}\left(\begin{array}{c} 1 - b_i^2\left(w_{i1} - {}^n R\right)^2 - b_i^2\left(w_{i2} - {}^n r\right)^2 + \frac{1}{2}\left(\frac{1}{{}^n R_1^2} + \frac{1}{{}^n R_2^2}\right) \times \\ \times\left(2\,{}^n c\left(w_{i3} - {}^n c\right) + (1 - {}^n c^2) - 2(1 - {}^n c^2)b_i^2\left(w_{i3} - {}^n c\right)^2\right) \end{array}\right) \tag{31}$$

The vibrational SE was then solved for the first five levels for different numbers of PES points, different numbers of neurons, and also for $\zeta=1$, 2, and 3 (Eq. (26)). Energies were initialized by randomly choosing a number in the window $E_{exact} \pm 10$ cm^{-1}, where E_{exact} are taken from Ref. [223]. The results showed that it is possible to obtain $\left\langle \sqrt{R}/E \right\rangle < 1\%$ and $\Delta E < 2$ cm^{-1} with 6000 PES samples and only 50 Gaussian neurons. With 1500 PES points and as few as 15 composite ($\zeta=3$) neurons, $\sqrt{R}$ was less than 2% of the energy and the error in energy was about 3 cm^{-1}. [69] That is, an accuracy generally sufficient for species identification can be obtained from a number of potential points that can easily be computed ab initio without building a surface, and with an extremely small basis.

In Ref. [70], we then applied the general formulation of the method, Eq. (22) to compute the spectrum of the water molecule directly from DFT data, which is the intended application. For electronic structure calculations, we used the SIESTA code [227], the PBE functional [228], and Troullier-Martins pseudopotentials [229] provided with SIESTA. The DZP basis with cutoff radius determined by $E_{shift}=0.01$ Ry was used [230]. These parameters provided a structural parameters as well as of normal mode frequencies (Table 1) close to experimental data [231] and previous calculations [232, 233], see Ref. [70] for more details. The PES was sampled at 1500 points as described in Section 0 with $V_{max}=20000$ cm^{-1}. As the exact PES is unknown, we used $V_{approx}(Q) = V_{unc}(Q) = \sum_{i=1}^{d} V_i^{slice}(Q_i)$ for sampling, as described above.

$V_{unc}(Q)$ is the uncoupled anharmonic potential, and its slices are shown in Figure 4 together with the corresponding harmonic slices. The anharmonicity is significant. As the goal is to obtain the fundamental frequencies, four vibrational levels were computed: the zero point energy (ZPE) and three levels each with one quantum in each of the normal modes. With no exact solution to compare to, we produced 3000 ab initio samples and in each run, used 1500 points randomly selected from the full set of 3000. Energy levels obtained from different calculations agreed to within ± 0.5 cm^{-1}. Fits were also performed by minimizing R', Eq. (23) and comparing the fitted energies to E_k^{int}, Eq. (24), and the difference was of the order of 0.1 cm^{-1} in different fits.

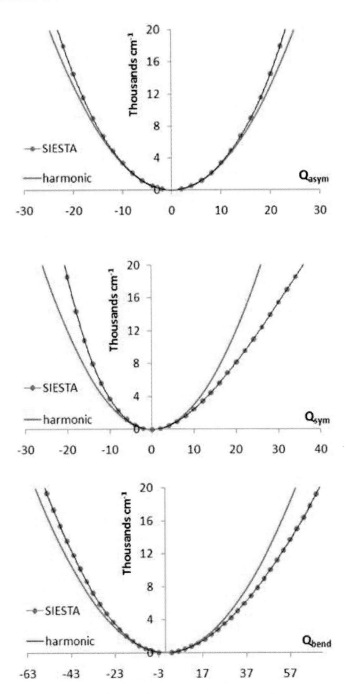

Figure 4. The slices along mass-weighted normal mode coordinates Q for water in vacuum. Solid lines are harmonic curves and lines with symbols are with DFT.

The solution on the anharmonic PES was found by using the parameter "walking" approach described in Section 0: we started from a harmonic potential built from the normal mode frequencies from Table 1, for which the exact energies are known, and slowly changed the potential first by using $V = \alpha V_{harm} + (1-\alpha) V_{unc}$, and then by using $V = \alpha V_{unc} + (1-\alpha) V_{coupl}$, where V_{coupl} is the coupled anharmonic potential defined by the DFT points. The parameter α was changed from 0 to 1 in 20 steps. This allowed us first to isolate the effect of anharmonicity before computing anharmonic and coupled solutions. Of course, it is possible to use $V = \alpha V_{harm} + (1-\alpha) V_{coupl}$. For the anharmonic uncoupled case, we obtained $\left\langle \sqrt{R}/E \right\rangle \approx 0.2\%$ with 30 neurons resulting in frequencies listed in the 5th column of Table 1.

The uncoupled and anharmonic energies were checked by solving one-dimensional Schrödinger equations on the slices with a Runge-Kutta [11] based method coded in Matlab. Levels computed with our method and from the Runge-Kutta method agree to within 0.5 cm^{-1}. For the anharmonic coupled case, i.e. the DFT potential, we obtained $\left\langle \sqrt{R}/E \right\rangle \approx 0.5\%$ with 35 neurons resulting in frequencies listed in the 8th column of Table 1. The theory/experiment ratios for frequencies given in brackets of column 8 are very similar for all modes and are <1, apparently due to an underestimation by DFT of the dissociation energies [234]. Our results are in agreement with the results of Ref. [137] obtained with the VSCF method and using a PES that included 2nd order coupling terms. In contrast, no PES function was constructed here, and there is no restriction on what coupling terms are allowed.

These tests confirmed that the method provides an accuracy of the order of 1 cm^{-1} by computing the vibrational spectrum directly from a small, computable ab initio, set of PES points and by using an extremely small basis set, for a small molecule.

E. A Molecule-surface System

In Ref. [70], the method of Section 0 was, for the first time, applied to a molecule-surface system: a water molecule on the Pt(111) surface. As argued in the introduction, this is an application where we believe the method holds most promise, because (i) its accuracy compares well with the accuracy of DFT calculations used to sample the PES and the accuracy of experimental IR spectra, (ii) PESs for molecule-surface systems are virtually non-existent, and (iii) only lowest vibrational levels are usually required for species identification. Also, errors introduced by approximations will largely cancel when computing spectral shifts upon adsorption.

The H_2O/Pt(111) system is important for catalysis, molecular electronics and biological applications [235] and has been the subject of considerable theoretical and experimental interest [108, 232, 236, 237]. Experimental measurements of IR spectra are difficult below about 0.1 ML [108, 238, 239], so the ability to compute the spectrum is important.

The Pt(111) surface was modeled with a three-layer slab, where the top layer was allowed to relax, and the bottom layers were fixed at the bulk positions. This thickness was shown to be adequate in the modeling of water-platinum interactions [236, 237, 240]. The slabs were separated by about 15.5 Å of vacuum. The 3x3 calculation cell used in the calculations is shown in Figure 5. It ensures negligible inter-cell interaction for the adsorbed water molecule and is repeated ad infinitum in all direction. The electronic structure method and parameters were the same as in Section 0. Converged results were obtained with the Brillouin zone sampled with Monkhorst-Pack 3x3x1 k-points. [241] The optimized geometry of $H_2O/Pt(111)$ (shown in Figure 5) and intramolecular modes of H_2O were in good agreement with other calculations [232, 236, 237, 242]. See Ref. [70] for more details.

As our purpose is to estimate the role of anharmonicity and coupling in frequency shifts upon adsorption, we considered only intramolecular vibrations of the water molecule. The coupling to frustrated translation and rotation DOFs is expected to be weak due to a large difference in frequencies [103, 111, 243]. The harmonic frequencies of the intramolecular modes are given in the 3rd column of Table 1.

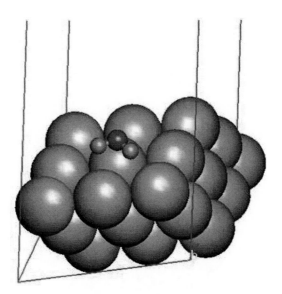

Figure 5. The calculation cell showing the optimized $H_2O/Pt(111)$. Grey -Pt, red – O, blue – H atoms.

The steps described in Section 0 were repeated for $H_2O/Pt(111)$. **Error! Reference source not found.** shows scans of the $H_2O/Pt(111)$ PES along the normal coordinates together with the harmonic curves. Clearly, binding to the substrate weakens intramolecular bonds resulting in lower potential curves. We therefore used V_{max}=18,000 cm^{-1}. For the uncoupled anharmonic PES, we obtain $\left\langle \sqrt{R}/E \right\rangle \approx 0.3\%$ with 30 basis functions, and for the fully coupled, ab initio potential $\left\langle \sqrt{R}/E \right\rangle \approx 0.7\%$ with 35 basis functions. The corresponding frequencies are listed in Table 1.

We can now compare the shift due to adsorption in the harmonic and harmonic approximations, given in the 4th and 10th columns of Table 1, respectively. Even though the numbers are of the same order of magnitude, the difference is several dozen cm^{-1} and is large both compared to typical instrumental precision [89, 108] and to typical differences in frequencies due to different adsorption configurations [99]. This confirms the importance of going beyond the uncoupled harmonic approximation for species assignment on surfaces.

Experimental IR measurements for low H_2O coverage report only one diffuse band for OH stretches at around 3400 cm^{-1}, [239] corresponding to a shift of 257-356 cm-1, which compares well to the present result. The HOH bend observed experimentally near 1630 cm^{-1} [239] is likely to come from interacting molecules [108], as it is higher than that of free H_2O. It is the advantage of theory that it can isolate the spectrum of a single molecule on an ideal surface.

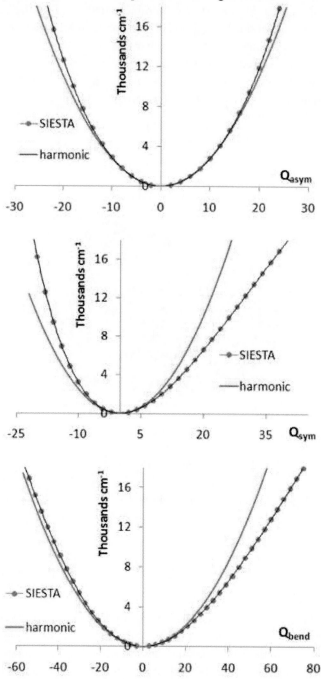

Figure 6. The slices along mass-weighted normal mode coordinates Q for internal modes of $H_2O/$ on Pt(111). Solid lines are harmonic curves and lines with symbols are with DFT.

4. Conclusion

Although well-developed methods exist for accurately computing vibrational spectra once a PES is known, it would be useful to have a procedure that enabled one to compute a spectrum from a (relatively) small set of potential values. These values could, themselves, be directly computed by solving the electronic Schrödinger equation. The number of potential values can only be small if the basis is. In this chapter, we reviewed mainstream methods and focus on a new procedure that does make it possible to compute a spectrum directly from points. This obviates the need to construct a PES which introduces extra cost and a source of error. Because we avoid fitting a PES, we map directly from electronic energies to vibrational energies and wavefunctions. The new approach combines the collocation method (obviating the need to do quadratures) with a small NN basis set (obviating the need for a PES). The NN basis set is very flexible and therefore small. Of course, adaptable basis functions must be better. Is the construction of the adaptable basis too expensive? Most of the cost comes from a high-dimensional non-linear optimization. This also carries the danger of local minima, as in the case of the standard NN approach. We have reduced the cost by combining linear optimization via a rectangular eigenvalue problem for some parameters with general non-linear optimization for others. We found that minima found by calculations with different initial parameters are of similar quality. Local minima with high values R are rare and easy to spot. With this approach, one is able to compute a small number of vibrational levels with a modest, of the order of 1 cm^{-1}, accuracy which is comparable to typical experimental uncertainties for vibrational frequencies of a molecule adsorbed on a surface. The calculations presented here were not costly. They were all done on a personal computer. If mass-weighted normal coordinates are used, the method can be applied with no modifications to any molecule.

The ideas we developed to improve the NN approach for the vibrational Schrödinger equation can certainly be used in other areas. One possible application is the solution of the single-electron Kohn-Sham (KS) equations in DFT [244]. These are three-dimensional equations which are repeatedly solved during the self-consistency loop [245] and the cost of their solution is a major component of the cost of DFT calculations, which is the limiting factor in rational design of materials [246-249]. Recently, neural networks have been used to solve the KS equations for the He and Li atoms with modest accuracy [213]. This is therefore another field where improvements in NN-based eigensolvers such as proposed here could have a significant impact.

5. Tables

Table 1. Frequencies of internal vibrations of H_2O in cm^{-1} and shifts due to adsorption for H_2O/Pt(111) for different levels of theory. The ratio of the theoretical to the experimental results for H_2O is given in brackets

H_2O Exper.	Harmonic			$V_{unc}(Q)$			$V_{coupl}(Q)$		
	H_2O	H_2O/Pt	Shift	H_2O	H_2O/Pt	Shift	H_2O	H_2O/Pt	Shift
3755.9	3789(1.01)	3480	-308	3875(1.03)	3561	-314	3600(0.96)	3216	-384
3657.1	3631(0.99)	3351	-280	3549(0.97)	3235	-314	3468(0.95)	3119	-349
1594.7	1578(0.99)	1536	-42	1575(0.99)	1510	-65	1525(0.96)	1452	-73

References

[1] Currant, A.; Hilbert, D. *Methods of Mathematical Physics*; John Wiley & Sons: New York, 1991.

[2] Papousek, D.; Aliev, M. R. *Molecular vibrational-rotational spectra*; Elsevier: Amsterdam, 1982.

[3] Carter, S.; Handy, N. C. *Comp. Phys. Rep.* 1986, 5, 115.

[4] Tennyson, J. *Comp. Phys. Rep.* 1986, 4, 1.

[5] Bacic, Z.; Light, J. C. *Annu. Rev. Phys. Chem.* 1989, 40, 469.

[6] Kosloff, R. *Annu. Rev. Phys. Chem.* 1994, 45, 145.

[7] Bowman, J. M.; Carrington Jr., T.; Meyer, H.-D. *Mol. Phys.* 2008, 106, 2145.

[8] Carrington Jr., T. Rovibrational Energy Level Calculations for Molecules. In *Encyclopedia of Computational Chemistry*, Schleyer, P. R., Allinger, N. L., Clark, T., Gasteiger, J., Kollman, P. A., Schaefer III, H. F., Schreiner, P. R., Eds.; John Wiley and Sons: New York, 1998; Vol. 5, pp 3157-3166.

[9] Schatz, G. C. *Rev. Mod. Phys.* 1989, 61, 669-688.

[10] Truhlar, D. G.; Steckler, R.; Gordon, M. S. *Chem. Rev.* 1987, 87, 217-236.

[11] Press, W. H.; Teukolsky, S. A.; Vetterling, W. T.; Flannery, B. P. Numerical recipes in Fortran 77: *The art of scientific computing*; 2 ed.; Cambridge University Press: Cambridge, 1992; Vol. 1.

[12] Hartree, D. R. *The Calculation of Atomic Structure*; John Wiley: New York, 1957.

[13] Cooley, J. W. *Math. Comput.* 1961, 15, 363.

[14] Hu, X.-G.; Ho, T.-S.; Rabitz, H. *Comp. Phys. Comm.* 1998, 113, 168.

[15] Wang, X.-G.; Carrington Jr., T. *J. Chem. Phys.* 2001, 115, 9781.

[16] Wang, X.-G.; Carrington Jr., T. *J. Chem. Phys.* 2008, 129, 234102.

[17] Light, J. C.; Carrington Jr., T. *Adv. Chem. Phys.* 2000, 114, 263.

[18] Bramley, M. J.; Carrington Jr., T. *J. Chem. Phys.* 1993, 99, 8519-8541.

[19] Yu, H.-G.; Muckerman, J. T. *J. Mol. Spectrosc.* 2002, 214, 11.

[20] Czako, G.; Furtenbacher, T.; Csaszr, A. G.; Szalay, V. *Mol. Phys.* 2004, 102, 2411.

[21] Wang, X.-G.; Carrington Jr., T. *J. Chem. Phys.* 2002, 117, 6923-6934.

[22] Bramley, M. J.; Carrington Jr., T. *J. Chem. Phys.* 1994, 101, 8494.

[23] Bramley, M. J.; Handy, N. C. *J. Chem. Phys.* 1993, 98, 1378.

[24] Bowman, J. M.; Gazdy, B. *J. Chem. Phys.* 1991, 94, 454.

[25] Luckhaus, D. *J. Chem. Phys.* 2000, 113, 1329.

[26] Mladenovic, M. *Spectrochimica Acta A* 2002, 58, 809.

[27] Manzhos, S.; Yamashita, K.; Carrington Jr., T. Extracting Functional Dependence from Sparse Data Using Dimensionality Reduction: Application to Potential Energy Surface Construction. In *Coping with Complexity: Model Reduction and Data Analysis*, Gorban, A. N., Roose, D., Eds.; Springer: Berlin, 2011; pp 133-149.

[28] Manzhos, S.; Yamashita, K. *Surf. Sci.* 2010, 604, 555-561.

[29] Manzhos, S.; Carrington Jr., T. *J. Chem. Phys.* 2006, 125, 084109.

[30] Hirst, D. M. *Potential energy surfaces*; Taylor and Francis: London, 1985.

[31] Collins, M. A. *Theor. Chem. Acc.* 2002, 108, 313-324.

[32] Carter, E. A. *Science* 2008, 321, 800.

[33] Hussain, A.; Curulla Ferre, D.; Gracia, J.; Nieuwenhuys, B. E.; Niemantsverdriet, J. W. *Surf. Sci.* 2009, 603, 2734.

[34] Sousa, S. F.; Fernandes, P. A.; Ramos, M. J. *J. Phys. Chem. A* 2007, 111, 10439.

[35] Manzhos, S.; Carrington Jr., T. *J. Chem. Phys.* 2008, 129, 224104.

[36] Sharma, A. R.; Braams, B. J.; Carter, S.; Shepler, B. C.; Bowman, J. M. *J. Chem. Phys.* 2009, 130, 174301.

[37] Malshe, M.; Raff, L. M.; Rockey, M. G.; Hagan, M. T.; Agrawal, P. A.; Komanduri, R. *J. Chem. Phys.* 2007, 127, 134105.

[38] Malshe, M.; Narulkar, R.; Raff, L. M.; Hagan, M. T.; Bukkapatnam, S.; Agrawal, P. A.; Komanduri, R. *J. Chem. Phys.* 2009, 130, 184102.

[39] Alis, O. F.; Rabitz, H. *J. Math. Chem.* 2001, 29, 127-142.

[40] Rabitz, H.; Alis, O. F.; Shorter, J.; Shim, K. *Comp. Phys. Comm.* 1999, 117, 11-20.

[41] Rabitz, H.; Alis, O. F. *J. Math. Chem.* 1999, 25, 197-233.

[42] Li, G.; Rosenthal, C.; Rabitz, H. *J. Phys. Chem. A* 2001, 105, 7765-7777.

[43] Carter, S.; Culik, S. J.; Bowman, J. M. *J. Chem. Phys.* 1997, 107, 10458-10469.

[44] Carter, S.; Bowman, J. M. *J. Phys. Chem. A* 2000, 104, 2355-2361.

[45] Carter, S.; Handy, N. C. *Chem. Phys. Lett.* 2002, 352, 1-7.

[46] Carter, S.; Bowman, J. M.; Harding, L. B. *Spectrochimica Acta A* 1997, 53, 1179-1188.

[47] Jaeckle, A.; Meyer, H.-D. *J. Chem. Phys.* 1998, 109, 3772-3779.

[48] Jaeckle, A.; Meyer, H.-D. *J. Chem. Phys.* 1996, 104, 7974-7984.

[49] Jaeckle, A.; Meyer, H.-D. *J. Chem. Phys.* 1995, 102, 5605-5615.

[50] Manzhos, S.; Carrington Jr., T. *J. Chem. Phys.* 2006, 125, 194105.

[51] Sugawara, M. *Comp. Phys. Comm.* 2001, 140, 366-380.

[52] Nakanishi, H.; Sugawara, M. *Chem. Phys. Lett.* 2000, 327, 429-438.

[53] Lagaris, I. E.; Likas, A.; Fotiadis, D. I. *Comp. Phys. Comm.* 1997, 104, 1-14.

[54] Diver, D. A. *J. Phys. A: Math. Gen.* 1993, 26, 3503.

[55] Zeiri, Y.; Fattal, E.; Kosloff, R. *J. Chem. Phys.* 1995, 102, 1859.

[56] Chaudhury, P.; Bhattacharyya, S. P. *Chem. Phys. Lett.* 1998, 296, 51.

[57] Darsey, J. A.; Noid, D. W.; Upadhyaya, B. R. *Chem. Phys. Lett.* 1991, 177, 189-194.

[58] Androsiuk, J.; Kulak, L.; Sienicki, K. *Chem. Phys.* 1993, 173, 377-383.

[59] Mushinski, A.; Nightingale, M. P. *J. Chem. Phys.* 1994, 101, 8831.

[60] van Milligen, B. Ph.; Tribaldos, V.; Jimenez, J. A. *Phys. Rev. Lett.* 1995, 75, 3594.

[61] Yang, W.; Peet, A. C. *Chem. Phys. Lett.* 1988, 153, 98.

[62] Finlayson, B. A. *The Method of Weighted Residuals and Variational Principles*; Academic Press: New York, 1972.

[63] Collatz, L. *The Numerical Treatment of Differential Equations*; Springer: Berlin, 1960.

[64] Yang, W.; Peet, A. C. *J. Chem. Phys.* 1990, 92, 522.

[65] Peet, A. C.; Yang, W. *J. Chem. Phys.* 1989, 90, 1746.

[66] Peet, A. C.; Yang, W. *J. Chem. Phys.* 1989, 91, 6598.

[67] Yang, W.; Peet, A. C.; Miller, W. H. *J. Chem. Phys.* 1989, 91, 7537.

[68] Manzhos, S.; Carrington Jr., T. *Can. J. Chem.* 2009, 87, 864-871.

[69] Manzhos, S.; Yamashita, K.; Carrington, T. *Chemical Physics Letters* 2009, 474, 217-221.

[70] Manzhos, S.; Yamashita, K.; Carrington Jr., T. *Surf. Sci.* 2010, submitted.

[71] Hassoun, M. H. *Fundamentals of artificial neural networks*; MIT Press: Cambridge, MA, 1995.

[72]   Sumpter, B. G.; Getino, C.; Noid, D. W. *Annu. Rev. Phys. Chem.* 1994, 45, 439-481.
[73]   Widrow, B.; Lehr, M. A. *Proc. IEEE* 1990, 78, 1415-1442.
[74]   Geremia, J. M.; Rabitz, H. *J. Chem. Phys.* 2001, 115, 8899-8912.
[75]   Geremia, J. M.; Rabitz, H.; Rosenthal, C. *J. Chem. Phys.* 2001, 114, 9325-9336.
[76]   Hornik, K. *Neural Networks* 1991, 4, 251-257.
[77]   Hornik, K.; Stinchcombe, M.; White, H. *Neural Networks* 1990, 3, 551-560.
[78]   Hornik, K. *Neural Networks* 1989, 2, 359-366.
[79]   Funahashi, K. *Neural Networks* 1989, 2, 183-192.
[80]   Kurkova, V. *Neural Networks* 1992, 5, 501-506.
[81]   Scarselli, F.; Tsoi, A. C. *Neural Networks* 1998, 11, 15-37.
[82]   Barron, A. *IEEE Trans. Information Theory* 1993, 39, 930.
[83]   Chui, C.; LI, X. *J. Approximation Theory* 1992, 70, 131.
[84]   Mhaskar, H.; Miccheli, C. *IBM J. Res. Development* 1994, 38, 277.
[85]   Mhaskar, H.; Miccheli, C. *Adv. Appl. Math.* 1992, 13, 350.
[86]   Boutry, G.; Elad, M.; Golub, G. H.; Milanfar, P. *SIAM J. Matrix Anal. Appl.* 2005, 27, 582-601.
[87]   Mulero-Martinez, J. I. *Neurocomputing* 2007, 70, 1439.
[88]   Meng, S.; Xu, L. F.; Wang, E. G.; Gao, S. *Phys. Rev. Lett.* 2002, 89, 176104.
[89]   Wang, Y.; Woell, C. *Surf. Sci.* 2009, 603, 1589.
[90]   Hirschmugl, C. J. *Surf. Sci.* 2002, 500, 577-604.
[91]   Qiu, H.; Idriss, H.; Wang, Y.; Woll, C. *J. Phys. Chem. C* 2008, 112, 9828-9834.
[92]   Rasko, J.; Solymosi, F. *J. Phys. Chem.* 1994, 98, 7147-7152.
[93]   Ramis, G.; Busca, G.; Lorenzelli, V. *Mater. Chem. Phys.* 1991, 29, 425-435.
[94]   Monder, D. S.; Karan, K. *J. Phys. Chem. C* 2010, 114, 22597-22602.
[95]   van Grootel, P. W.; Hensen, E. J. M.; van Santen, R. A. *Surf. Sci.* 2009, 603, 3275.
[96]   Haubrich, J.; Becker, C.; Wandelt, K. *Surf. Sci.* 2009, 603, 1476.
[97]   Koch, H. P.; Bako, I.; Schennach, R. *Surf. Sci.* 2010, 604, 596.
[98]   Kiss, J.; Witt, A.; Meyer, B.; Marx, D. *J. Chem. Phys.* 2009, 130, 184706.
[99]   Breedon, M.; Spencer, M. J. S.; Yarovsky, I. *J. Phys. Chem. C* 2010, 114, 16603.
[100] Strunk, J.; Kaehler, K.; Xia, X.; Muhler, M. *Surf. Sci.* 2009, 603, 1776.
[101] Boccuzzi, F.; Borello, E.; Zecchina, A.; Bossi, A.; Camia, M. *J. Catal.* 1978, 51, 150.
[102] Santana, J. A.; Ishikawa, Y. *Chem. Phys. Lett.* 2009, 478, 110.
[103] Bahel, A.; Bacic, Z. *J. Chem. Phys.* 1999, 111, 11164.
[104] He, H.; Zapol, P.; Curtiss, L. *J. Phys. Chem. C* 2010, 114, 21474-21481.
[105] Alexiadis, A.; Kassinos, S. *Chem. Rev.* 2008, 108, 5014-5034.
[106] Setyowati, K.; Piao, M. J.; Chen, J.; Liu, H. *Appl. Phys. Lett.* 2008, 92, 043105.
[107] Georgieva, I.; Trendafilova, N.; Creaven, B. S.; Walsh, M.; Noble, A.; McCann, M. *Chem. Phys.* 2009, 365, 69.
[108] Jacobi, K.; Beduerfti, K.; Wang, Y.; Ertl, E. *Surf. Sci.* 2001, 472, 9.
[109] Huang, P.; Carter, E. A. *Annu. Rev. Phys. Chem.* 2008, 59, 261.
[110] Park, S. C.; Bowman, J. M.; Jelski, D. A. *J. Chem. Phys.* 1996, 104, 2457-2460.
[111] Tremblay, J. C.; Beyvers, S.; Saalfrank, P. *J. Chem. Phys.* 2008, 128, 194709.
[112] Marquardt, R.; Cuvelier, F.; Olsen, R. A.; Baerends, E. J.; Tremblay, J. C.; Saalfrank, P. *J. Chem. Phys.* 2010, 132, 074108.
[113] Shemesh, D.; Mullin, J.; Gordon, M. S.; Gerber, R. B. *Chem. Phys.* 2008, 347, 218.
[114] Stewart, J. J. P. *J. Conput. Chem.* 1989, 10, 209-220.

[115] Brauer, B.; Chaban, G. M.; Gerber, R. B. *Phys. Chem. Chem. Phys.* 2004, 6, 2543.

[116] 1Kurten, T.; Biczysko, M.; Rajamaeki, T.; Laasonen, K.; Halonen, L. *J. Phys. Chem. B* 2005, 109, 8954.

[117] Sherrill, C. D. *J. Chem. Phys.* 2010, 132, 110902.

[118] MacDonald, J. K. L. *Phys. Rev.* 1933, 43, 830-833.

[119] Ritz, W. *J. Reine Angew. Math.* 1909, 135, 1-61.

[120] Boys, S. F. *Proc. Roy. Soc. A* 1969, 309, 195-208.

[121] Householder, A. S. *J. ACM* 1958, 5, 339-342.

[122] Francis, J. G. F. *Comput. J.* 1962, 4, 332-345.

[123] Francis, J. G. F. *Comput. J.* 1961, 4, 265-271.

[124] Mandelshtam, V. A.; Taylor, H. S. *J. Chem. Phys.* 1997, 106, 5085.

[125] Wall, M. R.; Neuhauser, D. *J. Chem. Phys.* 1995, 102, 8011.

[126] Chen, R.; Guo, H. *J. Chem. Phys.* 1998, 108, 6068.

[127] Mandelshtam, V. A.; Taylor, H. S. *J. Chem. Phys.* 1995, 102, 7390.

[128] McNichols, A.; Carrington Jr., T. *Chem. Phys. Lett.* 1993, 202, 464.

[129] Neuhauser, D. *J. Chem. Phys.* 1990, 93, 2611.

[130] Golub, G. H.; Loan, C. F. V. *Matrix Computations*; Johns Hopkins University Press: Baltimore, 1989.

[131] Cullum, J. K.; Willoughby, R. A. *Lanczos Algorithms for Large Symmetric Eigenvalue Computations*; Birkhauser: Boston, MA, 1985; Vol. 1.

[132] Manthe, U.; Koppel, H. *J. Chem. Phys.* 1990, 93, 345.

[133] Wang, X.-G.; Carrington Jr., T. *J. Chem. Phys.* 2003, 118, 6946.

[134] Wang, X.-G.; Carrington Jr., T. *J. Chem. Phys.* 2004, 121, 2937-2954.

[135] Davidson, E. *J. Comp. Phys.* 1975, 17, 87.

[136] Vendrell, O.; Gatti, F.; Lauvergnat, D.; Meyer, H.-D. *J. Chem. Phys.* 2007, 127, 184302.

[137] Njegic, B.; Gordon, M. S. *J. Chem. Phys.* 2006, 125, 224102.

[138] Rauhut, G. *J. Chem. Phys.* 2004, 121, 9313-9322.

[139] Carter, S.; Bowman, J. M.; Huang, X. *Int. Rev. Phys. Chem.* 2003, 22, 533.

[140] Beck, M. H.; Jaeckle, A.; Worth, G. A.; Meyer, H.-D. *Phys. Rep.* 2000, 324, 1.

[141] Meyer, H.-D.; Manthe, U.; Cederbaum, L. S. *Chem. Phys. Lett.* 1990, 165, 73.

[142] Manthe, U.; Meyer, H.-D.; Cederbaum, L. S. *J. Chem. Phys.* 1992, 97, 3199.

[143] Meyer, H.-D.; Worth, G. A. *Theor. Chem. Acc.* 2003, 109, 251.

[144] Avila, G.; Carrington Jr., T. *J. Chem. Phys.* 2009, 131, 174103.

[145] Light, J. C.; Hamilton, I. P.; Lill, J. V. *J. Chem. Phys.* 1985, 82, 1400.

[146] Bacic, Z.; Light, J. C. *J. Chem. Phys.* 1986, 85, 4594.

[147] Bacic, Z.; Light, J. C. *J. Chem. Phys.* 1987, 86, 3065.

[148] Wang, X.-G.; Carrington Jr., T. *J. Chem. Phys.* 2009, 130, 094101.

[149] Wang, X.-G.; Carrington Jr., T. *J. Chem. Phys.* 2008, 128, 194109.

[150] Carrington Jr., T.; Light, J. C. *Adv. Chem. Phys.* 2000, 114, 263.

[151] Bowman, J. M.; Braams, B. J.; Carter, S.; Chen, C.; Czako, G.; Fu, B.; Huang, X.; Kamarchik, E.; Sharma, A. R.; Shepler, B. C.; Wang, Y.; Xie, Z. *J. Phys. Chem. Lett.* 2010, 1, 1866-1874.

[152] Manzhos, S.; Wang, X.-G.; Dawes, R.; Carrington Jr., T. *J. Phys. Chem. A* 2006, 110, 5295-5304.

[153] Manzhos, S.; Carrington Jr., T. *J. Chem. Phys.* 2007, 127, 014103.

[154] Manzhos, S.; Yamashita, K.; Carrington Jr., T. *Comp. Phys. Comm.* 2009, 180, 2002.

[155] Czako, G.; Shepler, B.; Braams, B. J.; Bowman, J. M. *J. Chem. Phys.* 2009, 130, 084301.

[156] Thompson, K. C.; Jordan, M. J. T.; Collins, M. A. *J. Chem. Phys.* 1998, 108, 8302-8316.

[157] Crespos, C.; Collins, M. A.; Pijper, E.; Kroes, G. J. *J. Chem. Phys.* 2004, 120, 2392.

[158] Ramazani, S.; Frankcmbe, T. J.; Andersson, S.; Collins, M. A. *J. Chem. Phys.* 2009, 130, 244302.

[159] Zhang, D. H.; Collins, M. A.; Lee, S.-Y. *Science* 2000, 290, 961.

[160] Maisuradze, G. G.; Thompson, D. L.; Wagner, A. F.; Minkoff, M. *J. Chem. Phys.* 2003, 119, 10002-10014.

[161] Guo, Y.; Kawano, A.; Thompson, D. L.; Wagner, A. F.; Minkoff, M. *J. Chem. Phys.* 2004, 121, 5091-5097.

[162] Dawes, R.; Passalacqua, A.; Wagner, A. F.; Sewell, T. D.; Minkoff, M.; Thompson, D. L. *J. Chem. Phys.* 2009, 130, 144107.

[163] Dawes, R.; Thompson, D. L.; Wagner, A. F.; Minkoff, M. *J. Chem. Phys.* 2008, 128, 084107.

[164] Frishman, A. M.; Hoffman, D. K.; Rakauskas, R. J.; Kouri, D. *J. Chem. Phys. Lett.* 1996, 252, 62.

[165] Hoffman, D. K.; Frishman, A.; Kouri, D. *J. Chem. Phys. Lett.* 1996, 262, 393.

[166] Frishman, A. M.; Hoffman, D. K.; Kouri, D. J. *J. Chem. Phys.* 1997, 107, 804.

[167] Szalay, V. *J. Chem. Phys.* 1999, 119, 8804-8818.

[168] Hollebeek, T.; Ho, T.-S.; Rabitz, H. *Annu. Rev. Phys. Chem.* 1999, 50, 537-570.

[169] Ho, T.-S.; Rabitz, H. *J. Chem. Phys.* 2003, 119, 6433-6442.

[170] McLeod, M.; Carrington Jr., T. *Chem. Phys. Lett.* 2010, 501, 130.

[171] Wang, X.-G.; Carrington Jr., T. *J. Phys. Chem. A* 2001, 105, 2575.

[172] Dawes, R.; Carrington Jr., T. *J. Chem. Phys.* 2006, 124, 054102.

[173] Dawes, R.; Carrington Jr., T. *J. Chem. Phys.* 2005, 122, 134101.

[174] Cooper, J.; Carrington Jr., T. *J. Chem. Phys.* 2009, 130, 214110.

[175] Maynard, A. T.; Wyatt, R. E.; Iung, C. *J. Chem. Phys.* 1995, 103, 8372.

[176] Yu, H.-G. *J. Chem. Phys.* 2002, 117, 2030.

[177] Halonen, L.; Child, M. S. *J. Chem. Phys.* 1983, 79, 4355.

[178] Poirier, B. *J. Theor. Comput. Chem.* 2003, 2, 65.

[179] Lee, H.-S.; Light, J. C. *J. Chem. Phys.* 2003, 118, 3458.

[180] Zou, S.; Bowman, J. M.; Brown, A. *J. Chem. Phys.* 2003, 118, 10012.

[181] Tremblay, J. C.; Carrington Jr., T. *J. Chem. Phys.* 2006, 125, 094311.

[182] Wang, X.-G.; Carrington Jr., T. *J. Chem. Phys.* 2003, 119, 101-117.

[183] Liao, Y.; Fang, S.-C.; Nuttle, H. L. W. *Neural Networks* 2003, 16, 1019.

[184] Wu, W.; Nan, D.; Long, J.; Ma, Y. *Neural Networks* 2008, 21, 1462.

[185] Chesick, J. P. *J. Chem. Phys.* 1968, 49, 3772.

[186] Shore, B. W. *J. Chem. Phys.* 1973, 59, 6450.

[187] Davis, M. J.; Heller, E. J. *J. Chem. Phys.* 1979, 71, 3383.

[188] Hamilton, I. P.; Light, J. C. *J. Chem. Phys.* 1986, 84, 306.

[189] Husimi, K. *Proc. Phys. Math. Soc. Jpn.* 1940, 22, 264.

[190] Poirier, B.; Light, J. C. *J. Chem. Phys.* 2000, 113, 211.

[191] Friesner, R. *J. Chem. Phys.* 1987, 86, 3522.

[192] Friesner, R. *J. Chem. Phys.* 1986, 85, 1462.

[193] Friesner, R. *Chem. Phys. Lett.* 1985, 116, 39.

[194] Gottlieb, D.; Orszag, S. A. *Numerical Analysis of Spectral Methods: Theory and Applications*; SIAM: Philadelphia, 1977.

[195] Orszag, S. A. *Studies Appl. Math.* 1972, 51, 253.

[196] Enkovaara, J.; Rostgaard, C.; Mortensen, J. J.; Chen, J.; Dulak, M.; Ferrighi, L.; Gavnholt, J.; Glinsvad, C.; Haikola, V.; Hansen, H. A.; Kristoffersen, H. H.; Kuisma, M.; Larsen, A. H.; Lehtovaara, L.; Ljungberg, M.; Lopez-Acevedo, O.; Moses, P. G.; Ojanen, J.; Olsen, T.; Petzold, V.; Romero, N. A.; Stausholm-Moller, J.; Strange, M.; Tritsaris, G. A.; Vanin, M.; Walter, M.; Hammer, B.; Hakkinen, H.; Madsen, G. K. H.; Nieminen, R. M.; Norskov, J.; Puska, M.; Rantala, T. T.; Schiotz, J.; Thygesen, K. S.; Jacobsen, K. W. *J. Phys. : Condens. Matter* 2010, 22, 253202.

[197] Sumpter, B. G.; Noid, D. W. *Chem. Phys. Lett.* 1992, 192, 455-462.

[198] Lee, H. M.; Huynh, S.; Raff, L. M. *J. Chem. Phys.* 2009, 131, 014107.

[199] Malshe, M.; Narulkar, R.; Raff, L. M.; Hagan, M. T.; Bukkapatnam, S.; Komanduri, R. *J. Chem. Phys.* 2008, 129, 044111.

[200] Agrawal, P. A.; Raff, L. M.; Hagan, M. T.; Komanduri, R. *J. Chem. Phys.* 2006, 124, 134306.

[201] Doughan, D. I.; Raff, L. M.; Rockey, M. G.; Hagan, M. T.; Agrawal, P. A.; Komanduri, R. *J. Chem. Phys.* 2006, 124, 054321.

[202] Raff, L. M.; Malshe, M.; Hagan, M. T.; Doughan, D. I.; Rockey, M. G.; Komanduri, R. *J. Chem. Phys.* 2005, 122, 084104.

[203] Zupan, J.; Gasteiger, J. *Analytica Chimica Acta* 1991, 248, 1-30.

[204] Gassner, H.; Probst, M.; Lauenstein, A.; Hermansson, K. *J. Phys. Chem. A* 1998, 102, 4596-4605.

[205] Lorenz, S.; Gross, A.; Scheffler, M. *Chem. Phys. Lett.* 2004, 395, 210-215.

[206] Prudente, F. V.; Acioli, P. H.; Soares Neto, J. J. *J. Chem. Phys.* 1998, 109, 8801-8808.

[207] Prudente, F. V.; Soares Neto, J. *J. Chem. Phys. Lett.* 1998, 287, 585-589.

[208] Brown, D. F. R.; Gibbs, M. N.; Clary, D. C. *J. Chem. Phys.* 1996, 105, 7597-7604.

[209] Blank, T. S.; Brown, S. D.; Calhoun, A. W.; Doren, D. J. *J. Chem. Phys.* 1995, 103, 4129-4137.

[210] Witkoskie, J. B.; Doren, D. J. *J. Chem. Theory Comput.* 2005, 1, 14-23.

[211] Maiorov, V.; Pinkus, A. *Neurocomputing* 1999, 25, 81-91.

[212] Gorban, A. N. *Appl. Math. Lett.* 1998, 11, 45-49.

[213] Caetano, C.; Reis Jr., J. L.; Amorim, J.; Ruv Lemes, M.; Dal Pino Jr., A. *Int. J. Quantum Chem.* 2010, in print.

[214] Watson, J. K. G. *Mol. Phys.* 1968, 15, 479-490.

[215] Yagi, K.; Karasawa, H.; Hirata, S.; Hirao, K. *Chem. Phys. Chem.* 2009, 10, 1442.

[216] Chaban, G. M.; Jung, J. O.; Gerber, R. B. *J. Chem. Phys.* 1999, 111, 1823.

[217] Yagi, K.; Hirao, K.; Taketsugu, T.; Schmidt, M. W.; Gordon, M. S. *J. Chem. Phys.* 2004, 121, 1383-1389.

[218] Matsunaga, N.; Chaban, G. M.; Gerber, R. B. *J. Chem. Phys.* 2002, 117, 3541-3547.

[219] Chu, D.; Golub, G. H. *SIAM J. Matrix Anal. Appl.* 2006, 28, 770-787.

[220] Lecumberri, P.; Gomez, M.; Carlosena, A. *SIAM J. Matrix Anal. Appl.* 2008, 30, 41-55.

[221] MatLab R2009B, version 7.9.0.529; *MathWorks*,Inc.: 2009

[222] Sobol, I. M. *USSR Comput. Maths. Math. Phys.* 1967, 7, 86.

[223] Garashchuk, S.; Light, J. C. *J. Chem. Phys.* 2001, 114, 3929-3939.
[224] Jensen, P. *J. Mol. Spectrosc.* 1989, 133, 438.
[225] Choi, S. E.; Light, J. C. *J. Chem. Phys.* 1992, 97, 7031.
[226] Smith, F. *Phys. Rev. Lett.* 1980, 45, 1157-1160.
[227] Soler, J. M.; Artacho, E.; Dale, J. D.; Garcia, A.; Junquera, J.; Ordejon, P.; Sanchez-Portal, D. *J. Phys.: Condens. Matter* 2002, 14, 2745.
[228] Perdew, J. P.; Burke, K.; Ernzerhoff, M. *Phys. Rev. Lett.* 1996, 77, 3865.
[229] Troullier, N.; Martins, J. L. *Phys. Rev. B* 1991, 43, 1993.
[230] Artacho, E.; Anglada, E.; Dieguez, O.; Gale, J. D.; Garcia, A.; Junquera, J.; Martin, R. M.; Ordejon, P.; Pruneda, J. M.; Sanchez-Portal, D.; Soler, J. M. *J. Phys.: Condens. Matter* 2008, 20, 064208.
[231] Russell, D. J. I. *NIST Computational Chemistry Comparison and Benchmark Database*, NIST Standard Reference Database Number 101; 2005.
[232] Meng, S.; Wang, E. G.; Gao, S. *Phys. Rev. B* 2004, 69, 195404.
[233] Parreira, R. L. T.; Caramori, G. F.; Galembeck, S. E.; Huguenin, F. *J. Phys. Chem. A* 2008, 112, 11731.
[234] Cohen, A. J.; Mori-Sanchez, P.; Yang, W. *Science* 2008, 321, 792.
[235] Henderson, M. A. *Surf. Sci. Rep.* 2002, 46, 1.
[236] Feibelman, P. J.; Bartelt, N. C.; Nie, S.; Thuermer, K. *J. Chem. Phys.* 2010, 133, 154703.
[237] Arnadottir, L.; Stuve, E. M.; Jonsson, H. *Surf. Sci.* 2010, 604, 1978.
[238] Sexton, B. A. *Surf. Sci.* 1980, 94, 435-445.
[239] Baumann, P.; Pirug, G.; Reuter, D.; Bonzel, H. P. *Surf. Sci.* 1995, 335, 186-196.
[240] Martinez de la Hoz, J. M.; Leon-Quintero, D. F.; Hirunsit, P.; Balbuena, P. B. *Chem. Phys. Lett.* 2010, 498, 328.
[241] Monkhorst, H.; Pack, J. *Phys. Rev. B* 1976, 13, 5188.
[242] *Handbook of Chemistry and Physics*; CRC Press, Inc.: New York, 1998.
[243] Tully, J. C.; Gomez, M.; Head-Gordon, M. *J. Vac. Sci. Technol. A* 1993, 11, 1914.
[244] Parr, R. G.; Weitao, Y. *Density-Functional Theory of Atoms and Molecule*; Oxford University Press: Oxford, UK, 1994.
[245] Kohn, W.; Sham, L. *J. Phys. Rev.* 1965, 140, A1133.
[246] Norskov, J. K.; Bligaard, T.; Rossmeisl, J.; Christensen, C. H. *Nature Chemistry* 2009, 1, 37-46.
[247] Farrusseng, D. *Surf. Sci. Rep.* 2008, 63, 487-513.
[248] Thomas, J. M. *J. Chem. Phys.* 2008, 128, 182502.
[249] Christensen, C. H.; Norskov, J. K. *J. Chem. Phys.* 2008, 128, 182503.

In: Computer Physics
Editors: B.S. Doherty and A.N. Molloy, pp. 267-295

ISBN: 978-1-61324-790-7
© 2012 Nova Science Publishers, Inc.

Chapter 3

MULTISYMPLECTIC INTEGRATORS FOR COUPLED NONLINEAR PARTIAL DIFFERENTIAL EQUATIONS

B. Karasözen [1] **and A. Aydın**[2]

[1]Department of Mathematics & Institute of Middle East Technical University,
06531 Ankara, Turkey
[2]Department of Mathematics, Atılım University, 06836 Ankara, Turkey

Abstract

The numerical solution of nonlinear partial differential equations (PDEs) using symplectic geometric integrators has been the subject of many studies in recent years. Many nonlinear partial differential equations can be formulated as an infinite dimensional Hamiltonian system. After semi-discretization in the space variable, a system of Hamiltonian ordinary differential equations (ODEs) is obtained, for which various symplectic integrators can be applied. Numerical results show that symplectic schemes have superior performance, especially in long time simulations. The concept of multisymplectic PDEs and multisymplectic schemes can be viewed as the generalization of symplectic schemes. In the last decade, many multisymplectic methods have been proposed and applied to nonlinear PDEs, like to nonlinear wave equation, nonlinear Schrödinger equation, Korteweg de Vries equation, Dirac equation, Maxwell equation and sine-Gordon equation. In this review article, recent results of multisymplectic integration on the coupled nonlinear PDEs, the coupled nonlinear Schrödinger equation, the modified complex Korteweg de Vries equation and the Zakharov system will be given. The numerical results are discussed with respect to the stability of the schemes, accuracy of the solutions, conservation of the energy and momentum, preservation of dispersion relations.

PACS 02.70.Bf; 42.65.Sf

Keywords: coupled nonlinear Schrödinger equation; complex Korteweg de Vries equation, periodic waves; solitons, dispersion, multisymplectic integrators, splitting methods.

1. Introduction

When solving differential equations numerically it is important to preserve as much of the qualitative solution behavior as possible. Numerical methods which preserve the geometric features of ordinary and partial differential equations have became very popular. These methods are known as geometric integrators. The symplectic integrators for Hamiltonian ordinary differential equations (ODEs) has been thoroughly analyzed and the concept of symplectic time integration has also been extended to Hamiltonian partial differential equations (PDEs) (see for example [1], and [2]). Numerical results show that symplectic schemes have superior performance, especially in long time simulations.

Hamiltonian PDEs arise as models in nonlinear optics, solid mechanics, meteorology, oceanography, electromagnetism and quantum field theory. Many nonlinear partial differential equations like nonlinear wave equation, nonlinear Schrödinger equation, Korteweg de Vries equation, nonlinear Dirac equation, Maxwell equation and sine-Gordon equation can be formulated as an infinite dimensional Hamiltonian system. After semi-discretization in the space variable a system of Hamiltonian ODEs is obtained. Various symplectic integrators can be applied to the resulting ODE system. When generalizing from Hamiltonian ODEs to Hamiltonian PDEs, the phase space is parameterized by time and space. During the symplectic integration of semi–discretized PDEs, the space coordinate is treated passively; the spatial variations are fixed. However for many problems it is advantageous to discretize both in space and time simultaneously. In such situations multisymplectic methods allow for a combined treatment of spatial and temporal discretizations. Multisymplectic Hamiltonian formulations rely on local conservation laws and, therefore it is well suited for long time numerical preservation of the local energy and momentum. Because of these, the multisymplectic integration of multisymplectic PDEs can be viewed as the generalization of symplectic integrators for ODEs.

Multisymplectic methods have been proposed and investigated for some important Hamiltonian PDEs. Many well known nonlinear PDEs like the nonlinear Schrödinger equation, the Korteweg de Vries equation, the sine-Gordon equation can be written in multisymplectic form. The concept of multisymplectic integration was developed in ([3, 4]). The most used multisymplectic integrator is the Preissman box scheme, which is a generalization of the mid-point method in space and time variables. It was shown that the Preismann scheme exactly preserves a multisymplectic conservation law and and the dispersion relation. The local energy and momentum are in general not conserved exactly. Using Taylor series expansions, one obtains a modified multisymplectic PDE and modified conservation laws that are preserved to higher order. Numerical experiments with the multisymplectic integrators applied to nonlinear PDEs with solitary and traveling wave solutions show that the local and global energy and momentum are conserved well in long time integration. These results are summarized in the survey paper [5] which provides an introduction of the symplectic and mulisymplectic discretization methods for Hamiltonian PDEs.

In this review article we focus on the multisymplectic integration of coupled nonlinear PDEs. Among the nonlinear coupled PDEs, the coupled nonlinear Schrödinger equation (CNLS) plays an important role. The CNLS equation with second-order dispersion and cubic nonlinearity was first derived for two interacting nonlinear packets in a dispersive and conservative systems. Analytical solutions can be obtained only for a few special integrable cases, like for the Manakov model where the selfmodulation and wave-wave interaction coefficients are equal. For non-integrable cases, numerical methods have to be used in order to understand different nonlinear phenomena that arise by the interaction of stable and unstable wave packets in the CNLS system. There have been a great deal of numerical methods used to solve the CNLS equation, symplectic and multisymplectic methods [6, 7, 8], finite difference methods [9, 10, 11, 12, 13] and the discontinuous Galerkin method [14]. Most of the multisymplectic integrators are implicit and they are variations of the multisymplectic Preissman scheme. Recently, other multisymplectic schemes such as the Euler-box scheme and explicit multisymplectic integrators are developed based on the Lobatto IIIA-IIIB methods for the partitioned ODE systems. Besides this, the multisymplectic splitting methods has the advantage, that they decompose the original multisymplectic system into sub-multisymplectic systems which are easier to solve than the original one. Multisymplectic integrators are also applied to other coupled nonlinear PDEs like the complex modified Korteveg de Vries(CMKdV) equation, and to the Zakharov system for Langmuir waves in plasma physics. In all these works it was shown that the numerical results for different solitary wave solutions (elastic and inelastic collisions, fusion of two solitons), traveling waves and waves with periodic solutions confirm the excellent long time behavior of the multisymplectic integrators by preserving global energy, momentum and mass.

The article is organized as follows. In the rest of this Chapter, we give the concept of multisymplectic PDE and the multisymplectic discretization. In the following Chapters, multisymplectic integration of CNLS equation, strongly coupled CNLS equation, CMKdV equation and the Zakharov systems is outlined with numerical results about the solitary and traveling wave solutions. The last Chapter deals with the preservation of the dispersion relation under multisymplectic discretization.

Many partial differential equations (PDEs) can be written as a system of equations with only first–order derivatives in time and space (see for example [3, 15]). Let $\mathbf{M}, \mathbf{K} \in \mathbb{R}^{d \times d}$ be two skew symmetric matrices and let $S : \mathbb{R}^d \longrightarrow \mathbb{R}$ be any smooth function. Then, a PDE of the form

$$\mathbf{M}\mathbf{z}_t + \mathbf{K}\mathbf{z}_x = \nabla_{\mathbf{z}} S(\mathbf{z}), \qquad \mathbf{z} \in \mathbb{R}^d \tag{1}$$

where $\nabla_{\mathbf{z}}$ is the standard gradient in $\mathbb{R}^d$, is called a multisymplectic PDE [16]. For the system (1) there exits the following two forms

$$\omega(U,V) = <\mathbf{M}U, V) = V^T \mathbf{M} U \quad \text{and} \quad \kappa(U,V) = <\mathbf{K}U, V) = V^T \mathbf{K} U, \quad U, V \in \mathbb{R}^d \tag{2}$$

where ω defines a symplectic structure on $\mathbb{R}^m$ ($m = \text{rank}(\mathbf{M}) \leq d$) associated with the time variable, and κ defines a symplectic structure on $\mathbb{R}^k$ ($k = \text{rank}(\mathbf{K}) \leq d$) associated with the space variable. An important aspect of a multisymplectic structure is the multisymplectic conservation law

$$\omega_t + \kappa_x = 0. \tag{3}$$

Let $U, V \longrightarrow \mathbb{R}^d$ be any two solutions of the variational equation associated with (1)

$$\mathbf{M}\, d\mathbf{z}_t + \mathbf{K}\, d\mathbf{z}_x = \mathbf{D}_{\mathbf{z}\mathbf{z}}(\mathbf{S}\mathbf{z}\, d\mathbf{z}). \tag{4}$$

Then

$$\omega_t = <\mathbf{M}U_t, V> + <\mathbf{M}U, V_t>, \quad \kappa_x = <\mathbf{K}U_x, V> + <\mathbf{K}U, V_x> \tag{5}$$

noting that $\mathbf{D}_{\mathbf{z}\mathbf{z}}\mathbf{S}(\mathbf{z})$ is a symmetric matrix, one obtains the multisymplectic conservation law (3)

The multisymplectic conservation law (3) is equivalent to

$$\omega_t \, (\mathbf{dz} \wedge \mathbf{Mdz}) + \kappa_x \, (\mathbf{dz} \wedge \mathbf{Kdz}) = 0 \tag{6}$$

using the wedge product notation.

Multisymplectic integrators are approximations to (1) which conserve the discrete version of (3) associated with the discretization of (1). Conservation of multisymplecticity (3) is analogous to preservation of the two-form, $\omega_t = 0$, for Hamiltonian ODEs. Let

$$z = (p_1, \cdots, p_N, q_1, \cdots, q_N), \quad \mathbf{M} \equiv \mathbf{J} = \begin{pmatrix} \mathbf{0} & -\mathbf{I}_N \\ \mathbf{I}_N & \mathbf{0} \end{pmatrix},$$

where all p_j, q_j are spatially independent, then $\partial_x \mathbf{dz} \equiv 0$ leads to $\partial_x \kappa = \partial_x \, (\mathbf{dz} \wedge \mathbf{Kdz}) \equiv 0$ and (6) reduces to

$$\omega_t = \partial_t \, (\mathbf{dz} \wedge \mathbf{Jdz}) = \mathbf{dp} \wedge \mathbf{dq} = 0.$$

As symplectic integrators are discretizations preserving the two-form ω, multisymplectic integrators are approximations to (1) which also conserve a discretization of the multisymplectic conservation law (3). As the symplectic integrators conserve the Hamiltonian extremely well over very long times, multisymplectic schemes conserve the related energy and momentum conservation laws very well.

When the Hamiltonian $S(\mathbf{z})$ is independent of x and t, each independent variable gives rise to a conservation law namely energy and momentum conservations [3, 16]. Conservation of energy and momentum are associated with translation invariance in time and space, respectively. Multiplying (1) with $\mathbf{z}^T$ from the left provides the energy conservation law (ECL)

$$\partial_t E(z) + \partial_x F(z) = 0, \tag{7}$$

while multiplying (1) with $\mathbf{z}_x^T$ from the left yields the momentum conservation law (MCL)

$$\partial_t I(z) + \partial_x G(z) = 0, \tag{8}$$

where

$$\begin{aligned} E(\mathbf{z}) &= S(\mathbf{z}) - \frac{1}{2}\kappa(\mathbf{z}_x, \mathbf{z}), & F(\mathbf{z}) &= \frac{1}{2}\kappa(\mathbf{z}_t, \mathbf{z}), \\ G(\mathbf{z}) &= S(\mathbf{z}) - \frac{1}{2}\omega(\mathbf{z}_t, \mathbf{z}), & I(\mathbf{z}) &= \frac{1}{2}\omega(\mathbf{z}_x, \mathbf{z}), \end{aligned} \tag{9}$$

where ω and κ ar defined in (2). Under periodic boundary conditions, integration of $E(z)$ and $I(z)$ over the spatial domain leads to the global conserved quantities

$$\frac{d}{dt}\mathcal{E}(\mathbf{z}) = 0, \qquad \frac{d}{dt}\mathcal{I}(\mathbf{z}) = 0 \tag{10}$$

where $\mathcal{E}(\mathbf{z}) = \int_0^L E(\mathbf{z})dx$ and $\mathcal{I}(\mathbf{z}) = \int_0^L I(\mathbf{z})dx$.

Multisymplecticity is a geometric property of a partial differential equation, which is destroyed using standard integrators, such as fourth-order RungeKutta method. The multisymplecticness property should be preserved by discretization of the underlying PDE. Based on this idea, Bridges and Reich [3] and Marsden et al. [4] introduced the concept of the multisymplectic integrator from different aspects. A standard method for constructing a multisymplectic integrator is to apply a well known symplectic method to each independent variable. We now give two examples for this approaches. These methods are designed to preserve the discrete version of the multisymplectic conservation law (3). Moreover these schemes locally conserve energy (7) and momentum (8) remarkably well, though not exactly [17]. One the most popular multisymplectic integrator is the Preissman scheme, which corresponds to mid-point discretization in space and time variables [3]. Higher order extensions of Preissman scheme are constructed as Gauss–Legendre Runge–Kutta methods in [17]. The Preissman scheme for (1) is given by

$$
\mathbf{M} \left(\frac{\mathbf{z}_{j+\frac{1}{2}}^{n+1} - \mathbf{z}_{j+\frac{1}{2}}^{n}}{\Delta t} \right) + \mathbf{K} \left(\frac{\mathbf{z}_{j+1}^{n+\frac{1}{2}} - \mathbf{z}_{j}^{n+\frac{1}{2}}}{\Delta x} \right) = \nabla_z S \left(\mathbf{z}_{j+\frac{1}{2}}^{n+\frac{1}{2}} \right), \tag{11}
$$

with

$$
\mathbf{z}_{j+\frac{1}{2}}^{n} = \frac{\mathbf{z}_{j}^{n} + \mathbf{z}_{j+1}^{n}}{2}, \quad \mathbf{z}_{j}^{n+\frac{1}{2}} = \frac{\mathbf{z}_{j}^{n} + \mathbf{z}_{j}^{n+1}}{2}, \quad \mathbf{z}_{j+\frac{1}{2}}^{n+\frac{1}{2}} = \frac{\mathbf{z}_{j}^{n} + \mathbf{z}_{j+1}^{n} + \mathbf{z}_{j}^{n+1} + \mathbf{z}_{j+1}^{n+1}}{4},
$$

where Δt, Δx are time and space step sizes respectively and $\mathbf{z}_{j}^{n}$ is an approximation to $\mathbf{z}(n\Delta t, j\Delta x)$. The Preissman scheme preserves the discrete form of the multisymplecticity (3) exactly, i.e.

$$
\left(\frac{\omega_{j+\frac{1}{2}}^{n+1} - \omega_{j+\frac{1}{2}}^{n}}{\Delta t} \right) + \left(\frac{\kappa_{j+1}^{n+\frac{1}{2}} - \kappa_{j}^{n+\frac{1}{2}}}{\Delta x} \right) = 0 \tag{12}
$$

and is second order accurate in the time and space variables [3]. In terms of differencing and averaging finite difference operators

$$
\begin{aligned}
D_x \psi_j^n &= \frac{\psi_{j+1}^n - \psi_j^n}{\Delta x}, \quad L_x \psi_j^n = \frac{\psi_{j+1}^n + \psi_j^n}{2} \\
D_t \psi_j^n &= \frac{\psi_j^{n+1} - \psi_j^n}{\Delta t}, \quad L_t \psi_j^n = \frac{\psi_j^{n+1} + \psi_j^n}{2}
\end{aligned} \tag{13}
$$

the Preissmann box scheme (11) can be written as

$$
\mathbf{M} D_x L_t \mathbf{z} + \mathbf{K} D_t L_x \mathbf{z} = \nabla S(L_t L_x \mathbf{z}) \tag{14}
$$

with the discrete multisymplectic conservation law

$$
d\mathbf{z} \wedge \mathbf{M} D_x L_t d\mathbf{z} + d\mathbf{z} \wedge \mathbf{K} D_x L_t d\mathbf{z} = 0.
$$

Usually the accuracy of a multisymplectic scheme is tested by looking at the energy (7) and momentum (8) conservation properties. For the Preissman scheme (11), the residual in the local energy conservation is defined as

$$
R_E = \left(\frac{E_{j+\frac{1}{2}}^{n+1} - E_{j+\frac{1}{2}}^{n}}{\Delta t} \right) + \left(\frac{F_{j+1}^{n+\frac{1}{2}} - F_{j}^{n+\frac{1}{2}}}{\Delta x} \right), \tag{15}
$$

and the residual in the local energy conservation is defined as

$$R_M = \left(\frac{I_{j+\frac{1}{2}}^{n+1} - I_{j+\frac{1}{2}}^{n}}{\Delta t} \right) + \left(\frac{G_{j+1}^{n+\frac{1}{2}} - G_j^{n+\frac{1}{2}}}{\Delta x} \right). \tag{16}$$

The global energy G_E and global momentum G_M errors for the Preissman scheme (11) is given by

$$G_E = \Delta x \sum_{j=1}^{N} (E_j^n - E^0), \qquad G_M = \Delta x \sum_{j=1}^{N} (I_j^n - I^0) \tag{17}$$

where E^0 and I^0 are the initial energy and momentum respectively.

Moore and Reich [18] suggested another multisymplectic integrator which is the Euler box scheme

$$\mathbf{M}_+ \left(\frac{\mathbf{z}_j^{n+1} - \mathbf{z}_j^n}{\Delta t} \right) + \mathbf{M}_- \left(\frac{\mathbf{z}_j^n - \mathbf{z}_j^{n-1}}{\Delta t} \right)$$

$$+ \mathbf{K}_+ \left(\frac{\mathbf{z}_{j+1}^n - \mathbf{z}_j^n}{\Delta x} \right) + \mathbf{K}_- \left(\frac{\mathbf{z}_j^n - \mathbf{z}_{j-1}^n}{\Delta x} \right) = \nabla_{\mathbf{z}} S(\mathbf{z}_j^n) \quad (18)$$

with the discrete multisymplectic conservation law

$$\left(\frac{\omega_j^{n+1} - \omega_j^n}{\Delta t} \right) + \left(\frac{\kappa_{j+1}^n - \kappa_j^n}{\Delta x} \right) = 0$$

for

$$\omega_j^n = \mathbf{dz}_j^{n-1} \wedge \mathbf{M}_+ \mathbf{dz}_j^n, \qquad \text{and} \qquad \kappa_j^n = \mathbf{dz}_{j-1}^n \wedge \mathbf{K}_+ \mathbf{dz}_j^n$$

where $\mathbf{M} = \mathbf{M}_+ + \mathbf{M}_-$ and $\mathbf{K} = \mathbf{K}_+ + \mathbf{K}_-$.

Apart from multisymplectic finite difference discretization, Bridges and Reich [19] suggested the idea of multisymplectic spectral discretization on Fourier space for the Zakharov–Kuznetsov and shallow water equations. Then the multisymplectic spectral discretization and Fourier pseudospectral discretization are investigated for the nonlinear Schrödinger equation and Klein–Gordon equation (see [20, 21, 22, 23]) .

2. The Coupled Nonlinear Schrödinger Equation

The nonlinear Schrödinger equation (NLS) describes the dynamics of slowly varying wave packets in nonlinear optics and fluid dynamics. If there are two or more modes the coupled nonlinear Schrödinger (CNLS) system would be the relevant model [24]. The two coupled CNLS equations are given by

$$i\frac{\partial \psi_1}{\partial t} + \alpha_1 \frac{\partial^2 \psi_1}{\partial x^2} + (\sigma_1 |\psi_1|^2 + v_{12} |\psi_2|^2)\psi_1 = 0,$$

$$i\frac{\partial \psi_2}{\partial t} + \alpha_2 \frac{\partial^2 \psi_2}{\partial x^2} + (\sigma_2 |\psi_2|^2 + v_{21} |\psi_1|^2)\psi_2 = 0, \tag{19}$$

where $\psi_1(x,t)$ and $\psi_2(x,t)$ are complex amplitudes or 'envelopes' of two wave packets, i is the imaginary number, x and t are the space and time variables respectively. The CNLS system occurs as a model in nonlinear optics [25, 26] and geophysical fluid dynamics [27, 28]. The parameters α_j are the dispersion coefficients, and σ_j the Landau constants describe the self-modulation of the wave packets respectively. The cross-modulations of the wave packets are described by the wave-wave interaction coefficients v_{12} and v_{21}. [12, 29].

Analytical solutions of CLNS equations are available only for a few special integrable cases, like the Manakov model [25] where the self- modulation and wave-wave interaction coefficients are equal, i.e. $\alpha_1 = \alpha_2 = 1$, $\sigma_1 = \sigma_2 = v_{12} = v_{21}$. Other integrable cases which can be solved by inverse scattering method [30] are $\alpha_1 = \alpha_2$, $\sigma_1 = \sigma_2 = v_{12} = v_{21}$ and $\alpha_1 = -\alpha_2$, $\sigma_1 = \sigma_2 = -v_{12} = -v_{21}$.

We consider the CNLS equation (19) with $v_{12} = v_{21} = v$ because this choice of the parameters allows a multisymplectic formulation of the CNLS system.

By decomposing the complex functions ψ_1, ψ_2 of (19) into real and imaginary parts

$$\psi_1(x,t) = q_1(x,t) + iq_2(x,t), \qquad \psi_2(x,t) = q_3(x,t) + iq_4(x,t)$$

the CNLS systems (19) can be written as a system of real-valued equations

$$
\begin{aligned}
\frac{\partial q_1}{\partial t} + \alpha_1 \frac{\partial^2 q_2}{\partial x^2} + (\sigma_1(q_1^2 + q_2^2) + v(q_3^2 + q_4^2))q_2 &= 0, \\
\frac{\partial q_2}{\partial t} - \alpha_1 \frac{\partial^2 q_1}{\partial x^2} - (\sigma_1(q_1^2 + q_2^2) + v(q_3^2 + q_4^2))q_1 &= 0, \\
\frac{\partial q_3}{\partial t} + \alpha_2 \frac{\partial^2 q_4}{\partial x^2} + (v(q_1^2 + q_2^2) + \sigma_2(q_3^2 + q_4^2))q_4 &= 0, \\
\frac{\partial q_4}{\partial t} - \alpha_2 \frac{\partial^2 q_3}{\partial x^2} - (v(q_1^2 + q_2^2) + \sigma_2(q_3^2 + q_4^2))q_3 &= 0.
\end{aligned}
\tag{20}
$$

These equations represent an infinite-dimensional Hamiltonian system in the phase space $z = (q_1, q_2, q_3, q_4)^T$

$$z_t = \mathbf{J}^{-1} \frac{\delta \mathcal{H}}{\delta z}, \qquad \mathbf{J} = \begin{pmatrix} 0 & -1 & 0 & 0 \\ 1 & 0 & 0 & 0 \\ 0 & 0 & 0 & -1 \\ 0 & 0 & 1 & 0 \end{pmatrix} \tag{21}$$

where the Hamiltonian is

$$\mathcal{H}(z) = \int \left\{ W - \frac{\alpha_1}{2}\left[\left(\frac{\partial q_1}{\partial x}\right)^2 + \left(\frac{\partial q_2}{\partial x}\right)^2\right] - \frac{\alpha_2}{2}\left[\left(\frac{\partial q_3}{\partial x}\right)^2 + \left(\frac{\partial q_4}{\partial x}\right)^2\right] \right\} dx, \tag{22}$$

with $W = \frac{1}{4}(\sigma_1(q_1^2 + q_2^2)^2 + \sigma_2(q_3^2 + q_4^2)^2) + \frac{v}{2}(q_1^2 + q_2^2)(q_3^2 + q_4^2)$.

By splitting the Hamiltonian (22) into linear and nonlinear parts as $\mathcal{H} = \mathcal{H}_{Lin} + \mathcal{H}_{Non}$ with

$$\mathcal{H}_{Lin} = -\int \left\{ \frac{\alpha_1}{2}\left[\left(\frac{\partial q_1}{\partial x}\right)^2 + \left(\frac{\partial q_2}{\partial x}\right)^2\right] + \frac{\alpha_2}{2}\left[\left(\frac{\partial q_3}{\partial x}\right)^2 + \left(\frac{\partial q_4}{\partial x}\right)^2\right] \right\} dx, \tag{23}$$

and

$$\mathcal{H}_{Non} = \int \left\{ \frac{1}{4}(\sigma_1(q_1^2 + q_2^2)^2 + \sigma_2(q_3^2 + q_4^2)^2) + \frac{v}{2}(q_1^2 + q_2^2)(q_3^2 + q_4^2) \right\} dx \qquad (24)$$

we can derive a semi–implicit symplectic scheme for the CNLS equation (19). A standard discretization of the linear subproblem corresponding to (23) by means of the method of lines consists of discretization first in space and then in time. After the discretization in space using the central difference approximation for the second order derivatives, we get the semi-discretized linear subproblem

$$\frac{dq_{1j}}{dt} = -\alpha_1 \frac{q_{2j-1} - 2q_{2j} + q_{2j+1}}{2\Delta x^2}, \quad \frac{dq_{2j}}{dt} = \alpha_1 \frac{q_{1j-1} - 2q_{1j} + q_{1j+1}}{2\Delta x^2},$$

$$\frac{dq_{3j}}{dt} = -\alpha_2 \frac{q_{4j-1} - 2q_{4j} + q_{4j+1}}{2\Delta x^2}, \quad \frac{dq_{4j}}{dt} = \alpha_2 \frac{q_{3j-1} - 2q_{3j} + q_{3j+1}}{2\Delta x^2}. \qquad (25)$$

This system of ODEs can be formulated as a finite dimensional Hamiltonian system as

$$\frac{dZ}{dt} = J^{-1} \nabla H_{lin}(Z), \qquad \text{with} \quad Z := (Q_1, Q_2, Q_3, Q_4)^T \qquad (26)$$

where $Q_j := (q_{j1}, q_{j2}, \cdots, q_{jN}), j = 1, \cdots, 4$ and

$$J := \begin{pmatrix} 0 & -I & 0 & 0 \\ I & 0 & 0 & 0 \\ 0 & 0 & 0 & -I \\ 0 & 0 & I & 0 \end{pmatrix}.$$

The semidiscretized nonlinear subproblem corresponding to (24) is

$$\frac{\partial q_{1j}}{\partial t} = - \left[\sigma_1 \left(q_{1j}^2 + q_{2j}^2 \right) + v \left(q_{3j}^2 + q_{4j}^2 \right) \right] q_{2j},$$

$$\frac{\partial q_{2j}}{\partial t} = \left[\sigma_1 \left(q_{1j}^2 + q_{2j}^2 \right) + v \left(q_{3j}^2 + q_{4j}^2 \right) \right] q_{1j},$$

$$\frac{\partial q_{3j}}{\partial t} = - \left[v \left(q_{1j}^2 + q_{2j}^2 \right) + \sigma_2 \left(q_{3j}^2 + q_{4j}^2 \right) \right] q_{4j}, \qquad (27)$$

$$\frac{\partial q_{4j}}{\partial t} = \left[v \left(q_{1j}^2 + q_{2j}^2 \right) + \sigma_2 \left(q_{3j}^2 + q_{4j}^2 \right) \right] q_{3j}.$$

The systems (26) and (27) are solved by the symplectic implicit mid–point rule. After applying the implicit mid-point rule to the linear part (26)

$$\frac{Z^{n+1} - Z^n}{\Delta t} = J^{-1} \nabla H_{lin} \left(\frac{Z^{n+1} + Z^n}{2} \right), \qquad (28)$$

the discrete system of linear equations can be written as

$$\begin{pmatrix} I & A_1 & 0 & 0 \\ -A_1 & I & 0 & 0 \\ 0 & 0 & I & A_2 \\ 0 & 0 & -A_2 & I \end{pmatrix} \begin{pmatrix} Q_1^{n+1} \\ Q_2^{n+1} \\ Q_3^{n+1} \\ Q_4^{n+1} \end{pmatrix} = \begin{pmatrix} I & -A_1 & 0 & 0 \\ A_1 & I & 0 & 0 \\ 0 & 0 & I & -A_2 \\ 0 & 0 & A_2 & I \end{pmatrix} \begin{pmatrix} Q_1^n \\ Q_2^n \\ Q_3^n \\ Q_4^n \end{pmatrix} \qquad (29)$$

where

$$A_j = \alpha_j \frac{\Delta t}{2 \Delta x^2} \begin{pmatrix} -2 & 1 & & & & 1 \\ 1 & -2 & 1 & & & \\ & & \ddots & & & \\ & & & 1 & -2 & 1 \\ 1 & & & & 1 & -2 \end{pmatrix}, \qquad j = 1, 2.$$

However, the nonlinear subsystem (27) requires a nonlinear solver.

The solutions of (26) and (27) by the mid-point rule can be composed by the second order symmetric integrator [1, 31]

$$\varphi_2(\Delta t) = e^{\frac{\Delta t}{2} H_{non}} \circ e^{\Delta t H_{lin}} \circ e^{\frac{\Delta t}{2} H_{non}} \tag{30}$$

which results a symplectic integrator for the CNLS system (19). Higher order compositions can be obtained by suitable composition of the second order integrator [1, 31].

2.1.   Multisymplectic Integration of the CNLS Equation

Introducing the canonical momenta p_j , $j = 1, \cdots, 4$

$$p_1 + i p_2 = \alpha_1 \frac{\partial \psi_1}{\partial x}, \qquad p_3 + i p_4 = \alpha_2 \frac{\partial \psi_2}{\partial x}$$

the CNLS system can now be formulated as a multisymplectic form (1) with the state variable $z = (q_1, q_2, q_3, q_4, p_1, p_2, p_3, p_4)^T$, skew–symmetric matrices

$$\mathbf{M} = \begin{pmatrix} -\mathbf{J} & \mathbf{0} \\ \mathbf{0} & \mathbf{0} \end{pmatrix}, \quad \mathbf{K} = \begin{pmatrix} \mathbf{0} & -\mathbf{I} \\ \mathbf{I} & \mathbf{0} \end{pmatrix}, \tag{31}$$

and the hamiltonian function

$$S(z) = W + \frac{1}{2\alpha_1} \left(p_1^2 + p_2^2 \right) + \frac{1}{2\alpha_2} \left(p_3^2 + p_4^2 \right)$$

with

$$W = \frac{\sigma_1}{4} \left(q_1^2 + q_2^2 \right)^2 + \frac{\sigma_2}{4} \left(q_3^2 + q_4^2 \right)^2 + \frac{v}{2} \left(q_1^2 + q_2^2 \right) \left(q_3^2 + q_4^2 \right). \tag{32}$$

In (31) J as defined in (21) and the matrices $\mathbf{0}$, $\mathbf{I}$ denote the 4×4 zero and identity matrices respectively. The local energy conservation law (7) is given with

$$\begin{aligned} E(z) &= W - \frac{1}{2\alpha_1} \left(p_1^2 + p_2^2 \right) - \frac{1}{2\alpha_2} \left(p_3^2 + p_4^2 \right), \\ F(z) &= p_1 \frac{\partial q_1}{\partial t} + p_2 \frac{\partial q_2}{\partial t} + p_3 \frac{\partial q_3}{\partial t} + p_4 \frac{\partial q_4}{\partial t}, \end{aligned} \tag{33}$$

where W is defined in (32) and the local momentum conservation law (8) is given with

$$\begin{aligned} I(z) &= \frac{1}{2} \left(\frac{1}{\alpha_1} (q_1 p_2 - q_2 p_1) + \frac{1}{\alpha_2} (q_3 p_4 - q_4 p_3) \right), \\ G(z) &= S(z) - \frac{1}{2} \left(q_1 \frac{\partial q_2}{\partial t} - q_2 \frac{\partial q_1}{\partial t} + q_3 \frac{\partial q_4}{\partial t} - q_4 \frac{\partial q_3}{\partial t} \right). \end{aligned} \tag{34}$$

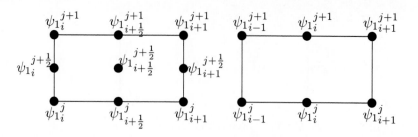

Figure 1. Preissman scheme (left) and six-point difference scheme (right) for the CNLS equation.

Under periodic boundary conditions the CNLS equation (19) has also mass conservations

$$C_1 = \int_0^L |\psi_1|^2 dx, \quad C_2 = \int_0^L |\psi_2|^2 dx. \tag{35}$$

We can write the Preissman scheme (11) for the CNLS equation (19) by using the state variables $z = (q_1, q_2, q_3, q_4, p_1, p_2, p_3, p_4)^T$ and the skew-symmetric matrices $\mathbf{M}$ and $\mathbf{K}$ in (31). However, for the numerical solution of the CNLS system (19) we need only the values of ψ_1 and ψ_2. Therefore from the Preissman scheme for the CNLS equation we eliminate the the canonical momenta p_j, $j = 1, \cdots, 4$ and obtain the following new scheme [8, 32]

$$i \left[\frac{(\psi_{1m-1}^{n+1} + 2\psi_{1m}^{n+1} + \psi_{1m+1}^{n+1}) - (\psi_{1m-1}^n + 2\psi_{1m}^n + \psi_{1m+1}^n)}{4\Delta t} \right]$$
$$+ \alpha_1 \frac{(\psi_{1m-1}^{n+1} - 2\psi_{1m}^{n+1} + \psi_{1m+1}^{n+1}) + (\psi_{1m-1}^n - 2\psi_{1m}^n + \psi_{1m+1}^n)}{2\Delta x^2} \tag{36}$$
$$+ \left(\sigma_1 \left| \widetilde{\psi_1} \right|^2 + \nu \left| \widetilde{\psi_2} \right|^2 \right) \widetilde{\psi_1} + \left(\sigma_1 \left| \widehat{\psi_1} \right|^2 + \nu \left| \widehat{\psi_2} \right|^2 \right) \widehat{\psi_1} = 0,$$

$$i \left[\frac{(\psi_{2m-1}^{n+1} + 2\psi_{2m}^{n+1} + \psi_{2m+1}^{n+1}) - (\psi_{2m-1}^n + 2\psi_{2m}^n + \psi_{2m+1}^n)}{4\Delta t} \right]$$
$$+ \alpha_2 \frac{(\psi_{2m-1}^{n+1} - 2\psi_{2m}^{n+1} + \psi_{2m+1}^{n+1}) + (\psi_{2m-1}^n - 2\psi_{2m}^n + \psi_{2m+1}^n)}{2\Delta x^2} \tag{37}$$
$$+ \left(\nu \left| \widetilde{\psi_1} \right|^2 + \sigma_2 \left| \widetilde{\psi_2} \right|^2 \right) \widetilde{\psi_2} + \left(\nu \left| \widehat{\psi_1} \right|^2 + \sigma_2 \left| \widehat{\psi_2} \right|^2 \right) \widehat{\psi_2} = 0.$$

where $\widetilde{\psi} = \psi_{m-1/2}^{n+1/2}$ and $\widehat{\psi} = \psi_{m+1/2}^{n+1/2}$. The scheme (36)-(37) is called the multisymplectic six-point scheme (MS6). The scheme couples two time levels in contrast to Preissman scheme, which involves three time levels (see Fig. 1). By using (33), the residual (15) for the energy conservation law for the multisymplectic six-point integrator (36)-(37) can be obtained with

$$
E_{j+\frac{1}{2}}^n = \frac{\sigma_1}{4}\left[(q_{1j+\frac{1}{2}}^n)^2 + (q_{2j+\frac{1}{2}}^n)^2\right]^2 + \frac{\sigma_2}{4}\left[(q_{3j+\frac{1}{2}}^n)^2 + (q_{4j+\frac{1}{2}}^n)^2\right]^2
$$
$$
- \frac{1}{2\alpha_1}\left[(p_{1j+\frac{1}{2}}^n)^2 + (p_{2j+\frac{1}{2}}^n)^2\right] - \frac{1}{2\alpha_2}\left[(p_{3j+\frac{1}{2}}^n)^2 + (p_{4j+\frac{1}{2}}^n)^2\right]
$$
$$
+ \frac{v}{2}\left[(q_{1j+\frac{1}{2}}^n)^2 + (q_{2j+\frac{1}{2}}^n)^2\right]\left[(q_{3j+\frac{1}{2}}^n)^2 + (q_{4j+\frac{1}{2}}^n)^2\right],
$$
$$
F_j^{n+\frac{1}{2}} = p_{1j}^{n+\frac{1}{2}}\left(\frac{q_{1j}^{n+1} - q_{1j}^n}{\Delta t}\right) + p_{2j}^{n+\frac{1}{2}}\left(\frac{q_{2j}^{n+1} - q_{2j}^n}{\Delta t}\right) +
$$
$$
p_{3j}^{n+\frac{1}{2}}\left(\frac{q_{3j}^{n+1} - q_{3j}^n}{\Delta t}\right) + p_{4j}^{n+\frac{1}{2}}\left(\frac{q_{4j}^{n+1} - q_{4j}^n}{\Delta t}\right).
$$

The residual (16) for the momentum conservation can be obtained analogously.

Figure 2 shows the numerical solution of the CNLS equation (19) with $v = 1$ (elliptic polarization) by using the multisymplectic six-point scheme (36)-(37) with the initial conditions

$$
\psi_1(x,0) = \psi_{10}\left(1 - \epsilon\cos\left(\frac{x}{2}\right)\right), \qquad \psi_2(x,0) = \psi_{20}\left(1 - \epsilon\cos\left(\frac{x}{2}\right)\right) \tag{38}
$$

where the parameter $\epsilon \ll 1$ describes the strength of the perturbation [12]. The amplitude of the wave, energy and momentum errors are shown for $0 \le x \le 8\pi$ and $0 \le t \le 100$ for the step sizes $\Delta x = 8\pi/128$ and $\Delta t = 0.05$ in space and time respectively. From the Fig.2 we see that there are two peaks, which are called as two–hump states, in the space interval $0 \le x \le 8\pi$. The amplitudes ψ_1 and ψ_2 oscillates between the near-uniform state and the two-hump state which is indeed the Fermi-Pasta-Ulam recurrence phenomenon [33]. The local energy and momentum dont't grow with time; they are well preserved. The enery and momentum errors errors are concentrated in the regions of the solutions where there are two peaks. The global errors are also well preserved by the multisymplectic integrator (36)-(37) (see [34].

Numerical solution of the CNLS equation for the linear polarization with the parameters $v_{12} = v_{21} = 2/3$ and circular polarization and $v_{12} = v_{21} = 2$ (circular polarization) are computed in [34].

3. The Strongly Coupled Nonlinear Schrödinger Equation

The strongly coupled nonlinear Schrödinger (SCNLS) equation was investigated in several articles[11, 13, 35, 36]

$$
\begin{aligned}
iu_t + \beta u_{xx} + \left[\alpha_1|u|^2 + (\alpha_1 + 2\alpha_2)|v|^2\right]u + \gamma u + \Gamma v &= 0, \\
iv_t + \beta v_{xx} + \left[\alpha_1|v|^2 + (\alpha_1 + 2\alpha_2)|u|^2\right]v + \gamma v + \Gamma u &= 0,
\end{aligned} \tag{39}
$$

with the initial conditions

$$
u(x,0) = u_0(x), \quad v(x,0) = v_0(x)
$$

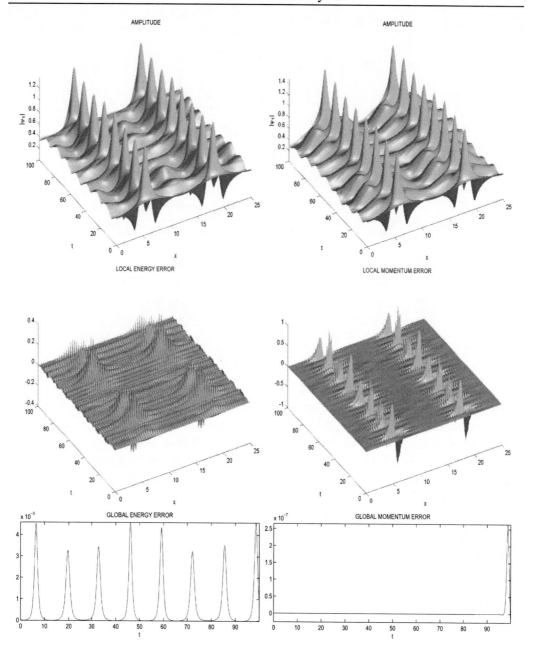

Figure 2. **Elliptic Polarization,** $v = 1$: Surface of waves and the errors of the CNLS equation (19) with $\psi_{10} = 0.5$, $\psi_{20} = 0.5$, $\epsilon = 0.1$.

and the periodic boundary conditions

$$u(x_l, t) = u(x_r, t), \quad v(x_l, t) = v(x_r, t), \quad t > 0,$$

where $u(x, t)$ and $v(x, t)$ are complex-valued functions of the spatial coordinate x and time t and β, α_1, α_2, γ, and Γ are real constants.

The SCNLS equation arises as a system of PDEs in many areas like in nonlinear optics, solid and fluid mechanics and in biology. The parameter β denotes the group velocity dispersion and the term proportional to α_1 describes the self-focusing of a signal for pulses in birefringent media. The cross-modulation parameter $\alpha = \alpha_1 + 2\alpha_2$ describes how each component of the solution is influenced by the other component. The constant γ is the ambient potential (normalized birefringent) and Γ is the linear coupling parameter (the linear birefringent). The weak coupling is proportional to Γ, which arises in addition to the weak coupling through the nonlinear terms [35].

The SCNLS equation (39) has mass and energy as conserved quantities [13, 35],

$$\frac{\partial}{\partial t} M(t) = \frac{\partial}{\partial t} \int_{x_l}^{x_r} (|u|^2 + |v|^2) \, dx = 0. \tag{40}$$

$$\begin{aligned}
\frac{\partial}{\partial t} E(t) &= \frac{\partial}{\partial t} \frac{1}{2} \int_{x_l}^{x_r} [-\beta(|u_x|^2 + |v_x|^2) + \frac{\alpha_1}{2}(|u|^4 + |v|^4) \\
&+ (\alpha_1 + 2\alpha_2)|u|^2|v|^2 + \gamma(|u|^2 + |v|^2) + 2\Gamma \cdot Re\{\overline{u}v\}] dx = 0.
\end{aligned} \tag{41}$$

The mass $M(t)$ and the energy $E(t)$ are not conserved individually for u and v when the coupling parameter $\Gamma \neq 0$.

3.1. Multisymplectic Integration of the SCNLS Equation

The SCNLS equation (39) can be written as a system of real–valued equations by decomposing the complex functions u and v into real and imaginary parts $u = p + iq$, $v = \mu + i\xi$

$$\begin{aligned}
-q_t &+ \beta\, p_{xx} &+ (\alpha_1(p^2 + q^2) + \alpha(\mu^2 + \xi^2))p &+ \gamma p &+ \Gamma\mu &= 0, \\
p_t &+ \beta\, q_{xx} &+ (\alpha_1(p^2 + q^2) + \alpha(\mu^2 + \xi^2))q &+ \gamma q &+ \Gamma\xi &= 0, \\
-\xi_t &+ \beta\, \mu_{xx} &+ (\alpha_1(\mu^2 + \xi^2) + \alpha(p^2 + q^2))\mu &+ \gamma\mu &+ \Gamma p &= 0, \\
\mu_t &+ \beta\, \xi_{xx} &+ (\alpha_1(\mu^2 + \xi^2) + \alpha(p^2 + q^2))\xi &+ \gamma\xi &+ \Gamma q &= 0.
\end{aligned} \tag{42}$$

Introducing new variables $\beta\, u_x = b + i\, a$, $\beta\, v_x = d + i\, c$ it can be written as a multisymplectic PDE (1) with the state variable $\mathbf{z} = (p, \mu, q, \xi, b, d, a, c)^T$

$$\mathbf{M} = \begin{pmatrix} \mathbf{0} & -\mathbf{I_2} & \mathbf{0} & \mathbf{0} \\ \mathbf{I_2} & \mathbf{0} & \mathbf{0} & \mathbf{0} \\ \mathbf{0} & \mathbf{0} & \mathbf{0} & \mathbf{0} \\ \mathbf{0} & \mathbf{0} & \mathbf{0} & \mathbf{0} \end{pmatrix} \quad \mathbf{K} = \begin{pmatrix} \mathbf{0} & \mathbf{0} & \mathbf{I_2} & \mathbf{0} \\ \mathbf{0} & \mathbf{0} & \mathbf{0} & \mathbf{I_2} \\ -\mathbf{I_2} & \mathbf{0} & \mathbf{0} & \mathbf{0} \\ \mathbf{0} & -\mathbf{I_2} & \mathbf{0} & \mathbf{0} \end{pmatrix} \tag{43}$$

where $\mathbf{0}$, $\mathbf{I_2}$ are 2×2 zero and identity matrices respectively. The Hamiltonian function $S(\mathbf{z}) : R^8 \to R$, is given by

$$S(\mathbf{z}) = -\frac{\alpha_1}{4}(p^2 + q^2)^2 - \frac{\alpha_1}{4}(\mu^2 + \xi^2)^2 - \left(\frac{\alpha_1 + 2\alpha_2}{2}\right)(p^2 + q^2)(\mu^2 + \xi^2)$$

$$-\frac{\gamma}{2}(p^2 + q^2 + \mu^2 + \xi^2) - \Gamma(p\mu + q\xi) - \frac{1}{2\beta}(a^2 + b^2 + c^2 + d^2). \tag{44}$$

The discretization of a multisymplectic PDE (1) in space and time with partitioned Runge–Kutta methods gives rise to a system of equations that formally satisfy a discrete multi-symplectic conservation law [37, 38]. When the same partitioning of the variables is used in both space and time discretization, the resulting scheme is, in general, fully implicit. Ryland and Mclachlan gave in [38] sufficient conditions on a multisymplectic PDE for a Lobatto IIIA-IIIB discretization in space to give rise to explicit ODEs. The variables are partitioned independently in space and time.

Theorem 3.1 ([38]). *We consider a multisymplectic PDE (1), with the matrices* $\mathbf{M}$ *and* $\mathbf{K}$:

$$
\mathbf{M} = \begin{pmatrix} & -\mathbf{I}_{\frac{1}{2}(d_1+d_2)} & \\ \mathbf{I}_{\frac{1}{2}(d_1+d_2)} & & \\ & & \mathbf{0}_{d_1} \end{pmatrix}, \quad \mathbf{K} = \begin{pmatrix} & & \mathbf{I}_{d_1} \\ & \mathbf{0}_{d_2} & \\ -\mathbf{I}_{d_1} & & \end{pmatrix} \tag{45}
$$

where $d_1 = n - rank(\mathbf{M})$, $d_2 = n - 2d_1 \le d_1$, $\mathbf{I}_d$ *and* $\mathbf{0}_d$ *are the* $d \times d$ *identity and zero matrices respectively. Let the variable* $\mathbf{z}$ *be partitioned into two parts* $\mathbf{z}^{(1)} \in \mathbb{R}^{d_1+d_2}$ *and* $\mathbf{z}^{(2)} \in \mathbb{R}^{d_1}$, *where we denote the first* d_1 *component of* $\mathbf{z}^{(1)}$ *by* $\mathbf{q}$, *the last* d_2 *components of* $\mathbf{z}^{(1)}$ *by* $\mathbf{v}$, *and the components of* $\mathbf{z}^{(2)}$ *by* $\mathbf{p}$ *such that the PDE may be written as*

$$
\begin{pmatrix} & -\mathbf{I}_{\frac{1}{2}(d_1+d_2)} & \\ \mathbf{I}_{\frac{1}{2}(d_1+d_2)} & & \\ & & \mathbf{0}_{d_1} \end{pmatrix} \begin{pmatrix} \mathbf{q} \\ \mathbf{v} \\ \mathbf{p} \end{pmatrix}_t
$$

$$
+ \begin{pmatrix} & & \mathbf{I}_{d_1} \\ & \mathbf{0}_{d_2} & \\ -\mathbf{I}_{d_1} & & \end{pmatrix} \begin{pmatrix} \mathbf{q} \\ \mathbf{v} \\ \mathbf{p} \end{pmatrix}_x = \begin{pmatrix} \nabla_{\mathbf{q}} S(\mathbf{z}) \\ \nabla_{\mathbf{v}} S(\mathbf{z}) \\ \nabla_{\mathbf{p}} S(\mathbf{z}) \end{pmatrix}. \tag{46}
$$

If the function $S(\mathbf{z})$ *can be written in the form*

$$
S(\mathbf{z}) = T(\mathbf{p}) + V(\mathbf{q}) + \widehat{V}(\overline{\mathbf{v}}) \tag{47}
$$

where $T(\mathbf{p}) = \frac{1}{2}\mathbf{p}^T \Lambda \mathbf{p}$ *and* $\widehat{V}(\mathbf{v}) = \frac{1}{2}\mathbf{v}^T \alpha \mathbf{v}$ *such that* $|\Lambda| \ne 0$ *and* $|\alpha| \ne 0$, *then applying an* r-*stage Lobatto IIIA-IIIB PRK discretization in space to the PDE lead to a set of explicit local ODEs in time in the stage variables associated with* $\mathbf{q}$.

The Comparison of the matrices $\mathbf{M}$ and $\mathbf{K}$ in (43) of the SCNLS equation (39) with (45) reveals that $d_1 = 4$ and $d_2 = 0$ with $\mathbf{z}^{(1)} = \mathbf{q} = \{p, \mu, q, \xi\}$ and $\mathbf{z}^{(2)} = \mathbf{p} = \{b, d, a, c\}$. $S(\mathbf{z})$ can be written as (47) with

$$
V(\mathbf{q}) = -\frac{\alpha_1}{4}(p^2 + q^2)^2 - \frac{\alpha_1}{4}(\mu^2 + \xi^2)^2 - \left(\frac{\alpha^2}{2}\right)(p^2 + q^2)(\mu^2 + \xi^2)
$$

$$
-\frac{\gamma}{2}(p^2 + q^2 + \mu^2 + \xi^2) - \Gamma(p\mu + q\xi)
$$

and $T(\mathbf{p}) = \dfrac{1}{2}\mathbf{p}^T \Lambda \mathbf{p}$ where

$$
\Lambda = \begin{pmatrix} -1/\beta & 0 & 0 & 0 \\ 0 & -1/\beta & 0 & 0 \\ 0 & 0 & -1/\beta & 0 \\ 0 & 0 & 0 & -1/\beta \end{pmatrix} \quad \text{and} \quad \mathbf{p} = \begin{pmatrix} b \\ d \\ a \\ c \end{pmatrix}.
$$

Because $d_2 = 0$, there are no terms with $\hat{v}$. Thus the SCNLS equation (39) satisfies the requirements of the Theorem 3.1. The multisymplectic formulation of the system (39) using the matrices (43) and the Hamiltonian function (44) is then discretized in the spatial variable by the 2–stage Lobatto IIIA–IIIB PRK method, which corresponds to the table

$$
\text{IIIA}: \quad
\begin{array}{c|cc}
0 & 0 & 0 \\
1 & 1/2 & 1/2 \\
\hline
 & 1/2 & 1/2
\end{array}
\quad , \qquad
\text{IIIB}: \quad
\begin{array}{c|cc}
0 & 1/2 & 0 \\
1 & 1/2 & 0 \\
\hline
 & 1/2 & 1/2
\end{array}
\quad .
\tag{48}
$$

We obtain the explicit local ODEs in time for the stage variables associated with $\mathbf{q} = \{p, \mu, q, \xi\}$

$$
\begin{aligned}
\partial_t q_i &= \beta \frac{p_{i-1} - 2p_i + p_{i+1}}{\Delta x^2} + (\alpha_1(p_i^2 + q_i^2) + \alpha(\mu_i^2 + \xi_i^2))p_i + \gamma p_i + \Gamma \mu_i, \\
\partial_t p_i &= -\beta \frac{q_{i-1} - 2q_i + q_{i+1}}{\Delta x^2} - (\alpha_1(p_i^2 + q_i^2) + \alpha(\mu_i^2 + \xi_i^2))q_i - \gamma q_i - \Gamma \xi_i, \\
\partial_t \xi_i &= \beta \frac{\mu_{i-1} - 2\mu_i + \mu_{i+1}}{\Delta x^2} + (\alpha_1(\mu_i^2 + \xi_i^2) + \alpha(p_i^2 + q_i^2))\mu_i + \gamma \mu_i + \Gamma p_i, \\
\partial_t \mu_i &= -\beta \frac{\xi_{i-1} - 2\xi_i + \xi_{i+1}}{\Delta x^2} - (\alpha_1(\mu_i^2 + \xi_i^2) + \alpha(p_i^2 + q_i^2))\xi_i - \gamma \xi_i - \Gamma q_i.
\end{aligned}
\tag{49}
$$

After elimination of the stage variables associated with $\mathbf{p} = \{b, d, a, c\}$ and rearranging the resulting equations, the Lobatto IIIA-IIIB semi-discretization in space corresponds to the replacement of p_{xx}, q_{xx}, μ_{xx} and ξ_{xx} in Eq. (42) by the central difference discretization.

In [39] a different partitioning $\mathbf{z}^{(3)} = (p, \mu, b, d)$ and $\mathbf{z}^{(4)} = (q, \xi, a, c)$ was chosen for the time discretization of (49). Application of the Lobatto IIIA method to the variables in $\mathbf{z}^{(3)}$ and Lobatto IIIB method to the variables in $\mathbf{z}^{(4)}$ gives an integrator which maps $(p_i^n, q_i^n, \xi_i^n, \mu_i^n)$ to $(p_i^{n+1}, q_i^{n+1}, \xi_i^{n+1}, \mu_i^{n+1})$. The resulting scheme is a nine-point, three time level multisymplectic integrator [39].

We present some numerical results from [39] for SCNLS equation in Figure 3 with $\beta = 1.0$, $\alpha_1 = 1.0$, $\gamma = 1.0$ for the initial condition

$$
\begin{aligned}
u(0, x) &= \sqrt{2} \operatorname{sech} \left(x + \tfrac{1}{2} D_0 \right) e^{iV_0 x/4} \\
v(0, x) &= \sqrt{2} \operatorname{sech} \left(x - \tfrac{1}{2} D_0 \right) e^{-iV_0 x/4},
\end{aligned}
\tag{50}
$$

where $D_0 = 25$ and $V_0 = 1.0$. The figure shows the surface of the wave and the global errors for $-30 \le x \le 30$ and $0 \le t \le 50$ with $N = 128$ number of space grid points and the time step size $\Delta t = 0.01$. Elastic and inelastic collisions are observed for the parameters $\alpha_2 = -1/6$, $\Gamma = 1.0$ and $\alpha_2 = -1/6$, $\Gamma = 0.0175$ respectively. Fusion of two solitons is obtained by choosing the parameters $\alpha_2 = -1/3$, $\Gamma = 0.0175$ [39]. In Fig. 3(a) the elastic interaction of two waves is shown; the waves emerge without any changes in their shapes. Fig. 3(b) shows an inelastic collision; after the interaction, the solitary waves leave dispersive oscillations; the shape and amplitudes of both waves are changed. In Fig. 3(c) the fusion of two solitons is shown. From the Fig. 3 we see that when the collision takes place, the global energy errors are corrupted and increased in all cases.

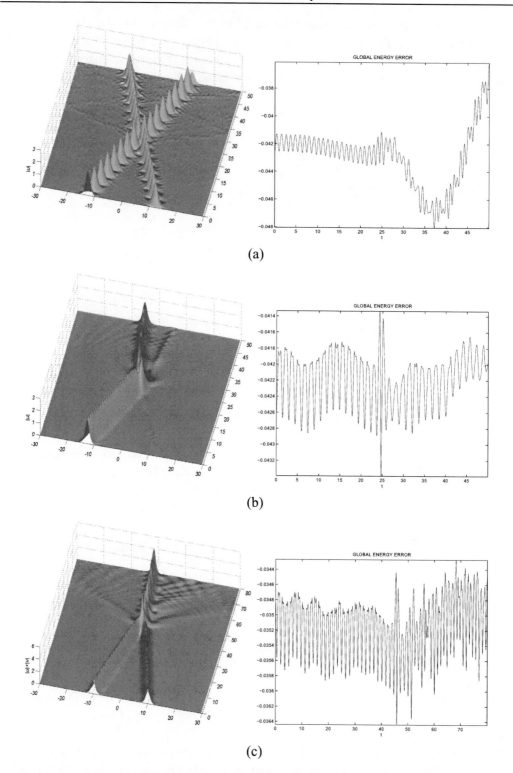

(a)

(b)

(c)

Figure 3. Surface of waves and global errors of the SCNLS equation (39). Top : Elastic collision with $\alpha_2 = -1/6$ and $\Gamma = 1.0$. Middle : Inelastic collision with $\alpha_2 = -1/6$ and $\Gamma = 0.0175$. Bottom : Fusion with $\alpha_2 = -1/3$ and $\Gamma = 0.0175$.

4. The Complex Modified Korteweg–de Vries Equation

In this Chapter we will investigate the application multisymplectic box schemes to the complex modified Korteweg-de Vries (CMKdV) equation [40, 41, 42]

$$\frac{\partial \psi}{\partial t} + \frac{\partial^3 \psi}{\partial x^3} + \alpha \frac{\partial \left(|\psi|^2 \psi\right)}{\partial x} = 0, \qquad -\infty < x < \infty, \quad t > 0, \tag{51}$$

where $\psi(x, t)$ is a complex-valued function of the spatial coordinate x and the time t, and α is a real parameter. CMKdV equation describes the nonlinear evolution of plasma waves and the propagation of transverse waves in a molecular chain model or in a generalized elastic solid [40, 41, 42]. We obtain the coupled pair of modified Korteweg-de Vries equations if $\psi(x, t)$ is decomposed into its real and imaginary parts, $\psi = u + iv$, $(i^2 = -1)$,

$$\begin{aligned} \frac{\partial u}{\partial t} + \frac{\partial^3 u}{\partial x^3} + \alpha \frac{\partial}{\partial x}[(u^2 + v^2)u] &= 0, \\ \frac{\partial v}{\partial t} + \frac{\partial^3 v}{\partial x^3} + \alpha \frac{\partial}{\partial x}[(u^2 + v^2)v] &= 0. \end{aligned} \tag{52}$$

These two coupled nonlinear equations describe the interaction of two orthogonally polarized transverse waves; where u and v represent y-polarized and z-polarized transverse waves, respectively, propagating in the x-direction in an xyz coordinate system.

Inverse scattering method has been used to obtain some analytical solutions, i.e., solitary wave solutions, of integrable nonlinear dispersive wave equations. The CMKdV equation is a non-integrable equation. However there are few analytical solutions corresponding to some special cases of the CMKdV equation, for example the linearly polarized solitary waves in form

$$\psi(x, t) = \sqrt{\frac{2c}{\alpha}} \operatorname{sech} \left[\sqrt{c}(x - x_0 - ct) \right] \exp(i\theta), \tag{53}$$

which represents a solitary wave located to the position $x = x_0$, moving to the right with velocity c and satisfying the boundary conditions $u \to 0$ as $x \to \pm\infty$.

The CMKdV equation posesses the following conserved quantities [40, 42]

$$I_1 = \int_{-\infty}^{\infty} \psi \, dx, \quad I_2 = \int_{-\infty}^{\infty} |\psi|^2 \, dx, \quad I_3 = \int_{-\infty}^{\infty} \left(\frac{\alpha}{2}|\psi|^4 - \left|\frac{\partial \psi}{\partial x}\right|^2 \right) dx. \tag{54}$$

4.1. Multisymplectic Integration of the CMKdV Equation

Introducing the new variables $\eta_1, \eta_2, \phi_1, \phi_2, w_1, w_2$ in (51) with

$$\begin{aligned} \psi_x(x, t) &= \eta_1(x, t) + i\eta_2(x, t), \\ -\tfrac{1}{2}\psi(x, t) &= \phi_{1x}(x, t) + i\phi_{2x}(x, t), \\ \tfrac{1}{2}w_1(x, t) &= -\phi_{1t}(x, t) + \eta_{1x}(x, t) + \alpha(u^2 + v^2)u, \\ \tfrac{1}{2}w_2(x, t) &= -\phi_{2t}(x, t) + \eta_{2x}(x, t) + \alpha(u^2 + v^2)v \end{aligned}$$

the CMKdV equation (51) can be rewritten as multisymplectic PDE system (1) with the state variable $\mathbf{z} = (u, v, \phi_1, \phi_2, \eta_1, \eta_2, w_1, w_2)^T$ and skew-symmetric matrices

$$
\mathbf{K} = \begin{pmatrix} \mathbf{J}_1 & \mathbf{0}_4 \\ \mathbf{0}_4 & \mathbf{0}_4 \end{pmatrix}, \quad \mathbf{L} = \begin{pmatrix} \mathbf{0}_4 & \mathbf{I}_4 \\ -\mathbf{I}_4 & \mathbf{0}_4 \end{pmatrix}, \quad \text{with} \quad \mathbf{J}_1 = \begin{pmatrix} \mathbf{0}_2 & -\mathbf{I}_2 \\ \mathbf{I}_2 & \mathbf{0}_2 \end{pmatrix} \tag{55}
$$

where $\mathbf{0}_2$, $\mathbf{I}_2$ are 2×2 and $\mathbf{0}_4$, $\mathbf{I}_4$ are the 4×4 zero and identity matrices respectively. The Hamiltonian function $S : R^8 \to R$, is given by

$$
S(\mathbf{z}) = \frac{1}{2}(uw_1 + vw_2) - \frac{1}{2}(\eta_1^2 + \eta_2^2) - \frac{\alpha}{4}(u^2 + v^2)^2. \tag{56}
$$

Application of the Preissmann scheme (14) to the multisymplectic formulation of (51) with (55) and (56) and eliminating the auxiliary variables $\phi_1, \phi_2, \eta_1, \eta_2, w_1, w_2$ we get

$$
D_t L_t L_x^3 \psi + D_x^3 L_t^2 \psi + \alpha D_x L_t L_x(|L_t L_x \psi| L_t L_x \psi) = 0. \tag{57}
$$

The scheme (57) can be written as

$$
\frac{1}{16\Delta t} \begin{bmatrix} 1 & 3 & 3 & 1 \\ 0 & 0 & 0 & 0 \\ -1 & -3 & -3 & -1 \end{bmatrix} \psi + \frac{1}{4\Delta x^3} \begin{bmatrix} -1 & 3 & -3 & 1 \\ -2 & 6 & -6 & 2 \\ -1 & 3 & -3 & 1 \end{bmatrix} \psi
$$

$$
+ \frac{\alpha}{4\Delta x} \begin{bmatrix} -1 & 0 & 1 \\ -1 & 0 & 1 \end{bmatrix} \left(\left| \frac{1}{4} \begin{bmatrix} 1 & 1 \\ 1 & 1 \end{bmatrix} \psi \right|^2 \frac{1}{4} \begin{bmatrix} 1 & 1 \\ 1 & 1 \end{bmatrix} \psi \right) = 0. \tag{58}
$$

The scheme (57) is the 12-point scheme since there it has four space levels and three time levels.

Note that every term in (57) contains a factor L_t but we can not eliminate the terms with L_t in (57) because L_t is not invertible. However, in the multisymplectic formulation of the equations (51), four terms contain no time derivatives, so they can be discretized by omitting the time averaging L_t to give $-D_x u = -L_x \eta_1$, $-D_x v = -L_x \eta_2$, $-D_x \phi_1 = \frac{1}{2} L_x u$ and $-D_x \phi_2 = \frac{1}{2} L_x v$. Then, discretization of the first four equations in the multisymplectic formulation of (51) and using the implicit midpoint rule in time and space and elimination of the auxiliary variables yields the eight-point scheme

$$
D_t L_x^3 \psi + D_x^3 L_t \psi + \alpha D_x L_x(|L_t L_x \psi| L_t L_x \psi) = 0. \tag{59}
$$

In the finite difference stencil format the eight-point scheme (59) is given by

$$
\frac{1}{8\Delta t} \begin{bmatrix} 1 & 3 & 3 & 1 \\ -1 & -3 & -3 & -1 \end{bmatrix} \psi + \frac{1}{2\Delta x^3} \begin{bmatrix} -1 & 3 & -3 & 1 \\ -1 & 3 & -3 & 1 \end{bmatrix} \psi
$$

$$
+ \frac{\alpha}{2\Delta x} \begin{bmatrix} -1 & 0 & 1 \end{bmatrix} \left(\left| \frac{1}{4} \begin{bmatrix} 1 & 1 \\ 1 & 1 \end{bmatrix} \psi \right|^2 \frac{1}{4} \begin{bmatrix} 1 & 1 \\ 1 & 1 \end{bmatrix} \psi \right) = 0. \tag{60}
$$

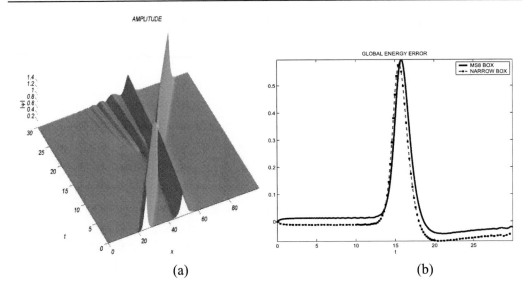

Figure 4. Interaction of two orthogonally traveling waves and the global errors for the CMKDV equation (51) with the initial condition (63).

Averaging the eight-point scheme (60) at time levels $n - 1$ and n gives the 12-point scheme (58). The scheme (60) is more efficient than the Preissmann scheme for the CMKDV equation (51) and equivalent to the 12-point scheme (57) up to the round-off error [43].

The box scheme (11) preserves the multisymplectic conservation [44]. It was obtained by applying finite volume method to the a KdV equation in [45] at the price of potentially abandoning multisymplecticity. In [46] a standard finite volume discretization was directly to a the CMKdV equation (52) following [45] and a variant on the box scheme (60) was obtained with a narrower stencil for the nonlinear term,

$$\frac{1}{2\Delta t}\begin{bmatrix} 1 & 1 \\ -1 & -1 \end{bmatrix}\psi + \frac{1}{2\Delta x^3}\begin{bmatrix} -1 & 3 & -3 & 1 \\ -1 & 3 & -3 & 1 \end{bmatrix}\psi$$

$$+\frac{\alpha}{\Delta x}\begin{bmatrix} -1 & 1 \end{bmatrix}\left(\left|\frac{1}{2}\begin{bmatrix} 1 \\ 1 \end{bmatrix}\psi\right|^2 \frac{1}{2}\begin{bmatrix} 1 \\ 1 \end{bmatrix}\psi\right) = 0. \qquad (61)$$

The scheme (61) is called as the narrow box scheme, since the approximations corresponding to the first derivatives ψ_t and $\left(|\psi|^2\psi\right)$ in (51) have narrower stencils than (60), whereas the term corresponding to the third derivative ψ_{xxx} is the same. Discretizing the CMKdV equation with two–stage Lobatto IIIA–IIIB and eliminating the auxiliary variables η_i, w_i and ϕ_i, $(i = 1, 2)$ gives the implicit ODE

$$M\partial_t\psi_j + D^3\psi_{j-1} + \alpha D\left(|\psi_j|^2 \psi\right) = 0. \qquad (62)$$

This is none other than the narrow box scheme (61). This shows that the narrow box scheme for the CMKdV equation (51) is multisymplectic [37, 46].

In the following we give some numerical results in [46]. We consider the CMKDV equation (51) with the initial condition [40]

$$\psi(x,0) = \sum_{j=1}^{2} \sqrt{\frac{2c_j}{\alpha}} \text{sech} \left[\sqrt{c_j}(x - x_j) \right] \exp(i\theta_j) \tag{63}$$

The propogation of the two single waves (63) with $\alpha = 2, c_1 = 2, c_2 = 0.5, x_1 = 25, x_2 = 50, \theta_1 = 0, \theta_2 = \pi/2$ is shown in Figure 4(a) for both integrators (59) and (61). The initial wave (63) consists of two solitary waves located at the positions $x_1 = 25$ and $x_2 = 50$ and moving to the right with the velocity $c_1 = 2$ and $c_2 = 0.5$ respectively. In Fig. 4(a) the interaction of these two $y-$ polarized solitary waves is shown for $\theta_1 = 0, \theta_2 = \pi/2$. The faster wave located at the position $x_1 = 25$ catches and interacts with the slower one and continues to move to the right without changing its profile. After the collision, there is a tail following the shorter one. The global energy errors are shown in Fig. 4(b).

4.2. Multisymplectic Splitting Method for the CMKdV Equation

Splitting methods are one of the most efficient methods for geometric integration of Hamiltonian ODEs (see for an overview [31]). Recently, splitting methods are applied to the multisymplectic PDEs [38]. The original multisymplectic equation is decomposed in such a way that each splitted equation is again multisymplectic. Composition of the splitted multisymplectic equations in a suitable yields a multisymplectic integrator for the original equation. A multisymplectic splitting method is proposed to solve a coupled nonlinear Schrödinger equation in [47] and a CMKDV equation in [46].

For the multisymplectic splitting method, we the multisymplectic PDE (1) and the split $\mathbf{L} = \sum_{j=1}^{N} \mathbf{L}^{(j)}$ and $S(\mathbf{z}) = \sum_{j=1}^{N} S^{(j)}(\mathbf{z})$ and obtain

$$\mathbf{K}\mathbf{z}_t + \mathbf{L}^{(j)}\mathbf{z}_x = \nabla_{\mathbf{z}} S^{(j)}(\mathbf{z}).$$

The flow of each subsystem satisfies a different multisymplectic conservation laws $\omega_t + \kappa_x^{(j)} = 0$ with $\kappa^{(j)} = 1/2\mathbf{L}^{(j)}d\mathbf{z} \wedge d\mathbf{z}$. The total symplecticity is conserved by each flow because the density ω in each conservation law is the same, which is known as symplectic splitting [37].

Now we introduce a linear–nonlinear splitting for the CMKdV equation (51) based on multisymplectic structure (1). The skew–symmetric matrix $\mathbf{L}$ can be written as the sum of two skew–symmetric matrices $\mathbf{L}^{(1)}$, $\mathbf{L}^{(2)}$ and the Hamiltonian function $S(\mathbf{z})$ can be split in two Hamiltonian functions $S^{(1)}(\mathbf{z})$, $S^{(2)}(\mathbf{z})$ respectively, both defining two different multisymplectic PDEs. Now we can write $\mathbf{L} = \mathbf{L}^{(1)} + \mathbf{L}^{(2)}$ where

$$\mathbf{L}^{(1)} = \begin{pmatrix} \mathbf{0}_4 & \mathbf{J}_2 \\ -\mathbf{J}_2 & \mathbf{0}_4 \end{pmatrix}, \quad \mathbf{L}^{(2)} = \begin{pmatrix} \mathbf{0}_4 & \mathbf{J}_3 \\ -\mathbf{J}_3 & \mathbf{0}_4 \end{pmatrix}$$

with

$$\mathbf{J}_2 = \begin{pmatrix} \mathbf{I}_2 & \mathbf{0}_2 \\ \mathbf{0}_2 & \frac{1}{2}\mathbf{I}_2 \end{pmatrix}, \quad \mathbf{J}_3 = \begin{pmatrix} \mathbf{0}_2 & \mathbf{0}_2 \\ \mathbf{0}_2 & \frac{1}{2}\mathbf{I}_2 \end{pmatrix},$$

and $S(\mathbf{z}) = S^{(1)} + S^{(2)}$ where

$$S^{(1)} = \frac{1}{4}(uw_1 + vw_2) - \frac{1}{2}(\eta_1^2 + \eta_2^2), \qquad S^{(2)} = \frac{1}{4}(uw_1 + vw_2) - \frac{\alpha}{4}(u^2 + v^2)^2.$$

We consider the time–space evolution generated by $\mathbf{L}^{(1)}$ and $S^{(1)}(\mathbf{z})$

$$\mathbf{K}\mathbf{z}_t + \mathbf{L}^{(1)}\mathbf{z}_x = \nabla_{\mathbf{z}} S^{(1)}(\mathbf{z}). \tag{64}$$

Eliminating the auxiliary variables $\phi_1, \phi_2, \eta_1, \eta_2, w_1, w_2$ we get the linear part of the CMKdV equation (51)

$$w_t + w_{xxx} = 0. \tag{65}$$

The time–space evolution generated by $\mathbf{L}^{(2)}$ and $S^{(2)}(\mathbf{z})$ is

$$\mathbf{K}\mathbf{z}_t + \mathbf{L}^{(2)}\mathbf{z}_x = \nabla_{\mathbf{z}} S^{(2)}(\mathbf{z}). \tag{66}$$

Again after eliminating the auxiliary variables $\phi_1, \phi_2, \eta_1, \eta_2, w_1, w_2$ the nonlinear part of the CMKdV equation (51) is obtained

$$w_t + \alpha \left(|w|^2 w\right)_x = 0. \tag{67}$$

Multisymplectic schemes can also be derived from the CMKdV equation in splitted form, which are obtained in a similar way to the eight-point scheme [46] We apply the Preissman scheme (11) to the linear system (64) and nonlinear system (66) and obtain

$$(aw_j^{n+1} + 3bw_{j+1}^{n+1} + 3aw_{j+2}^{n+1} + bw_{j+3}^{n+1}) - (bw_j^n + 3aw_{j+1}^n + 3bw_{j+2}^n + aw_{j+3}^n) = 0 \tag{68}$$

$$\begin{aligned}
w_{j+1}^{n+1} + w_j^{n+1} - w_{j+1}^n - w_j^n + e\Big[(w_{j+1}^{n+1} + w_{j+1}^n)|w_{j+1}^{n+1} + w_{j+1}^n|^2 \\
- (w_j^{n+1} + w_j^n)|w_j^{n+1} + w_j^n|^2\Big] = 0
\end{aligned} \tag{69}$$

where $e = \alpha \frac{\Delta t}{4\Delta x}$. The scheme (68) and (69) corresponds to the discretization of (65) and the discretization of (67) respectively. The solution of the multisymplectic two-term (linear–nonlinear) splitting is advanced in time according to the second–order compositions [48]

$$\exp\left(\frac{\Delta t}{2} L\right) \exp\left(\Delta t N\right) \exp\left(\frac{\Delta t}{2} L\right). \tag{70}$$

5. The Zakharov System

Another important coupled nonlinear PDE is the Zakharov system which describes the interaction between the Langmuir and ion acoustic waves in a plasma [49]

$$i\partial_t + \partial_{xx} + 2\phi\psi = 0, \partial_{tt}\psi - \partial_{xx}\psi + \partial_{xx}(|\phi|^2) = 0. \tag{71}$$

The complex function $\phi(x, t)$ is the envelope of the electric field, and the real function $\psi(x, t)$ denotes the fluctuation of the ion density about its equilibrium value. The Zakharov system (71) can be formulated as a multisymplectic PDE (1) by setting $\phi = u + iv$ and introducing new variables $\partial_u = p$, $\partial_v = q$, $\partial_t \phi = \partial_{xx} f$, $\partial_x = g$. After separating into real and complex variables i the equation (71) can be written as multisymplectic Hamiltonian PDE (1) with

$$
\mathbf{M} = \begin{pmatrix} 0 & -1 & 0 & 0 & 0 & 0 & 0 \\ 1 & 0 & 0 & 0 & 0 & 0 & 0 \\ 0 & 0 & 0 & 0 & 0 & 0 & 0 \\ 0 & 0 & 0 & 0 & 0 & 0 & 0 \\ 0 & 0 & 0 & 0 & 0 & 1 & 0 \\ 0 & 0 & 0 & 0 & -1 & 0 & 0 \\ 0 & 0 & 0 & 0 & 0 & 0 & 0 \end{pmatrix}, \quad \mathbf{K} = \begin{pmatrix} 0 & 0 & 1 & 0 & 0 & 0 & 0 \\ 0 & 0 & 0 & 1 & 0 & 0 & 0 \\ -1 & 0 & 0 & 0 & 0 & 0 & 0 \\ 0 & -1 & 0 & 0 & 0 & 0 & 0 \\ 0 & 0 & 0 & 0 & 0 & 0 & 0 \\ 0 & 0 & 0 & 0 & 0 & 0 & 1 \\ 0 & 0 & 0 & 0 & 0 & -1 & 0 \end{pmatrix}
$$

and $(S(z) = -\phi(u^2 + v^2) - \frac{1}{2}(p^2 + q^2)\frac{1}{2}\phi^2 - \frac{1}{2}g^2$ and $z = (u, v, p, \psi, f, g)^T$.

The Zakharov system was solved by symplectic and multisymplectic integrators in [49]

6. Dispersion Relations

In analyzing the behavior of the traveling wave and soliton solutions of nonlinear PDEs, the dispersion relations and group velocities play a fundamental role. Important properties like stability, sign of the group velocity, the existence of spurious solutions can be determined by considering the discrete dispersion relations. Dispersive properties of MS integrators were investigated for the KdV equation [43, 45], the linear wave equation [50, 51] and the sine-Gordon equation [51], the Boussinesq equation [52] and the nonlinear Schrödinger equation [53].

The dispersion relation of a PDE in multisymplectic form is calculated by considering the linearized PDE

$$
\mathbf{M}z_t + \mathbf{K}z_x = \mathcal{S}\mathbf{z} \tag{72}
$$

where $\mathcal{S}$ is a symmetric matrix such that $\mathcal{S}\mathbf{z}$ is the linear component of $\nabla_z S(\mathbf{z})$ [18].

The complex valued elementary single mode wave-like solutions can be written as

$$
\mathbf{z}(x, t) = \mathbf{a}e^{i(kx - \omega t)} \tag{73}
$$

where $\mathbf{a} = \mathbf{a}(k)$ denotes the amplitude, k the wave number and ω the frequency. Substituting this solution into the linear PDE leads to the linear system

$$
(-i\omega\mathbf{M} + ik\mathbf{K} - \mathcal{S})\mathbf{a} = 0.
$$

Since we seek a solution such that a is non-zero, k and ω must satisfy the dispersion relation

$$
\mathcal{D}(\omega, k) := \det(-i\omega\mathbf{M} + ik\mathbf{K} - \mathcal{S}) = 0. \tag{74}
$$

It is important to note that the matrix used in this calculation is self-adjoint which implies a real dispersion relation, meaning that there is no diffusion. In the case of dispersive equations, different waves have different wave speeds and the behavior of solutions depend on how the different waves interact with each other.

In this Chapter we derive dispersion relations for the CMKDV equation (51)[46]. For this we consider the linearization of the modified KdV equations (52) around the constant solutions $\bar{w} = (\bar{u}, \bar{v})^T$ and obtain the following linearized uncoupled modified KdV equations with $\tilde{z} = (\tilde{u}, \tilde{v})^T$

$$\frac{\partial \tilde{u}}{\partial t} + \frac{\partial^3 \tilde{u}}{\partial x^3} + \alpha\lambda_1 \frac{\partial \tilde{u}}{\partial x} = 0,$$
$$\frac{\partial \tilde{v}}{\partial t} + \frac{\partial^3 \tilde{v}}{\partial x^3} + \alpha\lambda_2 \frac{\partial \tilde{v}}{\partial x} = 0. \tag{75}$$

where $\lambda_1, \lambda_2 \in \mathbb{R}$. The linearized modified KdV equation (75) can be now recast in multisymplectic form (72) with

$$\mathcal{S} = \begin{pmatrix} -\alpha\lambda_1 & 0 & 0 & 0 & 0 & 0 & 1/2 & 0 \\ 0 & -\alpha\lambda_2 & 0 & 0 & 0 & 0 & 0 & 1/2 \\ 0 & 0 & 0 & 0 & 0 & 0 & 0 & 0 \\ 0 & 0 & 0 & 0 & 0 & 0 & 0 & 0 \\ 0 & 0 & 0 & 0 & -1 & 0 & 0 & 0 \\ 0 & 0 & 0 & 0 & 0 & -1 & 0 & 0 \\ 1/2 & 0 & 0 & 0 & 0 & 0 & 0 & 0 \\ 0 & 1/2 & 0 & 0 & 0 & 0 & 0 & 0 \end{pmatrix}.$$

Then, the continuous dispersion relation obtained from the determinant (74) yields the dispersion relations for $k \neq 0$

$$D_1(\omega, k) = \omega + k^3 - \alpha\lambda_1 k = 0, \qquad D_2(\omega, k) = \omega + k^3 - \alpha\lambda_2 k = 0. \tag{76}$$

We consider the dispersion relations (76) by setting $\lambda_1 = \lambda$, which yields

$$D(\omega, k) = \omega + k^3 - \alpha\lambda k = 0. \tag{77}$$

The nonlinear term k in the dispersion relation (77) shows that the CMKdV equation is dispersive, which means that solution tends to break up into oscillatory wave packets. Each wave travels with a phase velocity $v_p(k) = \omega/k = -k^2 + \alpha\lambda$. The group velocity $V_g(k) = d\omega/dk = -3k^2 + \alpha\lambda$ describes the velocities of different waves and it is usually a function of the wave number k. If the group velocity dispersion $d^2\omega/dk^2 = -6k$ is non-zero, waves with different wave numbers travel with different velocities and this results in the spatial spreading of the wave packet. In general the dispersion relation is polynomial in both ω and k and may have different numbers of solutions.

The continuous dispersion relation describes how the frequencies of a linearized PDE are related to the wave number k in space. The nonlinear PDE is linearized at some fixed points, therefore the behavior of the nonlinear pde can be predicted approximatively by the dispersion relation, which is valid around these points.

When we apply the eight-point scheme (59) to the linearized CMKdV equation (75) with $\tilde{u} = \psi$, we get the linearized multisymplectic eight-point scheme

$$\frac{(\psi_{j-1}^{n+1} + 3\psi_j^{n+1} + 3\psi_{j+1}^{n+1} + \psi_{j+2}^{n+1}) - (\psi_{j-1}^n + 3\psi_j^n + 3\psi_{j+1}^n + \psi_{j+2}^n)}{4\Delta t}$$
$$- \frac{(\psi_{j-1}^{n+1} - 3\psi_j^{n+1} + 3\psi_{j+1}^{n+1} - \psi_{j+2}^{n+1}) + (\psi_{j-1}^n - 3\psi_j^n + 3\psi_{j+1}^n - \psi_{j+2}^n)}{\Delta x^3}$$
$$- \frac{\lambda\alpha}{4\Delta x} \left[(\psi_{j-1}^{n+1} + \psi_j^{n+1} - \psi_{j+1}^{n+1} - \psi_{j+2}^{n+1}) + (\psi_{j-1}^n + \psi_j^n - \psi_{j+1}^n - \psi_{j+2}^n) \right] = 0 \tag{78}$$

which has a discrete general solution of the form

$$\psi_j^n = \hat{a}e^{i(kx_j - \omega t_n)} = \hat{a}e^{i(k\Delta x j - \omega \Delta t n)} = \hat{a}e^{i(\overline{k}j - \overline{\omega}n)} \tag{79}$$

where $\overline{k} = k\Delta x$ is the numerical wave number and $\overline{\omega} = \omega \Delta t$ is the numerical frequency such that $-\pi \leq \overline{k} \leq \pi$ and $-\pi \leq \overline{\omega} \leq \pi$. As in the continuous case, substituting the numerical plane wave solution (79) into the linearized equation (78) and simplifying we get the numerical dispersion relation

$$D_N(\overline{\omega}, \overline{k}) = \frac{2}{\Delta t} \tan\left(\frac{\overline{\omega}}{2}\right) + \left[\frac{2}{\Delta x} \tan\left(\frac{\overline{k}}{2}\right)\right]^3 - \alpha\lambda \left[\frac{2}{\Delta x} \tan\left(\frac{\overline{k}}{2}\right)\right] = 0. \tag{80}$$

The multisymplectic eight-point scheme preserves the form of the analytic dispersion relation.

Proposition 6.1. *The multisymplectic scheme (80) qualitatively preserve the dispersion relation of the linearized PDE. Specifically, there exist diffeomorphism Ψ_1 and Ψ_2 satisfying the exact dispersion relationship*

$$D_N(\overline{\omega}, \overline{k}) = D\left(\Psi_1(\overline{\omega}), \Psi_2(\overline{k})\right) = \Psi_1 + \Psi_2^3 - \alpha\lambda\Psi_2 = 0, \tag{81}$$

where

$$\left(\Psi_1(\overline{\omega}), \Psi_2(\overline{k})\right) = \left(\frac{2}{\Delta t} \tan\frac{\overline{\omega}}{2}, \frac{2}{\Delta x} \tan\frac{\overline{k}}{2}\right)$$

for $-\pi < \overline{k} < \pi$, and $-\pi < \overline{\omega} < \pi$.

The numerical group velocity corresponding to the eight-point multisymplectic scheme (59) is given by [46]

$$\frac{d\overline{\omega}}{d\overline{k}} = -\mu \frac{\left[1 + \tan^2\left(\frac{\overline{k}}{2}\right)\right]\left[3\left(\frac{2}{\Delta x} \tan\left(\frac{\overline{k}}{2}\right)\right)^2 - \alpha\lambda\right]}{1 + \left(\frac{\Delta t}{2}\right)^2 \left[-\left(\frac{2}{\Delta x} \tan\left(\frac{\overline{k}}{2}\right)\right)^3 + \alpha\lambda\frac{2}{\Delta x} \tan\left(\frac{\overline{k}}{2}\right)\right]^2}. \tag{82}$$

Numerical dispersion relation and the group velocities for the narrow box scheme (61) can obtained in [46].

In the following we give some numerical results in [46]. Figures 5–6 represent the analytical and numerical dispersion relations and the group velocities for three different values of the mesh ratio $\mu = \Delta t/\Delta x$ with $\mu = 0.1, 0.5, 2.0$ for $\alpha = 2$ and $\lambda = 1$. For convenience we write the analytic dispersion relation (76) as

$$D(\overline{\omega}, \overline{k}) = \overline{\omega} + \frac{1}{\Delta x^2}\mu\overline{k}^3 - \mu\alpha\lambda\overline{k} = 0. \tag{83}$$

In all computations we have chosen $\Delta x = 0.2$, $\alpha = 2$ and $\lambda = 1$. Each plot is shown only for $0 \leq \overline{k} \leq \pi$ and $-\pi \leq \overline{\omega} \leq 0$ since the dispersion relations are symmetric with respect to the origin.

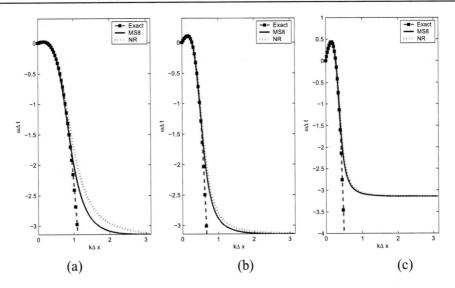

Figure 5. Dispersion relations for (a) $\mu = 0.1$, (b) $\mu = 0.5$, (c) $\mu = 2$.

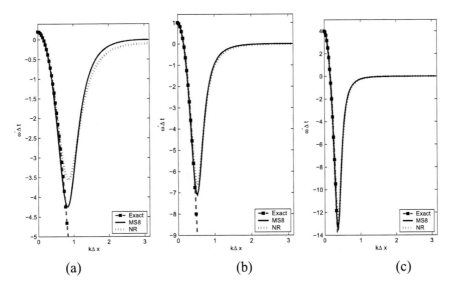

Figure 6. Group velocities for (a) $\mu = 0.1$, (b) $\mu = 0.5$, (c) $\mu = 2$.

From the Fig. 5 we can see the numerical dispersion is well preserved for different values of the Courant number $\mu = \frac{\Delta t}{\Delta x}$. There are no computational modes for each scheme because for every wave number there is only one frequency for each scheme. The numerical and exact dispersion relations are very close to each other. The dispersive properties of the schemes are the same. That is dispersion curves of the two schemes are above the analytic dispersion curves for each choice of μ. Fig. 6 represents the analytic and numerical group velocities $\overline{\omega}'(\overline{k})$. The figure shows that the group velocity curves of the eight-point scheme and the narrow box schemes are very close to the continuous one. Moreover the sign of the group velocity is preserved for both methods.

References

[1] E. Hairer, C. Lubich, and G. Wanner. *Geometric Numerical Integration Structure–Preserving Algorithms for Ordinary Differential Equations*. **31**. 2n edition, Springer, New York, 2006.

[2] B. Leimkuhler and S. Reich. *Simulating Hamiltonian Dynamics*, Cambridge University Press, 2005.

[3] T. J. Bridges, S. Reich. Multi–symplectic integrators: numerical schemes for Hamiltonian PDEs that conserve symplecticity, *Phys Lettets A*, **284**:184–93, 2001.

[4] J. Marsden, G. W. Partick, S. Shkoller. Multisymplectic geometry, variational integrators, and nonlinear PDEs. *Commun. Math. Phys.*, **199**:351–95, 1998.

[5] T. J. Bridges and S. Reich. Numerical methods for Hamiltonian PDEs *J. Phys. A: Math. Gen.*, **39**:5287-5320, 2006.

[6] T. Wanga, T. Nie, L. Zhanga, Analysis of a symplectic difference scheme for a coupled nonlinear Schrödinger system. *Journal of Computational and Applied Mathematics*, **231**:745-759, 2009.

[7] J.Q. Sun and X.Y. Gu Z.Q. Ma. Numerical study of the soliton waves of the coupled nonlinear Schrödinger system. *Physica D*, **196**:311-328, 2004.

[8] J.Q. Sun and M.Z. Qin. Multi–Symplectic methods for the coupled 1D nonlinear Schrödinger system. *Comp. Phys. Comm.*, **155**: 221-235, 2003.

[9] M. S. Ismail, T. R. Taha. Numerical simulation of coupled nonlinear Schrödinger equation, *Mathematics and Computers in Simulation*, **56**: 547–562, 2001.

[10] M. S. Ismail. A fourth-order explicit schemes for the coupled nonlinear Schrödinger equation. *Applied Mathematics and Computation*, **196**:273-284, 2008.

[11] M.D. Todorrov, C.I. Christov. Conservative numerical scheme in complex aithmetic for coupled nonlinear Schrödinger equaions. *Discrete and Continuous Dynamical Systems. Supplement*, **982**–992, 2007.

[12] S.C. Tsang and K.W. Chow. The evolution of periodic waves of the coupled nonlinear Schrödinger equations. *Math & Compt. Simul.*, **66**:551-564, 2004

[13] T. Wang, T. Nie, L. Zhang, F. Chen. Numerical simulation of nonlinearly coupled Schrödinger system: a linearly coupled finite difference scheme. *Mathematics and Computers in Simulation*, **79**:607–621, 2008.

[14] Y. Xu, C. W. Shu, Local discontinuous Galerkin methods for nonlinear Schroödinger equations. *Journal of Computational Physics*, **205**: 72–97, 2005.

[15] A.L. Islas, D.A. Karpeev, and C.M. Schober. Geometric integrators for the nonlinear Schrödinger equation. *Computational Physics*, **173**:116-148, 2001.

[16] T. J. Bridges. Multisymplectic structures and wave propagation. *Math. Proc. Cambridge Philos. Soc.*, **121**: 147–190, 1997.

[17] S. Reich. Multisymplectic Runge–Kutta collocation methods for Hamiltonian wave equations. *J. Comput. Phys.*, **156**: 1-27, 1999.

[18] B. Moore, S. Reich, Backward error analysis for multisymplectic integration methods. *Numer. Math.*, **95**: 625–652, 2003.

[19] T.J. Bridges, S. Reich, Multi–symplectic spectral discretizations for the Zakharov–Kuznetsov and shallow water equations. *Physica D*, **152**:491-504, 2001.

[20] J.-B. Chen and M.Z. Qin. Multisymplectic Fourier pseudospectral method for the nonlinear Schrodinger equation. *Electron. Trans. Numer. Anal.*, **12**:193–204, 2001.

[21] Jing-Bo Chen, Symplectic and Multisymplectic Fourier Pseudospectral Discretizations for the Klein-Gordon Equation, *Letters in Mathematical Physics*, 293-305, 2006.

[22] A.L. Islas and C.M. Schober. On the preservation of phase structure under multisymplectic discretization. *J. Computational Physics*, **197**:585–609, 2004.

[23] Jian Wang. Multisymplectic Fourier pseudospectral method for the nonlinear Schrödinger equations with wave operator. *Journal of Computational Mathematics*, **25**:31-48, 2007.

[24] M. J. Ablowitz, B. Prinari, A. D. Trubatch. *Discrete and continuous nonlinear Schrödinger systems*. London Mathematical Society Lecture Note Series, 2004.

[25] S.V. Manakov. On the theory of two–dimensional stationary self–focusing of electromagnetic waves. *Sov. Phys. JETP*, **38**:248, 1974.

[26] C.R. Menyuk. Stability of solitons in birefringent optical fibers. *Journal of Optics Society American B.* **5**:392, 1988.

[27] B. Tan. Collision interactions of envelope Rossby solitons in a barometric atmosphere. *J. Atmos. Sci.*, **53**:1604, 1996.

[28] B. Tan and S. Liu. Collision interactions of solitons in a barometric atmosphere. *J. Atmos. Sci.*, **52**:1501, 1995.

[29] B. Tan and J.P. Boyd. Stability and long time evolution the periodic solutions of the two coupled nonlinear Schrödinger equations. *Chaos Solitons and Fractals*, **12**:721-734, 2001.

[30] V.E. Zakharov and E.L. Schulman. To the integrability of the system of two coupled nonlinear Schrödinger equations. *Physica D*, **4**: 270, 1982.

[31] R. I. McLachlan and G R. W. Quispel. Splitting methods. *Acta Numerica*, **11**: 341-434 (2002)

[32] A. Aydın, B. Karasözen. Multisymplectic integration of coupled non-linear Schrödinger system with soliton solutions. *International Journal of Computer Mathematics* , **86**: 864–882, 2009.

[33] N.N. Akhmediev. Nonlinear physics–déjávu in optics. *Nature*, **413**: 267-268, 2001.

[34] A. Aydın, B. Karasözen. Symplectic and multisymplectic methods for coupled Nonlinear Schrödinger Equations with periodic solutions. *Computer Physics Communications*, **177**:566–583, 2007.

[35] W. J. Sonnier, C. I. Christov, Strong coupling of Schrödinger equations: conservative scheme approach, *Mathematics and Computers in Simulation*, **69**:514-525, 2005.

[36] T. Wang, L. Zhang, F. Chen Numerical analysis of a multi–symplectic scheme for a strongly coupled Schrödinger system. *Appl. Math. Comput.*, **203**:413–431, 2008.

[37] B. N. Ryland, R. I. McLachlan, and J. Frank. On the multisymplecticity of partitioned Runge-Kutta and splitting methods. *International Journal of Computer Mathematics*, **84**:847–869, 2007.

[38] B. N. Ryland, R. I. McLachlan. On the multisymplecticity of partitioned Runge–Kutta methods. *SIAM J. Sci. Comput.*, **30**: 1318–1340, 2008.

[39] A. Aydın, B. Karasözen. Lobatto IIIA-IIIB discretization of the strongly coupled nonlinear Schrödinger equation. to appear in *Journal of Computational and Applied Mathematics*, 2011.

[40] M. S. Ismail. Numerical solution of complex modified Korteweg-de Vries equation by collocation method. *Commun Nonlinear Sci Numer Simul.*, **14**: 749–759, 2009.

[41] M. S. Ismail. Numerical solution of complex modified Korteweg-de Vries equation by Petrov-Galerkin method. *Appl Math Comput.*, **202**:520-531, 2008.

[42] G. M. Muslu, H. A. Erbay. A split–step fourier method for the complex modified Korteweg–de Vries equation. *Comput. Math. Appl.* , **45**: 503–514, 2003.

[43] U. M. Ascher, R. I. McLachlan. Multisymplectic box schemes and the Korteweg-de Vries equation. *Appl Num Math.*, **48**:255–269, 2004.

[44] S. Reich. Finite volume methods for multisymplectic PDEs. *BIT*, **40**: 559–582, 2000.

[45] U. M. Ascher, R. I. McLachlan. On symplectic and multisymplectic schemes for the KdV equation. *Journal of Scientific Computing*, **25**:83–104 2005.

[46] A. Aydın, B. Karasözen. Multisymplectic box schemes for the complex modified Korteweg-de Vries Equation, *Journal of Mathematical Physics:* , **51**:083511, 2010.

[47] Y. Chen, H.Zhu, S.Song. Multi–symplectic splitting method for the coupled nonlinear Schrödinger equation. *Comput. Phys. Commun.*, **181**:1231–1241, (2010).

[48] H. Yoshida. Construction of higher order symplectic integrators. *Phys. Lett. A.* , **150**: 262-268, 1990.

[49] J. Wang. Multisymplectic numerical method for the Zakharov system. *Comp. Phys. Comm.*, **180**: 1063–1071, 2009.

[50] J. Frank, B.E. Moore, S. Reich. Linear PDEs and numerical methods that preserve a multisymplectic conservation law *SIAM. J. Sci. Comput.* , **28**: 260-277, 2006.

[51] C. M. Schober, T.H. Wlodarczyk. Dispersive properties of multisymplectic integrators. *J. Comput. Phys.* , **227**: 5090-5104, 2008.

[52] A. Aydın, B. Karasözen. Symplectic and multisymplectic Lobatto methods for the "good" Boussinesq equation. *J. Math Phys.*, **49**:083509, 2008.

[53] T.H. Wlodarczyk, Stability and Preservation Properties of Multisymplectic Integrators, PhD thesis, University of Central Florida, 2007.

In: Computer Physics
Editors: B.S. Doherty and A.N. Molloy, pp. 297-314

ISBN: 978-1-61324-790-7
© 2012 Nova Science Publishers, Inc.

Chapter 4

Grid Computing Multiple Shooting Algorithms for Extended Phase Space Sampling and Long Time Propagation in Molecular Dynamics

Vangelis Daskalakis[1,2,*] *Manos Giatromanolakis*[2], *Massimiliano Porrini*[2,†]
Stavros C. Farantos[1,2,‡] *and Osvaldo Gervasi*[3]

[1]Department of Chemistry, University of Crete, Heraklion 71003,
Crete, Greece
[2]Institute of Electronic Structure and Laser, Foundation for Research
and Technology-Hellas, Heraklion 71110, Crete, Greece
[3]Department of Mathematics and Computer Science, University of Perugia,
Perugia, Via Vanvitelli, 1-06123 Perugia, Italy

Abstract

Grid computing refers to a well established computational platform of geographically distributed computers that offer a seamless, integrated, computational and collaborative environment. It provides the means for solving highly demanded, in computer time and storage media problems of molecular dynamics. However, because of the rather high latency network, to exploit the unprecedented amount of computational resources of the Grid, it is necessary to develop new or to adapt old algorithms for investigating dynamical and statistical molecular behaviors at the desired temporal and spatial resolution. In this chapter we review methods that assist one to harness the current computational Grid infrastructure for carrying out extended samplings of phase space and integrating the classical mechanical equations of motion for long times. Packages that allow to automatically submit and propagate trajectories in the Grid and

---

*Current Address: Department of Environmental Science and Technology, Cyprus University of Technology, P.O. Box 50329, 3603 Lemesos, Cyprus.
†Current Address: Institute for Condensed Matter and Complex Systems, School of Physics & Astronomy, The University of Edinburgh, James Clerk Maxwell Building, The King's Buildings, Mayfield Road, Edinburgh EH9 3JZ
‡E-mail address: farantos@iesl.forth.gr

to check and store large amounts of intermediate data are described. We report our experience in employing the European production Grid infrastructure for investigating the dynamics and free energy hypersurfaces of enzymes such as Cytochrome c Oxidases. Time autocorrelation functions of dynamical variables yield vibrational spectra of the molecule and reveal the localization of energy in specific bonds in the active site of the enzyme. Dynamical calculations and free energy landscapes of the Cytochrome c Oxidase protein interacting with gases like O_2, CO and NO reveal the pathways for the molecules to penetrate in the cavities of the enzyme and how they reach the active site where the reactions take place. The discussed methods can be adopted in any intensive computational campaign, which involves the scheduling of a large number of long term running jobs.

PACS 33.20.Tp, 82.20.Wt, 83.10.Rs, 87.15.Aa.

Keywords: Grid Computing, EGI, multiple shooting algorithms, atomistic molecular simulations, protein dynamics and spectroscopy, free energy surfaces.

1. Introduction

Computational chemistry is voracious in computer power and one of the first disciplines which immediately adopts any innovation in computer sciences. In general, molecular computational sciences face two challenges in the twenty first century. The first has to do with the size of the systems; studies which cover molecules from a few atoms to nanostructures and the macroscopic states of matter are pursued. The second challenge is related to time and the dynamics of the systems; phenomena which last from femtoseconds to milliseconds are important in materials and biological sciences. Grid computing has been emerged as one solution in bridging the gaps in time and space. A Grid of computers is envisioned as a seamless, integrated, computational and collaborative environment embracing different categories of distributed systems. Following the classification introduced by Foster and Kesselman [1], computational Grids are categorised into five major classes of applications. In summary, these classes identify a specific context of applications such as supercomputing applications, high-throughput computing, on-demand computing, data-intensive computing and collaborative computing. This new paradigm of scientific computing is rapidly developing. Several Grids have been deployed around the world among which the European Grid Infrastructure (EGI) which is the successor of the Enabling Grids for E-sciencE (EGEE) project[1]. Reports on the status and evaluations of their performances have been published [2, 3].

Grid technology has promised to offer unprecedented computational resources, continuously increasing the number of CPUs (currently at about 114000) and improving the networking quality. Thus, it is not surprising that computational chemistry was immediately allured by the Grid and it has already developed applications for this new infrastructure [4, 5, 6]. Quantum chemistry electronic structure calculations were reported even during

---

[1]EGEE was a project funded by the European Union under contract INFSO-RI-508833. The EGEE infrastructure was built on the EU Research Network GEANT and provided interoperability with other Grids around the globe, including Israel, Russia, US and Asia, with the purpose of establishing a worldwide Grid infrastructure.

the first stages of Grid computing, then called *metacomputing*. The idea in metacomputing was to distribute the job among supercomputer centres, located even in different continents [7]. The evolution of these ideas led to computational Grids, and today's Grid is not only a software framework which provides layers of services to access and manage distributed hardware and software resources, but a network to manage "coordinated resource sharing and problem solving in dynamic, multi-institutional virtual organisation" as it is stated in Ref. [8]. This is, actually, the most excited aspect of Grid computing and it will pave the way to successful studies of complex systems.

The European Cooperation in the field of Scientific and Technical Research (COST), one of the longest-running European instruments supporting cooperation among scientists and researchers across Europe, addressed in the field of Chemistry and Molecular Sciences and Technologies (CMST) the metacomputing issues launched the D23 Action, called Metachem [9], and subsequently the Grid computing one D37 Action, called GridChem [10]. The authors were collaborating through the COST CMST Action D37 in order to coordinate with several European Research Laboratories the Grid activities necessary for exploiting new computational frontiers and for implementing a network to manage coordinated resource sharing and problem solving in dynamic, multi-institutional virtual organisations [8].

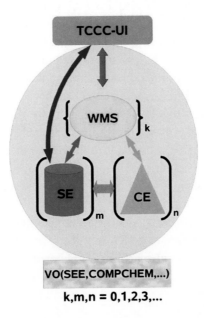

Figure 1. A schematic representation of a computational Grid. Users are members of the Virtual Organisations (VO) and TCCC-UI is the user interface host to the Grid. Executable programmes and input/output data are uploaded in the Storage Elements (SE), whereas jobs are submitted to the Workload Management System (WMS), which in turn sends them for execution to the Computing Elements (CE). Data can be read or written from the Computing Elements straight to the Storage Element or via the WMS.

1.1.   The EGI Production Grid

Among the various Grid implementations, we use intensively the European Grid Infrastructure, the largest e-infrastructure of the world, now functioning under the Dutch law. According to the EGI approach, the users, after becoming members of the appropriate Virtual Organisation (VO) (in our case the South-East Europe (SEE) and CompChem VOs have been used), can submit jobs via the User Interface (UI) host to the Grid resources as it is schematically shown in Figure 1. The selection of the most adequate site available on the Grid infrastructure to execute the submitted job is determined usually by the Workload Management System (WMS), even if the user can demand the execution on predetermined sites. The selection of the WMS is based on the information collected through a set of Grid Services from the Computing Elements (CEs) devoted to the management of the resources of the Grid site and the monitoring of their status. The EGI provides also the Storage Elements (SEs) for data management, that we use to safely store and retrieve files for the input/output of our jobs (Figure 1). An important task of the various VOs is to support their members in employing the Grid in their calculations, as well as encouraging several computational and collaborative endeavours. We access EGI via HellasGrid, the Greek National Grid Initiatives (NGI), which is a member of EGI and contributes with hardware to both, CompChem and SEE VOs.

Considering the complexity of molecular simulations, the execution of parallel jobs on several Grid resources is of great interest. However, in order to use successfully a worldwide distributed computing environment of thousands of heterogeneous processors in the Grid, communications among these processors should require the minimum time. At present, there are not many molecular simulation algorithms suitable for running in the Grid environment which can lead one to extract general strategies in writing codes for molecular applications. A practical rule is to allow each processor to work independently, even though calculations are repeated, and only if something important happens to one of them, then, they communicate. The common ingredient of the above mentioned applications in Internet Grid Computing is the computational algorithm employed, which guarantees minimum and asynchronous communications. Obviously, there are a few problems that can be solved with such algorithms. Up to now, most applications in Computational Chemistry require solutions of a large number of linear equations that can not be solved without significant communication among the computer nodes. On the other hand, parallelised codes written for high performance parallel local machines may be of no good use when thousands of computers are connected by high latency networks. Thus, new algorithms and new programming paradigms suitable for distributed computing should be developed in order to exploit Computational Grids [4].

1.2.   Protein Molecular Dynamics

Molecular dynamics (MD) are the methods for solving the many particle classical mechanical equations of motion for a molecular system [12]. The trajectories obtained provide structures (for free energy calculation) and time dependent data such as correlation functions or transport properties. The accumulation of data requires long time integration of the equations of motion in order to sample the energetically accessible coordinate space, a particularly time consuming problem for large molecules. However, for some problems we can invoke the ergodic hypothesis in a reduced dimension space and instead of integrating single trajectories for very long time, we sample many trajectories with different initial conditions whose evolution is followed for shorter multiple time intervals. The great advantage of replacing time averages by partially phase space averages is the parallelisation of the problem. We call this method *multiple shooting*, according to which we distribute the starting data and integrate the resulting trajectories in different CPUs of the Grid in multiple time intervals. It turns out, that biological molecules are suitable examples for applying this method, since they often show localisation of their motions in a subspace of phase space of the macromolecule [13, 14].

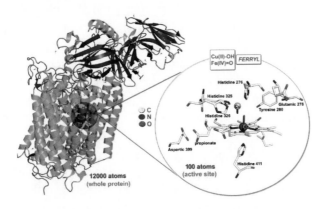

Figure 2. Crystal structure of the aa_3 Cytochrome c Oxidase from *Paracoccus denitrificans*. Heme a_3-Fe/Cu$_B$ of the active site is shown magnified on the right. In the proximal area Fe^{IV} (ferryl ion) is bonded to His411 aminoacid, whereas in the distal area there is a double bond to O^{-2} (oxo). Copper (Cu^{II}-OH) which coordinates three histidine residues, His276, His325, His326, is also present. Carbon appears grey, oxygen red and nitrogen blue.

Biochemists frequently associate biochemical action to specific sites in a protein and spectroscopic methods are useful to investigate their dynamical behaviours. Such a case is Cytochrome c Oxidases (CcO), the terminal enzymes in the respiratory chains, found in the inner mitochondrial membranes or in many bacteria and the last acceptor of electrons from oxidising processes involving nutrient molecules. Figure 2 depicts the enzyme of *Paracoccus denitrificans*. On the right, we can see a magnification of the *active site* where the heme-a_3 iron captures the dioxygen and the whole mechanism of its reduction to water molecules takes place. The active site shown consists of a heme group with the iron at the oxidation state of IV (ferryl), which bounds the oxo anion (O^{-2}). CcO facilitates the activation and by four electron reduction of oxygen molecule to water with the pump of four protons across the inner mitochondrial membrane attributing to the electrochemical gradient [15] that is used to drive the synthesis of ATP. CcO is particularly appealing for theoretical studies, because of the relative small size of its active site as well as there is a plethora of experimental results in the literature. Nevertheless, details of its molecular mechanism of action are still unclear being a matter of considerable debate for over three decades.

The role of CcO in proton translocation, as well as intermediate compounds and reaction mechanisms are given in a recent review article [16]. Proposed mechanisms for proton transfer across the membrane involve the protonation/deprotonation of water molecules or aminoacids, which yield local changes of the electrostatic potential. Some evidence has been given in the study of ba_3 oxidase and the so called Q-channel [17].

Both Cytochrome c Oxidases, aa_3 from *Paracoccus denitrificans* and ba_3 from *Thermus thermophilus* are currently under intensive investigations. For the first enzyme the motivation has come from questions with respect to the assignment of infrared and Raman spectra obtained in several laboratories [13]. In addition, several studies deal with ba_3-CO complex, as CO is highly polarisable and serves as a probe for the protein environment and cavities. CO molecule binds initially to Fe^{II} (ferrous) and after photodissociation and long time evolution rebinds to iron. Therefore, it is important to follow MD for very long times [18] and to explore its free energy surface (FES). Moreover, the impact on the dynamics and spectroscopy of protonated/deprotonated aminoacids is also a subject of investigation. We have constructed four different conformations of ferryl-oxo intermediate of CcO with combinations of protonated/deprotonated Asp399 residue (see Figure 2) and the A-propionate of heme-a_3, as well as protonated/deprotonated Glu278. We have studied the following cases: In structures A and C all carboxyl groups in equilibrium are protonated or deprotonated, respectively, while in B or D structures only one H^+ is available protonating either Asp399 or propionate, respectively. Glutamic acid 278 may appear as protonated or deprotonated, thus doubling the number of the possible conformations. It is not difficult to count the large amount of computer time needed to study these problems which can easily be enlarged by adding questions related to the role of temperature and dielectric constant variations. Such projects would require years with our local computational resources.

In the present chapter we report our experience in using the EGI (former EGEE) through SEE and CompChem VOs for studying molecular dynamics and spectroscopy of aa_3-CcO in a variety of protonated/deprotonated structures and the photodissociation of ba_3-CO complex. This has led us to develop a bundle of shell scripts that allow to automatically launch hundreds of trajectories and propagate them for an extended period of time by repeatedly submitting jobs (multiple shooting).

The rest of the chapter is organised as follows; in the next Section (2), we briefly describe the computational methods employed in molecular dynamics, but we explain in detail the scripts written to monitor thousands of jobs submitted to the Grid. Section 3 presents the results and Section 4 summarises and discusses future calculations and extensions.

2. Computational Methods

2.1. Molecular Dynamics

Molecular theories are basically formulated within the Born-Oppenheimer approximation, i.e. the adiabatic separation of the motions of nuclei from the motions of electrons. The latter are treated by solving the electronic Schrödinger equation, whereas the motions of the nuclei in the average field of electrons are studied with (semi)classical mechanics. Only small molecules are treated by quantum methods. Thus, molecular dynamics of large molecules mean integration in time of Newton or Hamilton equations of motion. Simulations are useful if realistic and accurate electronic potential functions (the Potential Energy

Surface (PES)) for the motions of nuclei are available. A tremendous effort has been invested to develop methods for solving the electronic Schrödinger equation (called *ab initio*), as well as to produce analytical functions which fit the ab initio calculations and experimental data [19]. Analyticity guarantees continuous first derivatives (forces) needed in the equations of motion.

In our investigations of Cytochrome *c* Oxidases we carry out Density Functional Theory calculations at a high level of theory (b3lyp/6-311g*) for a small part of the protein, the active site. Part of it is shown in Figure 2 and includes about 100 atoms compared to the $\sim$12000 atoms of the total protein. The DFT calculations provide the geometry, harmonic vibrational frequencies, charge distributions which together with the force field data bases available in the literature allow us to construct an analytical function for a reliable Potential Energy Surface of the active site of CcO and the whole enzyme. The PESs that we employ are constructed with Morse type functions for the stretches, harmonic potentials for the angles, cosine functions for the torsions, and Lennard-Jones and Coulomb intermolecular interactions [12]. The set of fitted parameters reasonably reproduces the theoretical and experimental data [13, 14, 18].

Molecular dynamics simulations are carried out with initial Cartesian coordinates from the two-subunit of aa_3 crystal structure of *Paracoccus denitrificans* (PDB code 1AR1) and the ba_3 crystal structure of *Thermus thermophilus* (PDB code 1XME). The A, B, C and D protonation cases of (A)propionate-Asp399 pair discussed above (Figure 2) and protonated/deprotonated Glu278 are constructed by adding or subtracting one or two protons. All crystallographic water molecules were retained in the structure (TIP3P force field). Simulations were performed at constant energy or temperature.

Two software packages have been employed for the MD simulations, Tinker (version 4.2) [20] and DL_POLY2 (version 2.18) [21] with Amber99 [22] force field. Tinker assists us to develop the empirical PES, to find equilibrium geometries and to carry out vibrational analysis. Trajectories are run with both Tinker and DL_POLY2. We have carefully transformed the parameters (Amber99) and potential functions of Tinker to those readable by DL_POLY, such as the same PES is employed in both MD codes. Tinker provides analytical first and second derivatives of the potential function, whereas DL_POLY only first derivatives. DL_POLY2 is a parallelised code based on the Replicated Data Algorithm [23]. Although, this feature of the code substantially accelerates the calculations in our local cluster, for the Grid we run it sequentially. As we discuss it later, at present, parallel codes with Message Passing Interface (MPI) are executed on specific clusters in the Grid and not all sites support the MPI.

Several parameters, especially for the heme groups, were fitted either to DFT data or extracted from CHARMM27 [24] force field and altered accordingly to fit our DFT results or the experimental data. The bond type parameter was set to Morse, and for the integration of equations of motion the time steps were varied in the range of [0.15-0.5] femtoseconds (fs). All aminoacids were free to move during the simulations, except one case in the aa_3 enzyme, where a force constant of 7 kcal/molÅ was applied to restrain the distance between the C_γ atom of Glu278 and the Δ-methyl carbon of heme-a_3 at its crystallographic value as proposed elsewhere [25]. For non-bonded interactions a cutoff distance of 12 Å with a smoothing window between 9.6-12 Å was applied. Structures were minimised and equilibrated before each simulation.

2.2.   The Grid Multiple Shooting Algorithm (GMDmult)

Comparing the Grid (a geographically distributed system) with a local cluster of computers we immediately notice the following drawbacks of the Grid:

1. inhomogeneous hardware (but highly homogeneous middleware),

2. larger network latencies,

3. limited control on the Grid Computing Elements (CE),

4. an upper limit for the execution time of each job defined on a local policy by the EGI/NGI site administrator,

5. limitation on the size of the compressed files to be uploaded in a WMS or Computing Element,

6. MPI parallel jobs run only on specific sites of the Grid.

Thus, failures in executions and aborted jobs in the Grid are not at all a surprise. To overcome the above drawbacks and to minimise the number of failures, we have adopted two simple rules to our applications: i) run jobs for short times: most of our jobs require 24-48 hours, and ii) the smaller is the amount of input/output data the less is the possibility to loose them. Thus, the algorithm we have developed is based on the principle to run short jobs, store intermediate results, and resubmit the jobs as many times as needed to achieve the predetermined total integration time.

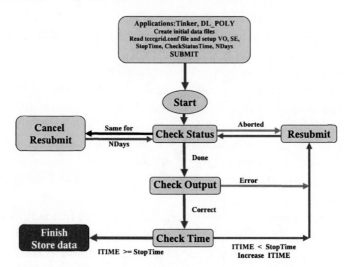

Figure 3. A flow-chart for Grid molecular dynamics multiple shooting (GMDmult) bundle of shell scripts that determine the flow of jobs in the Grid. For details see text. ITIME accumulates the integration time of the trajectory in each job.

A flow-chart of the Grid molecular dynamics multiple shooting (GMDmult) algorithm is shown in Figure 3. It is parametrised by several parameters that make it flexible enough to be used for different applications. The parameters are defined in a configuration file (tcc-cgrid.conf), which is read at the beginning of the script. In this file we define the Myproxy server to renew the proxy authorisation based on a X.509 digital certificate until the completion of the job, the Virtual Organisation, the Storage Element to be accessed, the application type (the program to be used: Tinker/DL_POLY) and the initial data. The executable programme and the input files are stored in the SE. There are three parameters related to the execution time of the job, **CheckStatusTime, NDays**, and **StopTime**.

For the present case, an application means to run a batch of trajectories using Tinker or DL_POLY software. Thus, each **application** consists of a number of tasks. For each **task** we integrate one trajectory with predetermined initial conditions and total integration time (**StopTime**) in multiple jobs. Each **job** integrates the equations of motion for a subinterval of time. Thus, we have to launch several jobs to complete one task (trajectory).

We first upload the executable programme and the input files for the tasks into SE, and then we submit the various jobs to the Grid, which means that different trajectories may be sent to different CE. Every **CheckStatusTime** interval we retrieve the status of the job. The return value of the status may be, *Scheduled/ Waiting/ Ready/ Running/ Cancel/ Done/ Aborted/ Cleared*. If the job remains in the same status for **NDays**, then, we cancel it and start a new one with the same initial conditions. Aborted and Cancelled jobs are considered to be in a failed status and they are resubmitted, whereas for Done jobs we store the outputs in the SE. Next, the outputs or some characteristic values are examined for their correctness. If an error is detected the job is resubmitted with the same initial conditions, otherwise we proceed and the integration time is examined. If the total integration time (**StopTime**) has been reached the task has been accomplished and the output is retrieved, otherwise a new job is launched for the next cycle (not necessarily in the same CE).

In our calculations on EGI, we run applications with batches of 50 to 500 trajectories, with a number of input and output files of about ten per job. It turns out that GMDmult scripts are robust enough to control data and evolution time for thousands of jobs.

3. Results and Discussion

Molecular dynamics simulations is the means for exploring dynamical phenomena of large molecules. Furthermore, it can be used for calculating statistical averages such as the topography of free energy along reaction paths. In both cases long time integration of the equations of motion for an effective phase space sampling is required. However, in cases where ergodicity can be invoked the same results are obtained by integrating for shorter time a large number of trajectories with different initial conditions that span the relevant regions of phase space. It is obvious that averaging over many trajectories is the preferred method for distributed computing.

3.1. Vibrational Spectra from Classical Trajectories

Among the observables which can be calculated as phase space averages are autocorrelation functions. For example, the classical mechanical approximation of the vibrational spectra is given by the Fourier transform of the survival probability function [26], $\Omega(t)$,

$$\Omega(t) = \frac{1}{(2\pi\hbar)^F} \int \rho_0(x,p)\rho[x(t),p(t)]dxdp, \qquad (1)$$

$$I_c(\omega) = \frac{1}{2\pi} \int e^{i\omega t}\Omega(t). \qquad (2)$$

ρ_0 is a distribution function of the initial coordinates and conjugate momenta for a system with F degrees of freedom. $\rho[x(t),p(t)]$ denotes the time evolution of this distribution according to the classical equations of motion. $\hbar$ is Planck's constant. The trajectories have to be integrated long enough to cover the characteristic times of interest for the studied system.

We have adopted this approach to compute distances and angles of Cytochrome c Oxidases as time series, as well as their power spectra [13, 18]. For the ferryl intermediate of CcO we are interested in the frequencies that involve the Fe=O bond which correspond to subpicoseconds vibrational periods. Thus, we do not have to integrate for very long time, but of course we have to average over many trajectories by sampling the initial phase space. In a previous study [13], we demonstrated that the motions of ferryl-oxo species stay localised in the active site, and therefore, we can also restrict averaging in the configuration space. Here, we check this assumption by comparing time averages versus phase space averages.

For the ferryl-oxo intermediate state of aa_3 enzyme and using several protonated/deprotonated conformations described above, time-bond distance series are recorded in every molecular dynamics simulation. Two different methods are used to derive the vibrational power spectra (Raman transition moments are not taken into account): (a) Fast Fourier Transform (FFT) over time-distance series from one trajectory (1.2 ns), using one CPU and (b) averages of FFTs of around one hundred trajectories of 52 picoseconds (ps) long, using the EGI. Similar calculations are repeated by substituting ^{16}O by ^{18}O bound on Fe and Cu (see Figure 2). Subtracting the two spectra of the two isotopically different species, we reveal the frequencies that are relevant to the chemical bonds which contain the oxygen atoms. Do not forget that for biomolecules, we deal with thousands of vibrational modes and the assignment of the spectra needs special care. Spectra at different temperatures and dielectric constants are also calculated.

In Figures 4 and 5 we compare the results of CcO from *Paracoccus denitrificans* of the 1.2 ns simulations (blue curves) with the averaged ones run in the Grid (red-dash curves). The integration time is considered to be adequate for convergence, since after 0.6 ns positions and intensities of the peaks in the spectra remain unchanged. The biophysical significance of the results and arguments about the localisation of the motions in the active site are yield in a recent article [13]. The main conclusions of this work are, first, the well separated band at about 773 cm^{-1}, that remains practically stable for all protonated/deprotonated states of CcO, and second, the rest of the spectrum covers the same frequency interval (780-860 cm^{-1}), approximately having peaks at the same frequencies (marked on the plot), although with variable intensities. These resolve previous questions with respect to the appearance of either one or two peaks with v(Fe=O) character in the experimental Raman spectra.

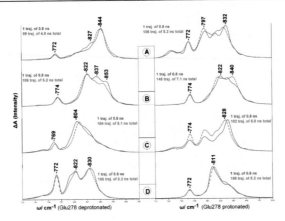

Figure 4. Calculated Fe=O bond power spectra of A, B, C and D protonated/deprotonated states of the enzyme aa_3 (see text) when Glu278 is deprotonated (left) or protonated (right). Simulations were carried out at constant energy which corresponds to an average temperature of 300 K. Blue curves denote the one trajectory results, and red-dash curves the averages of many trajectories run in the Grid.

In Figure 5 we depict the difference spectra obtained from ^{16}O minus ^{18}O isotopic substitution of Fe^{IV} and Cu_B^{II} oxygens. For most of them two positive bands can clearly be seen which cover a width of frequencies of approximately 50 cm^{-1}. This is in accord to the experimental observations [27, 28, 29, 30, 31]. Although, in these publications dominate lines are discussed, they do show broad bands of about 50 cm^{-1} width.

Comparing the two smoothed spectra in Figures 4 and 5 (blue and red-dash curves) for all protonated cases, the agreement in the positions, intensities and widths of the spectral bands is found satisfactory. All calculations were done with Tinker software. For the one-trajectory runs we used the smallest integration step of 0.15 fs. The calculations required months to complete each trajectory, and of course we had to occupy several CPUs for examining different conformations and temperatures. Contrary to that, calculations in the Grid required only weeks by employing GMDmult scripts to submit and control jobs.

We must say, that we spent a considerable amount of time to identify the subset of reliable EGI resources among the large set of available ones, and only using such sites the percentage of failed jobs decreased. In the following, we show representative statistical results from our latest runs. In Tables 1 and 2 we tabulate the number of failed and successful jobs launched in SEE and CompChem virtual organisations, using Tinker and DL_POLY packages in the calculations of both enzymes, aa_3 and ba_3.

In Table 1, we show the statistics for four applications: the A and C conformations of ferryl-oxo intermediate of CcO from *Paracoccus denitrificans* with deprotonated Glu278 (E278) aminoacid, and the AE-H and CE-H corresponding conformations with protonated Glu278. Calculations were performed with Tinker software. The total number of jobs to complete the application which involves the integration of each trajectory (task) for 52 ps, as well as the number of aborted, cancelled and succeeded jobs are shown. Ntr is the number of trajectories in each application, run-time means the total physical time in days needed to

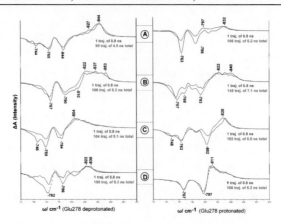

Figure 5. Difference spectra of isotopically substituted oxygen (^{16}O minus ^{18}O) for all protonated/deprotonated structures of CcO studied. Plots (left) for the deprotonated Glu278 and plots (right) for the protonated Glu278 aminoacid. Blue curves are for one trajectory results, while the red-dashed ones for averages of many trajectories run in the Grid.

get back the last successful job, data-size is the total number of megabytes of data circulated in the network during the run of the application, and finally CPU-time is the total CPU time gained from the Grid. The total size of data (input-output) circulated in the network and the total CPU-time used in the Grid for each application are obtained by multiplying data-size (14 MBytes) and CPU-time (900 min) of each job by the number of succeeded jobs.

Table 1. Batches of trajectories run on EGI for SEE VO and for the A, C, AE-H and CE-H conformations of CcO. (E-H) means protonated Glu278 aminoacid. The size of data and CPU-time required per job were 14 MBytes and 900 min, respectively. The total size of data (input-output) circulated in the network and the total CPU-time used in the Grid are obtained by multiplying the above magnitudes by the number of succeeded jobs. Ntr is the number of trajectories in each batch and run-time means the total physical time in days needed to get back the last successful job. CheckStatusTime = 30 min, StopTime = 52 ps, and NDays=2 days

	Ntr	run-time (days)	submitted jobs	aborted jobs	cancelled jobs	succeeded jobs(%)	data-size (MBytes)	total CPU-time (days)
AE-H	109	18	1810	273	6	84.59	21434	956.88
A	109	19	1784	253	3	85.65	21392	955.00
CE-H	109	19	1905	366	11	80.21	21378	954.38
C	109	19	1833	300	6	83.31	21392	955.00

In Table 2 results from the runs of DL_POLY at SEE and CompChem VOs and for the ba_3 enzyme are presented. The purpose of these runs was to explore the cavities around the active site using CO gas as probe. Photodissociating CO from Fe with different initial conditions is a process that can be tackled very efficiently on the Grid, as we can run thousands of parallel trajectories for 100 ps.

Table 2. Applications run for the ba_3 enzyme (∼13000 atoms) with DL_POLY2 MD software. The size of data and CPU-time required per job were 42 MBytes and 968 min, respectively. The total size of data (input-output) circulated in the network and the total CPU-time used in the Grid are obtained by multiplying the above magnitudes by the number of succeeded jobs. StopTime = 100 ps, CheckStatusTime = 60 min, and NDays = 2 days. *For this application the CPU-time per job was 1210 min

VO	Ntr	run-time (days)	submitted jobs	aborted jobs	cancelled jobs	succeeded jobs(%)	data-size (MBytes)	total CPU-time (days)
SEE	48	5	180	11	4	91.7	6930	110.9
SEE	41	5	194	15	21	81.4	6636	106.2
SEE	230	4	717	12	4	97.8	29442	471.2
SEE*	455	7	1122	134	33	85.1	40110	802.4
CompChem	235	8	1518	766	88	43.7	27888	446.4
CompChem	457	16	2549	351	200	78.4	83916	1343.1

The performance of an application seems to depend on the value of the parameter **CheckStatusTime**. Large values make the total physical run times longer and short times inflate the network with time checks. For the present applications a value of 60 minutes seems to be the optimum.

Finally, as it is expected, the tasks in each application are similar but their execution physical times differ depending on the hardware they are run and on the **CheckStatusTime** parameter. Therefore, we expect and get a distribution of several execution times in each application.

3.2.   Free Energy Surfaces from Thermodynamic Perturbation Calculations

Another application that we have carried out in the European production Grid is the calculation of the free energy surface (FES) of CcO interacting with several gases like, O_2, CO, NO and Xe. The FES portraits the difference in the free energy of protein with the ligand's centre of mass fixed at a specific site of the protein compared to the non interacting protein-ligand system.

In the perturbation theory the Hamiltonian of the protein-ligand system, H, consists of the sum of the unperturbed system Hamiltonian, H_0, plus the interaction term V,

$$H(p, q, p_r, r) = H_0(p, q) + K(p_r) + V(q, r), \tag{3}$$

where (p, q) denote the momenta and coordinates of the protein, p_r, r the momenta and coordinates of the ligand and $K(p_r)$ is the kinetic energy of the ligand. The dissociation

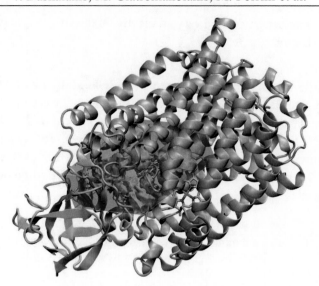

Figure 6. Three-dimensional view of the ba_3-CcO interacting with O_2 ligand. Embedded in the protein the volume explored by the thermodynamic perturbation calculations and the orange free energy iso-surface of -1.5 kcal/mol can be seen.

limit of protein-ligand system is defined

$$r \to \infty, \quad V(q, r_\infty) = 0. \tag{4}$$

Also, we assume that the internal degrees of freedom of the ligand molecule are kept fixed at their equilibrium values. Hence, r denotes the centre of mass coordinates and the Euler rotational angles. Then, the free energy as a function of r ($A(r)$) is approximated by the formula

$$A(r) = -k_B T \ln \left[\langle \exp[-V(q, r)/k_B T] \rangle_0 \right]. \tag{5}$$

k_B is the Boltzmann constant, T the temperature and $\langle \rangle_0$ denotes averaging over a set of microstates produced from a simulation of the unperturbed protein. Discretising the coordinates r we can distribute the evaluation of $A(r)$ over the thousand of CPUs of the Grid, thus, parallelising the calculations.

We explore the free energy landscapes of enzymes such as cytochrome c oxidases interacting with O_2 diatomic molecule. The free energy has been calculated as a function of the three Cartesian coordinates of the centre of mass of the ligand and averaging over the two Euler rotational angles of the diatom. Exploration of the topographical characteristics of the surfaces is done by employing the VMD graphics package [32]. A representative FES for ba_3-O_2 system is shown in Figure 6. We distinguish the cavities (low energy regions), that are common to both proteins, ba_3 and aa_3, DP (distal pocket), Xe1 and WP (water pocket), in spite of their differences in residues and metal atoms. The employment of Grid allows us to construct similar FES for other ligands as well as to examine the consequences of protonation or deprotonation of the propionates, water molecules, and aminoacids, which may affect the function of the active site [33]. Obviously, all these calculations can not be performed in reasonable time with our local computational resources.

The last decades a plethora of methods have been proposed to enhance the sampling of system configurations, and thus, to get more reliable free energy surfaces. These techniques are based on umbrella sampling, Jarzynski's identity equation, adaptive biasing force, and metadynamics [34]. Software for interfacing these methods with standard MD packages has been proposed [35]. Most of these methods are easily paralellized, and thus, our scripts can be used for distributing jobs in the Grid.

4.   Conclusion

Molecular dynamics calculations of Cytochrome c Oxidase enzymes from *Paracoccus denitrificans* (aa_3) and *Thermus thermophilus* (ba_3) have been carried out in the EGI (former EGEE) Grid through SEE and CompChem VOs. The proteins have a size of approximately 12000-13000 atoms and the software packages employed are Tinker [20] and DL_POLY2 [21]. The large number of Computing and Storage Elements available in this new infrastructure allowed us to obtain statistically reliable results [13, 18] in short project's times. This has been achieved thanks to a suite of shell scripts (GMDmult) parametrised in such a way that allows one to submit and resubmit thousands of jobs controlling MD integration times and input/output data. The total duration of the calculations, which counts years of CPU-time, and the size of data produced, unequivocally prove the superiority of the Grid, compared to custom local and distributed resources.

The Grid is the proper means for executing molecular dynamics jobs provided the problems we want to study can be formalised in such a way that parallel trajectories are run. Using the spectra of ferryl-oxo intermediate from *Paracoccus denitrificans* (aa_3) as an example, we demonstrate that, because of the localisation of the motions in the active site we may assume partial ergodicity in a subspace of the phase space, and thus, time averages can be replaced by averages over multiple trajectories. Research in the last decades has shown that energy localisation is very common in a variety of non-linear complex dynamical systems [36]. In this way we believe, that the Grid offers the computational environment that bridges the gaps in time and space by carrying out long time integration of the equations of motion for complex molecules, and sampling the phase space such as reliable averages can be obtained.

Currently, we run batches of trajectories of one nanosecond length with the DL_POLY software and 0.5 fs time step. With the available CPUs and for the size of proteins studied, that means one month of run-time. By running parallelly 1000 trajectories we achieve total integration times of microsecond scale, which is the current target simulation time [5]. Without doubt, we can increase the integration time of the trajectories by an order of magnitude if DL_POLY is run parallelly [23]. However, we found that the EGI is more efficient when it is used for sequential jobs than parallel, since the quality of the MPI support differs from site to site. Although a need for improvement is obvious, EGI is becoming a *planetary cluster of computers* which continuously grows, and computational groups have just started to realise and take advantage of its power.

Acknowledgments

This research was financially supported by the European Union ToK grant GRID-COMPCHEM (MTKD-CT-2005-029583). We acknowledge the EGEE Project funded by the European Commission through the program INFSO-RI-508833, as well as COST CMST D37 Action GridChem.

References

[1] Foster I.; Kesselman C.; Editors., *The Grid: Blueprint for a new Computing Infrastructure*, Morgan Kaufmann, 1999.

[2] European Network for Advanced Computing Technology for Science (ENACTS), http://www.epcc.ed.ac.uk/enacts/

[3] GN2-07-331v3:   DN5.0.2,7 - Report   on   Researchers'   Requirements: http://www.geant2.net/server/show/nav.1969

[4] Farantos S. C.; Stamatiadis S.; Nellari N.; Maric D. *Grid Enabling Technologies*, http://www.epcc.ed.ac.uk/wp-content/uploads/2007/02/ gridenabling.pdf

[5] Rodrigues C. I.; Hardy D. J.; Stone J. E.; Schulten K.; Hwu W.-M. W.,GPU acceleration of cut off pair potentials for molecular modelling applications In CF08: *Proceedings of the 2008 conference on Computing frontiers*, 2008, pp. 273282, New York, NY, USA, ACM.

[6] Ng M.-H.; Johnston S.; Wu B.; Murdock S. E.; Tai K.; Fangohr H.; Cox S. J.; Essex J. W.; Sansom M. S. P.; Jeffreys P., BioSimGrid: Grid-Enabled Biomolecular Simulation Data Storage and Analysis, *Future Generation Computer Systems* 2006, 22, 657-664.

[7] Luthi H. P.; Metacomputing, an emerging technology?, *Comp. Phys. Comm.*, 2000, 128, 326-332.

[8] Maozhen I.; Maozhen L.; Baker M. *The Grid: core technologies*, John Wiley & Sons, 2005.

[9] Laganà A.; *METACHEM: Metalaboratories for cooperative innovative computational chemical applications*, METACHEM workshop, Brussels, November 1999 http://costchemistry.epfl.ch/docs/D23/d23-main.htm

[10] *Grid Computing in Chemistry: GridChem* is the European Cooperation in the Field of Scientific and Technical Research (COST) action number D37. http://www.cost.esf.org/index.php?id=189&action_number=D37

[11] European Grid Infrastructure (EGI). http://www.egi.eu/

[12] Rapaport D. C., *The Art of Molecular Dynamics Simulation*, Cambridge University Press, 1995.

[13] Daskalakis V.; Farantos S. C.; Varotsis C., Assigning vibrational spectra of ferryl-oxo intermediates of Cytochrome c Oxidase by periodic orbits and molecular dynamics, *J. Am. Chem. Soc.* 2008, 130, 12385-12393.

[14] Daskalakis V.; Farantos S. C.; Guallar V.; Varotsis C., Vibrational Resonances and Cu_B displacement controlled by proton motion in Cytochrome c Oxidase, *J. Phys. Chem. B* 2009, 114, 1136-1143.

[15] Wikström M. K., Proton pump coupled to Cytochrome c Oxidase in mitochondria, *Nature* 1977, 266, 271-273.

[16] Kaila V. R. I.; Verkhovsky M. I.; Wikström M. K., Proton-Coupled Electron Transfer in Cytochrome Oxidase, *Chem. Rev.* 2010, 110, 7062-7081.

[17] Than M. E.; Soulimane T., ba_3-Cytochrome c Oxidase from Thermus thermophilus, In *Handbook of Metalloproteins*, Eds Messerschimdt A.; Huber R.; Poulos T.; K. Wieghardt K., John Wiley and Sons, Ltd., Chichester, UK, 2001, 363-378.

[18] Porrini M.; Daskalakis V.; Farantos S. C.; Varotsis C., Heme Cavity Dynamics of Photodissociated CO from ba_3-Cytochrome c Oxidase: the Role of Ring-D Propionate, *J. Phys. Chem. B* 2009, 113, 12129-12135.

[19] Murrell J. N.; Carter S.; Farantos S. C.; Huxley P.; Varandas A. J. C., *Molecular Potential Energy Functions*, John Wiley and Sons Ltd, 1984.

[20] Ponder J. W., *TINKER-Software tools for molecular design*, Version 4.2, 2004, Department of Biochemistry and Molecular Biophysics, Washington University School of Medicine.

[21] Smith W.; Forester T. R.; Todorov I. T.; Leslie M. , *The DL_POLY Molecular Simulation Package*, 2006, CSE Department, STFC Daresbury Laboratory.

[22] Wang J.; Cieplak P.; Kollman P. A., How well does a restrained electrostatic potential (RESP) model perform in calculating conformational energies of organic and biological molecules?, *J. Comput. Chem.* 2000, 21, 1049-1074.

[23] Smith W., Molecular Dynamics on distributed memory (MIMD) parallel computers, *Theor. Chim. Acta* 1993, 84, 385-398.

[24] Foloppe N.; MacKerell Jr. A. D., All-atom empirical force field for nucleic acids: I. Parameter optimization based on small molecule and condensed phase macromolecular target data, *J. Comput. Chem.* 2000, 21, 86-104.

[25] Wikström M. K.; Ribacka C.; Molin M.; Laakkonen L.; Verkhovsky M.; Puustinen A., Gating of proton and water transfer in the respiratory enzyme Cytochrome c Oxidase, *Proc. Natl. Acad. Sci. U.S.A.* 2005, 102, 10478-10481.

[26] Gomez Llorente J. M.; Taylor H. S., A classical trajectory study of the photodissociation spectrum of H_3^+, *J. Chem. Phys.* 1989, 90, 5406-5419; ibid. *Spectra in the chaotic region: A classical analysis for the sodium trimer*, 1989, 91, 953-962.

[27] Varotsis C.; Zhang Y.; Appelman E. H.; Babcock G. T., Resolution of the reaction sequence during the reduction of O_2 by Cytochrome Oxidase, *Proc. Natl. Acad. Sci. U.S.A.* 1993, 90, 237-241.

[28] Proshlyakov D. A.; Pressler M. A.; Babcock G. T., Dioxygen activation and bond cleavage by mixed-valence Cytochrome *c* Oxidase, *Proc. Natl. Acad. Sci. U.S.A.* 1998, 95, 8020-8025.

[29] Varotsis C.; Babcock G. T., Appearance of the $\nu(Fe^{IV} =O)$ vibration from a ferryl-oxo intermediate in the Cytochrome Oxidase/dioxygen reaction, *Biochemistry* 1990, 29, 7357-7362.

[30] Ogura T.; Hirota S.; Proshlyakov D. A.; Shinzawa-Itoh K.; Yoshikawa S.; Kitagawa T., Time-Resolved Resonance Raman Evidence for Tight Coupling between Electron Transfer and Proton Pumping of Cytochrome *c* Oxidase upon the Change from the Fe^V Oxidation Level to the Fe^{IV} Oxidation Level, *J. Am. Chem. Soc.* 1996, 118, 5443-5449.

[31] Han S.; Takahashi S.; Rousseau D. L., Time-Dependence of the Catalytic Intermediates in Cytochrome *c* Oxidase, *J. Biol. Chem.* 2000, 275, 1910-1919.

[32] Humphrey W.; Dalke A.; Schulten K., Visual Molecular Dynamics (VMD), *J. Mol. Graphics* 1996, 14, 33-38.

[33] Porrini M.; Daskalakis V.; Farantos S. C., to be published.

[34] Chipot C.; Pohorille A., Eds; Free Energy Calculations. Theory and Applications in *Chemistry and Biology*, Springer Verlag, Berlin, New York, 2007.

[35] Bonomi M.; Branduardi D.; Bussi G.; Camilloni C.; Provasi D.; Raiteri P.; Donadio D.; Marinelli F.; Pietrucci F.; Broglia R. A.; Parrinello M., PLUMED: A portable plugin for free-energy calculations with molecular dynamics, *Comp. Phys. Comm.* 2009, 180, 1961-1972.

[36] Farantos S. C.; Schinke R.; Guo H.; Joyeux M., Energy Localization in Molecules, Bifurcation Phenomena, and their Spectroscopic Signatures: The Global View, *Chem. Rev.* 2009, 109, 4248-4271.

In: Computer Physics
Editors: B.S. Doherty and A.N. Molloy, pp. 315-332

ISBN: 978-1-61324-790-7
© 2012 Nova Science Publishers, Inc.

Chapter 5

SIMULATION OF COLLISIONLESS PLASMA WITH THE VLASOV METHOD

Takayuki Umeda[*]
Solar-Terrestrial Environment Laboratory, Nagoya University, Aichi, Japan

Abstract

Numerical schemes for solving the Vlasov-Maxwell system of equations are presented. Our Universe is filled with collisionless plasma, which is a dielectric medium with nonlinear interactions between charged particles and electromagnetic fields. Thus computer simulations play an essential role in studies of such highly nonlinear systems. The full kinetics of collisionless plasma is described by the Vlasov-Maxwell equations. Since the Vlasov equation treats charged particles as position-velocity phase-space distribution functions in hyper dimensions, huge supercomputers and highly-scalable parallel codes are essential. Recently, a new parallel Vlasov-Maxwell solver is developed by adopting a stable but less-dissipative scheme for time integration of conservation laws, which has successfully achieved a high scalability on massively parallel supercomputers with multi-core scalar processors. The new code has applied to 2P3V (two dimensions for position and three dimensions for velocity) problems of cross-scale plasma processes such as magnetic reconnection, Kelvin-Helmholtz instability and the interaction between the solar wind and an asteroid.

PACS 52.65.-y, 52.65.Ff, 52.25.Dg, 52.30.Cv,

Keywords: plasma; electromagnetic wave; conservation law; multi-scale physics

AMS Subject Classification: 82D10, 83C50, 35L65

1. Introduction

No less than 99.9% of the matter in the visible Universe is in the plasma state. The plasma is a gas in which a certain portion of particles are ionized, which is considered to be the "fourth" state of the matter (solid, liquid, gas, and plasma). The Universe is filled with

---

[*]E-mail address: umeda@stelab.nagoya-u.ac.jp

plasma particles ejected from the upper atmosphere of stars. The stream of plasma is called the stellar wind, which also carries the intrinsic magnetic field of the stars.

Our solar system is filled with plasma particles ejected from the Sun as the solar wind. Neutral gases in the upper atmosphere of the Earth are ionized by a photoelectric effect due to absorption of energy from sunlight, and the most of these ionized particles are trapped by the intrinsic magnetic field of the Earth. The number density far above the Earth's ionosphere is $\sim 100 \text{cm}^{-3}$ or much less, and a typical mean-free path of solar-wind plasma is about 1AU[1]. The number density of plasma particles in space is so low that the system is regarded as collisionless. The word "space plasma" is generally equivalent to collisionless plasma.

The motion of plasma particles is affected by electromagnetic fields due to the electromagnetic Coulomb-Lorentz force. The change in the motion of plasma particles results in an electric current, which modifies the surrounding electromagnetic fields. Thus the plasma behaves as a dielectric medium with strong nonlinear interactions between plasma particles and electromagnetic fields. The computer simulation plays an essential role for studying nonlinear behavior of space plasma. The purpose of this chapter is to give a brief review on the numerical schemes for a first-principle simulation in space plasma based on the Vlasov-Maxwell equations.

2. Basic Equations

There are numerous types of self-consistent computer simulations on space plasma that treat charged particles according to several approximations. The macroscopic (fluid-scale) processes are commonly described by magneto-hydro-dynamic (MHD) or multi-fluid equations, while microscopic (particle-scale) processes are described by the kinetic models, i.e., the Maxwell equations (1) and either the Newton-Lorentz equations (2) or the Vlasov (collisionless Boltzmann) equation (3).

$$\left. \begin{aligned} \nabla \times \mathbf{B} &= \mu_0 \mathbf{J} + \frac{1}{c^2} \frac{\partial \mathbf{E}}{\partial t} \\ \nabla \times \mathbf{E} &= -\frac{\partial \mathbf{B}}{\partial t} \\ \nabla \cdot \mathbf{E} &= \frac{\rho}{\epsilon_0} \\ \nabla \cdot \mathbf{B} &= 0 \end{aligned} \right\} \tag{1}$$

$$\left. \begin{aligned} \frac{d\mathbf{r}_n}{dt} &= \mathbf{v}_n \\ \frac{d\mathbf{v}_n}{dt} &= \frac{q_n}{m_n} \left[\mathbf{E} + \mathbf{v}_n \times \mathbf{B} \right] \end{aligned} \right\} \tag{2}$$

$$\frac{\partial f_s}{\partial t} + \mathbf{v} \frac{\partial f_s}{\partial \mathbf{r}} + \frac{q_s}{m_s} \left[\mathbf{E} + \mathbf{v} \times \mathbf{B} \right] \frac{\partial f_s}{\partial \mathbf{v}} = 0 \tag{3}$$

where $\mathbf{E}$, $\mathbf{B}$, $\mathbf{J}$, ρ, μ_0, ϵ_0 and c represent the electric field, magnetic field, current density, charge density, magnetic permeability, dielectric constant and light speed, respectively. The quantities $\mathbf{r}$, $\mathbf{v}$, q and m in Eq.(2) are the position, velocity, charge, and mass of the n-th

---

[1] Astronomical Unit: the distance from the Sun to the Earth. $1 \text{AU} \sim 150,000,000 \text{km}$.

particle. The Vlasov equation (3) describes the development of the phase-space distribution functions by the electromagnetic (Coulomb-Lorentz) force $\mathbf{F} = q\left[\mathbf{E} + \mathbf{v} \times \mathbf{B}\right]$, with the collision term in the right hand side set to be zero. The distribution function $f_s(\mathbf{r}, \mathbf{v}, t)$ is defined in position-velocity phase space with the subscript s being the species of singly-charged particles (e.g., $s = i, e$ for ions and electrons, respectively). The Maxwell equations and either the Newton-Lorentz equations or the Vlasov equation are coupled with each other via the current density $\mathbf{J}$ that satisfies the continuity equation for charge

$$\frac{\partial \rho}{\partial t} + \nabla \cdot \mathbf{J} = 0 \tag{4}$$

These kinetic equations are regarded as the "first principle" of collisionless plasma.

The above equations can treat all physical processes in space plasma, from the fluid-scale (macroscopic) dynamics, such as the solar wind, solar flares, and global magnetospheres of stars and planets, to the particle-scale (microscopic) kinetics such as particle acceleration, particle diffusion, and wave-particle interactions. The Newton-Lorentz equation treats the motion of N-particles. This means that the computational load depends on the number of particles (N). The Vlasov equation treats the evolution of phase-space distributions in "hyper" dimensions (higher than four dimensions): three dimensions in configuration space and three dimensions in velocity space for an example.

One may understand that it is not easy to solve macroscopic processes by the first-principle kinetic models, and that fluid models are appropriate for solving the macro scale. The fluid equations are derived by taking the zeroth, first, and second moments of the Vlasov equations (3), where the zeroth, first, and second moments correspond to the conservation laws of the density, momentum, and energy, respectively.

$$\frac{\partial n_s}{\partial t} + \nabla \cdot (n_s \mathbf{u}_s) = 0 \tag{5}$$

$$\frac{\partial}{\partial t}(m_s n_s \mathbf{u}_s) + \nabla \cdot (m_s n_s \mathbf{u}_s \mathbf{u}_s + \mathbf{P}_s) - \rho_s \mathbf{E} - \mathbf{J}_s \times \mathbf{B} = 0 \tag{6}$$

$$\frac{\partial}{\partial t}\left(\frac{1}{2}m_s n_s |\mathbf{u}|_s^2 + \frac{3}{2}p_s\right) + \nabla \cdot \left(\frac{1}{2}m_s n_s |\mathbf{u}|_s^2 \mathbf{u}_s + \frac{3}{2}p_s \mathbf{u}_s + \mathbf{P}_s \cdot \mathbf{u}_s + \mathbf{h}_s\right) \tag{7}$$
$$-\mathbf{E} \cdot \mathbf{J}_s = 0$$

where $\mathbf{u}$ represents the average velocity, i.e., the flow velocity, $\mathbf{P}_s$ represents the pressure tensor, p_s represents a scalar pressure given by the trace of $\mathbf{P}_s$ ($p_s \equiv \frac{1}{3}\sum_{i=x,y,z} P_{i,i,s}$), and $\mathbf{h}_s$ represents the heat flux density of the s-th fluid. The Magneto-Hydro-Dynamic (MHD) equations are obtained by applying the single-fluid approximation to these equations.

When we consider low-frequency phenomena with a timescale much longer than electron plasma or cyclotron oscillations, the electromagnetic light-mode waves and electron dynamics can be neglected. Then, the displacement current in Eq.(1) becomes negligible, i.e., $\frac{\partial \mathbf{E}}{\partial t} = 0$, which is called the Darwin approximation [1]. In this case, the electric field is derived from the equation of motion for electron fluid.

$$\frac{\partial}{\partial t}(m_e n_e \mathbf{u}_e) + \nabla \cdot (m_e n_e \mathbf{u}_e \mathbf{u}_e) = -\nabla \cdot \mathbf{P}_e + \rho_e \mathbf{E} + \mathbf{J}_e \times \mathbf{B} \tag{8}$$

In the low-frequency phenomena, the left hand side in Eq.(8) can be set to be zero ($m_i \gg m_e \to 0$), and the electric field is obtained as

$$\mathbf{E} = -\mathbf{u}_e \times \mathbf{B} + \frac{\nabla \cdot \mathbf{P}_e}{\rho_e} = -\mathbf{u}_i \times \mathbf{B} + \frac{\mathbf{J} \times \mathbf{B}}{\rho_i} - \frac{\nabla \cdot \mathbf{P}_e}{\rho_i} \tag{9}$$

where the total current density is given by $\mathbf{J} = \mathbf{J}_i + \mathbf{J}_e$ and the quasi charge neutrality is assumed ($\rho_e + \rho_i \to 0$). Note that the second term in the right hand side of Eq.(9) (($\mathbf{J} \times \mathbf{B})/\rho_i$) is called the Hall term. The above equation is used in the Hall-MHD and hybrid models. Note that the hybrid model treat ions particles but electrons as a fluid by Eq.(9), in which only full kinetics of ions are included.

It is noted that the fluid approximations need resistivity, conductivity, adiabatic index, or diffusion coefficients as a closure model of higher-order moments that describes the time development of off-diagonal pressure terms and the heat flux. However, these coefficients are essentially due to first-principle kinetic processes that are eliminated in the framework of the fluid approximations. In this chapter, we especially focus on the first-principle Vlasov-Maxwell model where full kinetics of both ions and electrons are included by solving the Maxwell equations (1) together with the Vlasov equation (3) and the continuity equation for charge (4).

3.   Historical Development

Numerical procedures for solving kinetic equations of collisionless plasma have been studied for more than a half century since the advent of the von Neumann-type computers in late 1940's. As described in Sec.2., there are two approaches for solving kinetic equations of collisionless plasma. One is to solve the Newton-Lorentz equations (2), and the other is to solve the Vlasov equation (3). The former one is called the particle-in-cell (PIC) method, because plasma particles are treated as individual charged particles and they freely moves in grid cells of electromagnetic fields. The standard explicit PIC method is a well-developed numerical technique which has numerical procedures and concepts that are quite simple without approximation in the basic laws of collisionless plasma. The PIC method is very powerful and widely used for studying full kinetics in space plasma since 1960's. However, this method has a drawback in numerical noises, because a limitation on the number of particles per cell gives rise to strong numerical thermal fluctuations. There is also a difficulty in load-balancing for massively parallel computers. Since the PIC method uses both Eulerian variables (field variables that depend on both time and space) and Lagrangian variables (particle positions and velocities that depend only on time), the number of particles on each processor element (PE) becomes sometimes nonuniform.

The latter approach is called the Vlasov method, in which spatial and temporal developments of distribution functions defined in the position-velocity phase space are directly solved based on the Vlasov equation (3). The Vlasov method is considered to be an alternative to the PIC method, because this method is free from numerical noises. However, computational load of the Vlasov method is much heavier than that of the PIC method, and therefore the development of numerical schemes for the Vlasov method is much slower than those for the PIC method.

Ideas of standard methods for time integration of the Vlasov equation were established in 1970's. One of standard time-advance methods for Vlasov simulations is the spectral methods [2–5], in which phase-space distribution functions are transformed to spectral components in configuration and velocity spaces by the Fourier or Hermite transformation. Another standard time-advance method is based on a time-advance algorithm called the "splitting scheme" [6, 7]. In 1970's, computer simulations with two dimensions were familiar because the physical size of computer memory was so small (less than hundred MBs). Let us consider a simple 1P1V (one dimension in position and one dimension in velocity) phase space as in 1970's. By neglecting the effect of the magnetic field, the Vlasov-Maxwell equations are reduced as follows.

$$\frac{\partial E_x}{\partial x} = \frac{\rho}{\epsilon_0} \tag{10}$$

$$\frac{\partial f_s}{\partial t} + v_x \frac{\partial f_s}{\partial x} + \frac{q_s}{m_s} E_x \frac{\partial f_s}{\partial v_x} = 0 \tag{11}$$

With the operator splitting, the Vlasov equation (11) is separated into the following two advection equations.

$$\frac{\partial f_s}{\partial t} + v_x \frac{\partial f_s}{\partial x} = 0 \tag{12}$$

$$\frac{\partial f_s}{\partial t} + \frac{q_s}{m_s} E_x \frac{\partial f_s}{\partial v_x} = 0 \tag{13}$$

This means that the formal solutions to the Vlasov equation (11) is approximated by the solutions to the two advection equations [8].

$$f_s\left(x_i, v_{x,j}, t + \Delta t\right) = f_s\left(x_i - v_{x,j}\Delta t, v_{x,j} - \frac{q_s}{m_s}E_{x,i}\Delta t, t\right) \tag{14}$$

$$\sim f_s^*\left(x_i, v_{x,j} - \frac{q_s}{m_s}E_{x,i}\Delta t\right), \quad f_s^*\left(x_i, v_{x,j}\right) = f_s\left(x_i - v_{x,j}\Delta t, v_{x,j}, t\right) \tag{15}$$

Here, the phase-space distribution functions are descritized on a two-dimensional mesh in the $x - v_x$ plane, and the advection velocities v_x and $\frac{q_s}{m_s}E_x$ are functions of v_x and x, respectively. Equation (15) represents a shift of the distribution function in the x direction, and then the distribution function has to be shifted in the v_x direction. One may understand that the general solution to the two-dimensional advection equation (14) is well approximated by Eq.(15) when the advection velocities are constant for all x, v_x, and t. However, this approximation is valid for the Vlasov equation only in a short time scale (Δt), in which the profile of a phase-space distribution function does not move over one cell mesh (Δx and Δv). Note that the splitting scheme for the 1P1V phase space has been extended to hyper-dimensional phase space including the effect of the magnetic field [9, 10]. However, there have not been applications of hyper-dimensional Vlasov-Maxwell solvers to nonlinear plasma physics until 1990's because of the limitation of computer memory.

The idea of the splitting scheme is widely adopted in recent Vlasov solvers because of its simplicity of the algorithms and ease of programming. The time integration of the Vlasov equation is approximated by combinations of the one-dimensional numerical interpolation as seen in Eq.(15). Since the Vlasov solvers in the 20th century adopt the cubic spline interpolation as the one-dimensional numerical interpolation [6, 7], they are also called semi-Lagrangian Vlasov solvers where the solution can be approximated as a function of time (see e.g., Eqs.(14) and (15)). Recently, this idea is extended to the B-spline interpolation (e.g., Refs. [11,12]), the CIP (constrained interpolation profile) scheme [13–15], and conservative semi-Lagrangian schemes (e.g., Refs. [16–22]). Although numerical diffusion in conservative semi-Lagrangian schemes is somewhat higher than that in non-conservative semi-Lagrangian schemes, the conservative semi-Lagrangian schemes appear to be more efficient for Vlasov simulations of several classical problems of plasma physics [17, 19, 21, 23]. The inherent conservative property is important in plasma physics, because electromagnetic fields are affected by the current density that satisfies the charge conservation law. In addition, conservative schemes has an advantages in easiness for introducing slope limiters in the reconstruction to ensure specific properties such as positivity and monotonicity [22].

There are not many examples of hyper-dimensional Vlasov-Maxwell simulations for studies of nonlinear plasma physics. 2P2V Vlasov-Maxwell solvers have been applied to the interaction between laser (intense electromagnetic waves) and unmagnetized plasma since 1990's (e.g., Refs. [10, 24]). In recent days, hyper-dimensional (2P2V and 3P2V) Vlasov-Maxwell solvers are applied to magnetically confined plasma, such as tokamak plasma in thermonuclear fusion devices, with the guiding center, drift-kinetic and gyro-kinetic approximations (e.g., Refs. [25, 26]). For magnetized plasma, a 1P3V Vlasov-Maxwell solver is applied to wave-particle interactions (e.g., Refs [27]), where the velocity space is taken in the Cylindrical (polar) coordinate [28]. In these previous works, non-conservative semi-Lagrangian schemes were adopted in hyper-dimensional Vlasov-Maxwell solvers in 1990' and early 2000's, while recent hyper-dimensional Vlasov-Maxwell solvers adopt conservative schemes. Note that hyper-dimensional Vlasov-Maxwell solvers based on spectral schemes have also been proposed (e.g., Refs [29]). However, applications of these solvers are limited to linear instability problems. Only recent hyper-dimensional Vlasov-Maxwell solvers with conservative schemes have successfully applied to multi-dimensional problem on nonlinear processes in Geophysics [30–33].

4. Numerical Procedures

4.1. Time-advance Scheme

The Vlasov equation (3) consists of two advection equations with a constant advection velocity and a rotation equation by a centripetal force without diffusion terms. To simplify the numerical time-integration of the Vlasov equation, we adopt an operator splitting technique for magnetized plasma [6,10]. We have developed a modified version of the operator splitting, where the Vlasov equation splits into the following three equations [34],

$$\frac{\partial f_s}{\partial t} + \mathbf{v}\frac{\partial f_s}{\partial \mathbf{r}} = 0 \tag{16}$$

$$\frac{\partial f_s}{\partial t} + \frac{q_s}{m_s}\mathbf{E}\frac{\partial f_s}{\partial \mathbf{v}} = 0 \tag{17}$$

$$\frac{\partial f_s}{\partial t} + \frac{q_s}{m_s}\left[\mathbf{v} \times \mathbf{B}\right]\frac{\partial f_s}{\partial \mathbf{v}} = 0 \tag{18}$$

Equations (16) and (17) are scalar (linear) advection equations in which **v** and **E** are independent of **r** and **v**, respectively. We have developed a multidimensional conservative semi-Lagrangian scheme for solving these two advection equations [34], which is briefly reviewed in Subsection 4.2. Note that it is essential to use conservative schemes for satisfying the continuity equation for charge in the full electromagnetic method. With the multidimensional conservative semi-Lagrangian scheme, the solution to the advection equation in the configuration space (16) is given by

$$f_{i,j,k}^{t+\Delta t} \leftarrow f_{i,j,k}^{t} + \frac{\Delta t}{\Delta x}\left[U_{x,i+\frac{1}{2},j,k} - U_{x,i-\frac{1}{2},j,k}\right] + \frac{\Delta t}{\Delta y}\left[U_{y,i,j+\frac{1}{2},k} - U_{y,i,j-\frac{1}{2},k}\right] \quad (19)$$
$$+ \frac{\Delta t}{\Delta z}\left[U_{z,i,j,k+\frac{1}{2}} - U_{z,i,j,k-\frac{1}{2}}\right]$$

where $\mathbf{U} \equiv (U_x, U_y, U_z)$ represent numerical flux in the configuration space, and the subscript s is omitted. One can find that the above equation exactly satisfies the continuity equation for charge (4) with

$$\rho = \sum_s q_s \int f_s d\mathbf{v} \quad (20)$$

$$\mathbf{J} = \sum_s q_s \int \mathbf{U}_s d\mathbf{v} \quad (21)$$

In the present study, we adopt a positive, non-oscillatory and conservative scheme [21] for stable time-integration of advection equations, which is briefly reviewed in Subsection 4.3. Equation (18), by contrast, is a multi-dimensional rotation equation which follows a circular motion of a profile at constant angular speed by a centripetal force. For stable rotation of the profile on the Cartesian grid system, the "back-substitution" technique [35] is applied. In addition, Maxwell's equations are solved by the implicit Finite Difference Time Domain (FDTD) method on the Yee grid system [36], which is free from the Courant condition for electromagnetic light mode waves.

The Maxwell-Vlasov system is advanced by using the following sequences [34], which is consistent with the second-order leap-frog time-integration algorithm used in particle-in-cell simulations.

1. Shift phase-space distribution functions in the configuration space with the full time step Δt by using the multidimensional conservative scheme [34].

$$f_s^*(\mathbf{r}, \mathbf{v}) \leftarrow f_s^t(\mathbf{r} - \mathbf{v}\Delta t, \mathbf{v}) \quad (22)$$

2. Compute the current density by integrating the numerical flux **U** over the velocity **v** as Eq.(21).

3. Advance electromagnetic fields from t to $t + \Delta t$ by solving Maxwell's equations with the implicit FDTD method.

4. Shift phase-space distribution functions in the velocity space by an electric force with the half time step $\Delta t/2$ by using the multidimensional conservative scheme [34].

$$f_s^{**}(\mathbf{r}, \mathbf{v}) \leftarrow f_s^*\left(\mathbf{r}, \mathbf{v} - \frac{q_s}{m_s}\mathbf{E}^{t+\Delta t}\frac{\Delta t}{2}\right) \quad (23)$$

5. Rotate phase-space distribution functions in the velocity space by a magnetic force with the full time step Δt by using the back-substitution scheme [35].

$$f_s^{***}(\mathbf{r}, \mathbf{v}) \leftarrow f_s^{**}(\mathbf{r}, \mathbf{v}^*) \tag{24}$$

where

$$\mathbf{v}^* = \mathbf{v} - \frac{q_s}{m_s} \frac{\Delta t}{1 + \left[\frac{q_s \Delta t}{2m_s}|\mathbf{B}^{t+\Delta t}|\right]^2} \left\{ \left[\mathbf{v} \times \mathbf{B}^{t+\Delta t}\right] + \frac{q_s \Delta t}{2m_s}\left[\mathbf{v} \times \mathbf{B}^{t+\Delta t} \times \mathbf{B}^{t+\Delta t}\right] \right\}$$

6. Shift phase-space distribution functions in the velocity space by an electric force with the half time step $\Delta t/2$ by using the multidimensional conservative scheme [34].

$$f_s^{t+\Delta t}(\mathbf{r}, \mathbf{v}) \leftarrow f_s^{***}\left(\mathbf{r}, \mathbf{v} - \frac{q_s}{m_s}\mathbf{E}^{t+\Delta t}\frac{\Delta t}{2}\right) \tag{25}$$

The detailed descriptions of the numerical schemes are provided in Refs. [21, 34, 35].

In the present Vlasov code, there is a numerical constraint on the Courant condition for rotation in velocity space by magnetic fields. For stable rotation of distribution functions with the back-substitution scheme on the Cartesian grid system, we need to choose the timestep Δt such that

$$\Delta v_e > \frac{q_e}{m_e}|\mathbf{v}_{\text{max},e}|B_0\Delta t = \omega_{ce}|\mathbf{v}_{\text{max},e}|\Delta t \tag{26}$$

where B_0 represents an ambient magnetic field. This means that the timestep Δt must be taken to be small for strongly magnetized plasma.

4.2. Multi-dimensional Conservative Scheme

Let us consider that a numerical solution to Eq.(22) takes the following conservative form (see also Eq.(19)),

$$f_s^*(x_i, y_j, z_k, v_{x,l}, v_{y,m}, v_{z,n}) = f_s^t(x_i, y_j, z_k, v_{x,l}, v_{y,m}, v_{z,n}) \tag{27}$$
$$-\frac{\Delta t}{\Delta x}\left[U_{x,s}\left(x_{i+\frac{1}{2}}, y_j, z_k, v_{x,l}, v_{y,m}, v_{z,n}\right) - U_{x,s}\left(x_{i-\frac{1}{2}}, y_j, z_k, v_{x,l}, v_{y,m}, v_{z,n}\right)\right]$$
$$-\frac{\Delta t}{\Delta y}\left[U_{y,s}\left(x_i, y_{j+\frac{1}{2}}, z_k, v_{x,l}, v_{y,m}, v_{z,n}\right) - U_{y,s}\left(x_i, y_{j-\frac{1}{2}}, z_k, v_{x,l}, v_{y,m}, v_{z,n}\right)\right]$$
$$-\frac{\Delta t}{\Delta z}\left[U_{z,s}\left(x_i, y_j, z_{k+\frac{1}{2}}, v_{x,l}, v_{y,m}, v_{z,n}\right) - U_{z,s}\left(x_i, y_j, z_{k-\frac{1}{2}}, v_{x,l}, v_{y,m}, v_{z,n}\right)\right]$$

where U_x, U_y and U_z are numerical fluxes in the x, y and z directions, respectively. The subscripts i, j, k, l, m and n represent the grid numbers. Then, the continuity equation for charge (4) is automatically satisfied.

Although there is an arbitrary solution for the multi-dimensional numerical flux in Eq.(27), we approximate the numerical flux by using one-dimensional numerical flux in x,

y and z directions: $U_{x,s}^{1D}\left(x_{i+\frac{\sigma_i}{2}}, y_j, z_k\right)$, $U_{y,s}^{1D}\left(x_i, y_{j+\frac{\sigma_j}{2}}, z_k\right)$ and $U_{z,s}^{1D}\left(x_i, y_j, z_{k+\frac{\sigma_k}{2}}\right)$.

$$
\left.
\begin{aligned}
U_{x,s}\left(x_{i+\frac{\sigma_i}{2}}, y_{j+\sigma_j}, z_{k+\sigma_k}\right) &= \sigma_i \frac{\Delta t^2}{\Delta y \Delta z} \frac{U_{x,s}^{1D} U_{y,s}^{1D} U_{z,s}^{1D}}{3\left[f_s^t(x_i, y_j, z_k)\right]^2} \\
U_{x,s}\left(x_{i+\frac{\sigma_i}{2}}, y_{j+\sigma_j}, z_k\right) &= \sigma_i \frac{\Delta t}{\Delta y} \frac{U_{x,s}^{1D} U_{y,s}^{1D}}{2 f_s^t(x_i, y_j, z_k)} - U_{x,s}\left(x_{i+\frac{\sigma_i}{2}}, y_{j+\sigma_j}, z_{k+\sigma_k}\right) \\
U_{x,s}\left(x_{i+\frac{\sigma_i}{2}}, y_j, z_{k+\sigma_k}\right) &= \sigma_i \frac{\Delta t}{\Delta z} \frac{U_{x,s}^{1D} U_{z,s}^{1D}}{2 f_s^t(x_i, y_j, z_k)} - U_{x,s}\left(x_{i+\frac{\sigma_i}{2}}, y_{j+\sigma_j}, z_{k+\sigma_k}\right) \\
U_{x,s}\left(x_{i+\frac{\sigma_i}{2}}, y_j, z_k\right) &= U_{x,s}^{1D} - U_{x,s}\left(x_{i+\frac{\sigma_i}{2}}, y_{j+\sigma_j}, z_k\right) \\
&\quad - U_{x,s}\left(x_{i+\frac{\sigma_i}{2}}, y_j, z_{k+\sigma_k}\right) - U_{x,s}\left(x_{i+\frac{\sigma_i}{2}}, y_{j+\sigma_j}, z_{k+\sigma_k}\right)
\end{aligned}
\right\}
\tag{28}
$$

$$
\left.
\begin{aligned}
U_{y,s}\left(x_{i+\sigma_i}, y_{j+\frac{\sigma_j}{2}}, z_{k+\sigma_k}\right) &= \sigma_j \frac{\Delta t^2}{\Delta z \Delta x} \frac{U_{x,s}^{1D} U_{y,s}^{1D} U_{z,s}^{1D}}{3\left[f_s^t(x_i, y_j, z_k)\right]^2} \\
U_{y,s}\left(x_i, y_{j+\frac{\sigma_j}{2}}, z_{k+\sigma_k}\right) &= \sigma_j \frac{\Delta t}{\Delta z} \frac{U_{y,s}^{1D} U_{z,s}^{1D}}{2 f_s^t(x_i, y_j, z_k)} - U_{y,s}\left(x_{i+\sigma_i}, y_{j+\frac{\sigma_j}{2}}, z_{k+\sigma_k}\right) \\
U_{y,s}\left(x_{i+\sigma_i}, y_{j+\frac{\sigma_j}{2}}, z_k\right) &= \sigma_j \frac{\Delta t}{\Delta x} \frac{U_{x,s}^{1D} U_{y,s}^{1D}}{2 f_s^t(x_i, y_j, z_k)} - U_{y,s}\left(x_{i+\sigma_i}, y_{j+\frac{\sigma_j}{2}}, z_{k+\sigma_k}\right) \\
U_{y,s}\left(x_i, y_{j+\frac{\sigma_j}{2}}, z_k\right) &= U_{y,s}^{1D} - U_{y,s}\left(x_i, y_{j+\frac{\sigma_j}{2}}, z_{k+\sigma_k}\right) \\
&\quad - U_{y,s}\left(x_{i+\sigma_i}, y_{j+\frac{\sigma_j}{2}}, z_k\right) - U_{y,s}\left(x_{i+\sigma_i}, y_{j+\frac{\sigma_j}{2}}, z_{k+\sigma_k}\right)
\end{aligned}
\right\}
\tag{29}
$$

$$
\left.
\begin{aligned}
U_{z,s}\left(x_{i+\sigma_i}, y_{j+\sigma_j}, z_{k+\frac{\sigma_k}{2}}\right) &= \sigma_k \frac{\Delta t^2}{\Delta x \Delta y} \frac{U_{x,s}^{1D} U_{y,s}^{1D} U_{z,s}^{1D}}{3\left[f_s^t(x_i, y_j, z_k)\right]^2} \\
U_{z,s}\left(x_i, y_{j+\sigma_j}, z_{k+\frac{\sigma_k}{2}}\right) &= \sigma_k \frac{\Delta t}{\Delta y} \frac{U_{y,s}^{1D} U_{z,s}^{1D}}{2 f_s^t(x_i, y_j, z_k)} - U_{z,s}\left(x_{i+\sigma_i}, y_{j+\sigma_j}, z_{k+\frac{\sigma_k}{2}}\right) \\
U_{z,s}\left(x_{i+\sigma_i}, y_j, z_{k+\frac{\sigma_k}{2}}\right) &= \sigma_k \frac{\Delta t}{\Delta x} \frac{U_{z,s}^{1D} U_{x,s}^{1D}}{2 f_s^t(x_i, y_j, z_k)} - U_{z,s}\left(x_{i+\sigma_i}, y_{j+\sigma_j}, z_{k+\frac{\sigma_k}{2}}\right) \\
U_{z,s}\left(x_i, y_j, z_{k+\frac{\sigma_k}{2}}\right) &= U_{z,s}^{1D} - U_{z,s}\left(x_i, y_{j+\sigma_j}, z_{k+\frac{\sigma_k}{2}}\right) \\
&\quad - U_{z,s}\left(x_{i+\sigma_i}, y_j, z_{k+\frac{\sigma_k}{2}}\right) - U_{z,s}\left(x_{i+\sigma_i}, y_{j+\sigma_j}, z_{k+\frac{\sigma_k}{2}}\right)
\end{aligned}
\right\}
\tag{30}
$$

where $\sigma_i = \text{sign}(v_{x,l})$, $\sigma_j = \text{sign}(v_{y,m})$ and $\sigma_k = \text{sign}(v_{z,n})$. Note that $v_{x,l}$, $v_{y,m}$ and $v_{z,n}$ are omitted for simplicity. Here the one-dimensional numerical flux U^{1D} is computed by using a set of one-dimensional data. For an example, a numerical flux in the x direction, $U_{x,s}^{1D}\left(x_{i+\frac{\sigma_i}{2}}, y_j, z_k\right)$, is obtained as a function of $f(x_i, y_j, z_k)$, $f(x_{i-1}, y_j, z_k)$, $f(x_{i+1}, y_j, z_k)$, $f(x_{i-2}, y_j, z_k)$, $f(x_{i+2}, y_j, z_k)$, $\cdots$. The detailed derivation is given in Ref. [34].

4.3. One-dimensional Conservative Semi-Lagrangian Scheme

We can use an arbitrary conservative scheme to compute the one-dimensional numerical flux U^{1D}. In the present study, we use the positive and non-oscillatory scheme [21]. Let us consider numerical solutions to the one-dimensional advection equation

$$\frac{\partial f}{\partial t} + v\frac{\partial f}{\partial x} = 0. \tag{31}$$

The general solution of Eq.(31) is obtained as

$$f(t + \Delta t, i\Delta x) = f(t, i\Delta x - v\Delta t) \tag{32}$$

and a numerical solution with the 3rd-degree upwind-biased Lagrange polynomial interpolation is obtained as

$$f(x) = f_i + \frac{x - i\Delta x}{\Delta x}(f_i - f_{i-1})$$
$$+\frac{x - i\Delta x}{\Delta x}\left(1 + \frac{x - i\Delta x}{\Delta x}\right)\left(2 + \frac{x - i\Delta x}{\Delta x}\right)\frac{f_{i-1} - 2f_i + f_{i+1}}{6}$$
$$+\frac{x - i\Delta x}{\Delta x}\left(1 + \frac{x - i\Delta x}{\Delta x}\right)\left(1 - \frac{x - i\Delta x}{\Delta x}\right)\frac{f_{i-2} - 2f_{i-1} + f_i}{6}$$

$$\tag{33}$$

The conservative form of the above equation is

$$U_{i+\frac{1}{2}}(\nu) = \nu f_i + \nu(1 - \nu)(2 - \nu)\frac{f_{i+1} - f_i}{6} \tag{34}$$
$$+\nu(1 - \nu)(1 + \nu)\frac{f_i - f_{i-1}}{6}$$

where $\nu \equiv (i\Delta x - x)/\Delta x$ with

$$f_i^{t+\Delta t} = f_i^t + U_{i-\frac{1}{2}}(\nu) - U_{i+\frac{1}{2}}(\nu) \tag{35}$$

We introduce a flux limiter in Eq.(34) to suppress the numerical oscillation (e.g., Ref. [16]),

$$U_{i+\frac{1}{2}}(\nu) = \nu f_i + \nu(1 - \nu)(2 - \nu)\frac{L_i^{(+)}}{6} \tag{36}$$
$$+\nu(1 - \nu)(1 + \nu)\frac{L_i^{(-)}}{6}$$

where

$$L_i^{(+)} = \begin{cases} \min[2(f_i - f_{min}), (f_{i+1} - f_i)] & \text{if } f_{i+1} \geq f_i \\ \max[2(f_i - f_{max}), (f_{i+1} - f_i)] & \text{if } f_{i+1} < f_i \end{cases}$$

$$L_i^{(-)} = \begin{cases} \min[2(f_{max} - f_i), (f_i - f_{i-1})] & \text{if } f_i \geq f_{i-1} \\ \max[2(f_{min} - f_i), (f_i - f_{i-1})] & \text{if } f_i < f_{i-1} \end{cases}$$

with

$$f_{max} = \max[f_{max1}, f_{max2}]$$
$$f_{min} = \max[0, \min[f_{min1}, f_{min2}]]$$

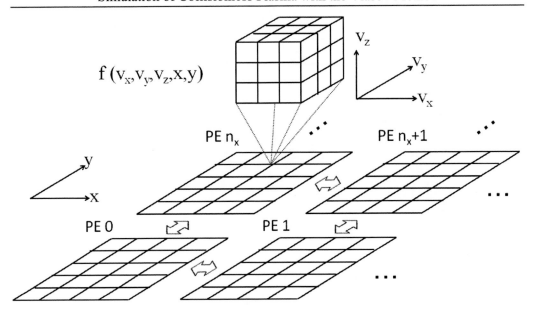

$f(v_x, v_y, v_z, x, y)$

Figure 1. Schematic illustration on the domain decomposition of the 2P3V Vlasov solver.

and

$$f_{max1}=\max\left[\max[f_{i-1}, f_i], \min[2f_{i-1} - f_{i-2}, 2f_i - f_{i+1}]\right]$$
$$f_{max2}=\max\left[\max[f_{i+1}, f_i], \min[2f_{i+1} - f_{i+2}, 2f_i - f_{i-1}]\right]$$
$$f_{min1}=\min\left[\min[f_{i-1}, f_i], \max[2f_{i-1} - f_{i-2}, 2f_i - f_{i+1}]\right]$$
$$f_{min2}=\min\left[\min[f_{i+1}, f_i], \max[2f_{i+1} - f_{i+2}, 2f_i - f_{i-1}]\right]$$

The detailed derivation is given in Ref. [21]. One can see than the present conservative scheme needs five grid points ($f_{i+2}, f_{i+1}, f_i, f_{i-1}, f_{i-2}$) to detect local extrema for computing the numerical flux (and six grid points are needed for interpolation). Note that the conservative scheme based on the third-degree polynomial is now extended to the fourth- and fifth-degree polynomials [37].

4.4. Parallel Implementation

The velocity distribution function has both configuration-space and velocity-space dimensions, and is defined as a hyper-dimensional array, as schematically illustrated in Figure 1. Thus we adopt the "domain decomposition" only in configuration space, where the distribution functions f and electromagnetic fields E_x, E_y, E_z, B_x, B_y, and B_z are decomposed over the configuration-space dimensions (see Figure 1). This involves the exchange of ghost values for the distribution function and electromagnetic field data along each PE boundary. Note that there is additional communication overhead in parallelizing over the velocity-space dimensions since a reduction operation is required to compute the charge and current densities (the zeroth and first moments) at a given point in configuration space as seen in Eqs.(20) and (21). However, the code allows thread parallelization over the velocity-space dimensions via OpenMP.

As seen in Sec.4.1., the Vlasov-Maxwell solver consists of three procedures:

(a) Shifting distribution function in configuration space.

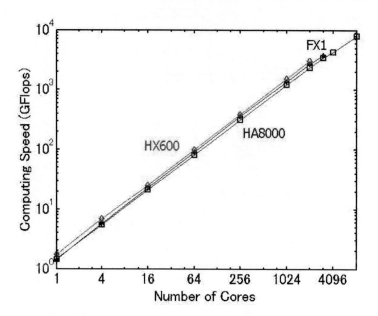

Figure 2. Computational speed of the 2P3V Vlasov solver as a function of the number of cores on various scalar-type supercomputers: Hitach HA8000 system (T2K) at University of Tokyo, Fujitsu FX1 and HX600 systems at Nagoya University.

(b) Shifting distribution function in velocity space.

(c) Updating electromagnetic fields.

Communications between PEs are required at the end of each procedure in order to exchange the ghost values at boundaries. The positive, non-oscillatory and conservative scheme [21] uses six grid points for numerical interpolation (see also Subsec.4.3.), and three ghost grids are exchanged by using the "`Mpi_Sendrecv()`" subroutine in the standard message passing interface (MPI) library for simplicity and portability. The electromagnetic field solver (c) needs additional communications because the convergence check is required in the implicit FDTD method. The "`Mpi_Allreduce()`" subroutine is called on each iteration for the field solver. As shown in Figure 2, the present parallel Vlasov-Maxwell solver achieved a high scalability on various scalar-type supercomputers in Japan.

5. Applications

5.1. Magnetic Reconnection

We adopt the similar initial condition to the Geospace Environment Modeling (GEM) reconnection challenge [38]. There is a Harris-type sheet equilibrium in the $x - y$ plane. We use $N_x \times N_y = 320 \times 160$ grid cells for configuration space, and $N_{v_x} \times N_{v_y} \times N_{v_z} = 40 \times 40 \times 40$ grid cells for velocity space. Thus the simulation domain is taken in the 5D phase space. The half thickness of the current sheet is chosen to be $l = 0.5d_{i0}$, where d_{i0} is the initial ion inertial length ($d_{i0} = c/\omega_{pi0}$) at the center of the current sheet. The ion

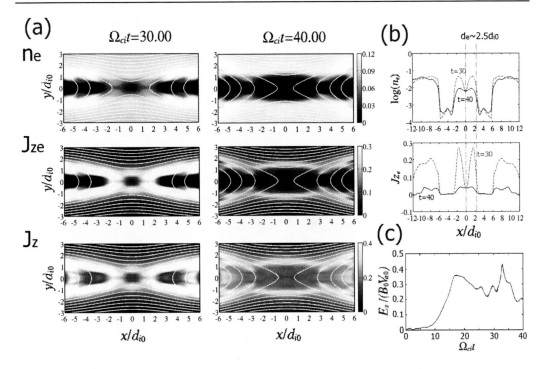

Figure 3. (a) Spatial profiles of the electron number density n_e, the out-of-plane electron current density J_{ze}, and the out-of-plane total current density J_z at $\Omega_{ci}t = 30$ and 40. (b) Cross-section of n_e and J_{ze} at $t = 30, 40$ and $y = 0$. (c) Time history of the reconnection rate.

cyclotron frequency is set as $\omega_{ci}/\omega_{pi} = 0.04$, and the speed of light is set as $c/v_{ti} = 100.0$. For computational efficiency we adopt a very small number of the ion-to-electron mass ratio, $m_i/m_e = 25$. The ion-to-electron temperature ratio is chosen to be $T_i/T_e = 7$. The grid spacing is set as $\Delta = 4.0v_{ti}/\omega_{pi}(= 0.04d_{i0} = 0.2d_{e0})$. Detailed descriptions on the initial setup for the Vlasov simulation are given in Ref. [31]. The boundary conditions are symmetric at the inner boundaries and open at the outer boundaries. That is, we adopt the one quarter model with the symmetry constraints.

Figure 3 shows the nonlinear development of magnetic reconnection. One can see from Figure 3c that the reconnection rate saturates at $\Omega_{ci}t = 18$ but keeps a high value due to the open boundary condition. The spatial profiles of the total current density at $\Omega_{ci}t = 30$ and 40 (Figure 3a) show that the current sheet around the X-point has multiple structures. It is clearly shown that the multiple structures of the current sheet is dominated by electrons, suggesting the existence of multiple electron diffusion regions (e.g., Refs. [39,40]). Figure 3b shows the cross-section of the electron number density and the out-of-plane electron current density at $t = 30, 40$ and $y = 0$. It is found that the electron number density around the X-point decreases to $\sim 10^{-2}n_{e0}$. Note that the Vlasov method would have an advantage in handling such a low-density region because of its noiselessness. In such a low-density region, the electron inertial length becomes comparable to the ion inertial length, and the spatial scale of the inner electron diffusion region is determined by the local electron inertial length. Note that the ion-to-electron mass ratio in the present simulation is very small, and therefore it is not easy to separate ion and electron scales. However, this result suggests influence of electron-scale processes on magnetic reconnection.

5.2. Kelvin-Helmholtz Instability

We adopt the same initial condition as the previous works (e.g., Refs. [31, 41]). There are low-density and high-density regions directed in the negative and positive x directions, respectively, which are sheared in the y direction. We use $N_x \times N_y = 512 \times 960$ grid cells for configuration space, and $N_{v_x} \times N_{v_y} = 80 \times 80$ grid cells for velocity space. Thus the simulation domain is taken in the 4D phase space. The half thickness of the shear layer is chosen to be $l = 4.0r_i$ ($r_i = v_{ti}/\omega_{ci}$). The velocity shear is chosen to be $u_0 = \sqrt{V_A^2 + V_S^2} = 7.28v_{ti}$, with the Alfven velocity $V_A = 7.0v_{ti}$ and the ion sound velocity $V_S = 2.0v_{ti}$. A number density ratio of low-density to high-density regions is set as 0.1. The ion cyclotron frequency is set as $\omega_{ci}/\omega_{pi} = 0.0875$, and the speed of light is set as $c/v_{ti} = 80.0$. For computational efficiency we adopt a very small number of the ion-to-electron mass ratio, $m_i/m_e = 16$. The ion-to-electron temperature ratio is chosen to be $T_i/T_e = 1$. The grid spacing is set as $\Delta = 2.0v_{ti}/\omega_{pi}(= 0.175r_i = 0.7r_e)$. The boundary conditions are periodic in the x direction and open in the y direction.

In Figure 4, we show spatial profiles of the total charge density. At $t = 100$ one large vortex is formed by the primary K-H instability, and at $t = 120$ several small vortices are formed by the secondary Rayleigh-Taylor instability [41]. We see that the charge separation in the primary vortex takes place on the spatial scale of the initial half thickness l, while the charge separation in the secondary vortices takes place on the spatial scale of ion cyclotron radius r_i. In Figure 4b, we plot a cross-section of the total charge density at $y/l = -6.7$ and $t = 140$. Note that the distance in Figure 4b is normalized by two different quantities, i.e., l (top) and r_i (bottom). It is found that the spatial scale of small vortices at $y/l = -6.7$ is $\sim 5r_i$ and that the charge separation takes place on the spatial scale smaller than r_i ($\sim 2r_e$). Note that the ion-to-electron mass ratio in the present simulation is very small, and therefore it is not easy to completely separate ion and electron scales. However, this result suggests existence of strong in-plane electrostatic fields on the spatial scale of electron cyclotron radius.

5.3. Interaction between Solar Wind and a Dielectric Body

We adopt the same initial and boundary condition as the previous work [33]. There exists an insulative sphere at $(x, y) = (0, 0)$, in which the charge accumulates at the surface. The system size of the simulation box is taken for $-10R_S \leq L_x \leq 30R_S$ and $-10R_S \leq L_y \leq 10R_S$, where R_S is the radius of the object. We use $N_x \times N_y = 400 \times 200$ grid cells for configuration space, and $N_{v_x} \times N_{v_y} \times N_{v_z} = 40 \times 40 \times 40$ grid cells for velocity space. Thus the simulation domain is taken in the 5D phase space. The solar wind velocity is set as $V_s = 12V_{ti}$, and the plasma beta in the solar wind is set as $\beta_i = \beta_e = 0.5$. The magnitude of the interplanetary magnetic field (IMF) is given such that $\omega_{ci}/\omega_{pi} = 0.01$. The radius of the body is set as $R_S = r_i = 10\Delta$. Electrons are assumed to be much heavier than the reality with the mass ratio $m_i/m_e = 100$, and the ions and electrons have the same temperature ($T_i/T_e = 1.0$). The IMF is taken in the out-of-plane (z) direction so that the

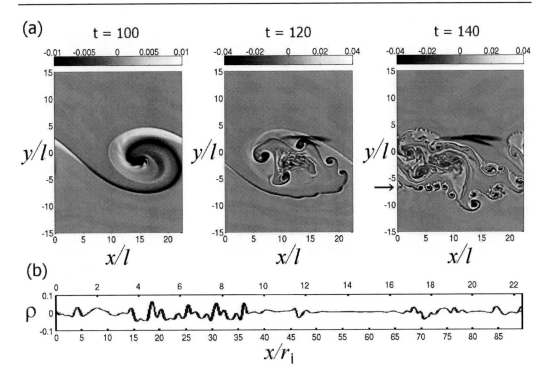

Figure 4. (a) Spatial profiles of the total charge density at different times. (b) Cross-section of the total charge density at $y/l = -6.7$ and $t = 140$ (as indicated by the arrow). The time is normalized by l/u_0.

effect of the finite cyclotron radius in the $x - y$ plane is included. Detailed descriptions on the initial setup for the Vlasov simulation are given in Ref. [42].

Figure 5 shows the the spatial profiles of the ion density N_i and the electric field δE_y component. The plasma void is formed on the nightside of the body while plasma particles accumulate on the dayside surface of the body. At both sides of the plasma void in the nightside, there exist electric fields called "wake" fields. It is found that the structure of the wake field becomes asymmetric, with a strong bipolar signature at the $+y$ side and a weak unipolar signature at the $-y$ side. In the present simulation with the out-of-plane IMF, the trajectory of the $\vec{E} \times \vec{B}$ drift motion of ions becomes a trochoid. In this case, ions with a large gyro radius can penetrate into the wake from the $+y$ side of the wake, where the electric force $\vec{F} = q\vec{E}$ and the density gradient ∇n are in the same direction, i.e., $\nabla n \cdot \vec{F} > 0$. On the other hand, ions with a large gyro radius are absorbed at the surface of the dielectric body and cannot penetrate into the wake from the $-y$ side of the wake where $\nabla n \cdot \vec{F} < 0$. This feature results in the asymmetric structure of the wake fields.

6. Conclusion

In this chapter, numerical schemes of electromagnetic Vlasov simulations for space plasma are briefly reviewed. A latest parallel Vlasov-Maxwell solver with a conservative scheme has successfully applied to magnetic reconnection [31], the K-H instability [32],

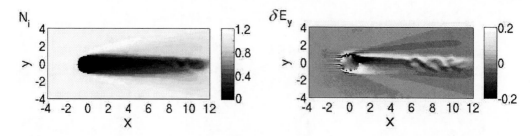

Figure 5. Spatial profiles of the ion density N_i and the electric field δE_y component. The density is normalized by n_0. The electric field is normalized by $|B_0 V_S|$. The distance is normalized by the radius of the object R_S.

and the interaction between solar/stellar winds and dielectric bodies [33]. Currently, 256-1024 cores are used for the parallel computations. However, future Peta-scale supercomputers with 100,000 cores are necessary for 3P3V simulations.

It should be noted, finally, that the numerical schemes for the Vlasov-Maxwell solver is still developing. That is, the numerical schemes presented in this chapter is NOT the final goal of Vlasov-Maxwell solvers; There are still many unsolved issues for the numerical integration of the Vlasov equations in basic numerical problems of multidimensional advection and rotation.

Acknowledgments

This work was supported by MEXT/JSPS under Grant-in-Aid for Young Scientists (B) No.21740352 and No.23740367. The computer simulations were performed on the Hitach HA8000 (T2K) supercomputer system at University of Tokyo and the Fujitsu FX1 and HX600 supercomputer systems at Nagoya University. The computer resources were provided as a computational joint research program at the STEL, Nagoya University, and as a JHPCN program at the Joint Usage/Research Center for Interdisciplinary Large-Scale Information Infrastructures.

The author is grateful to K. Togano, Y. Ito, K. Fukazawa, Y. Nariyuki, Y. Matsumoto, T. K. M. Nakamura, and T. Ogino for discussions.

References

[1] Darwin, C. G. *Philos. Mag.* 1920, 39, 537–551.

[2] Joyce, G.; Knorr, G.; Meier, H. K. *J. Comput. Phys.* 1971, 8, 53–63.

[3] Emery, M. H.; Joyce, G. *J. Comput. Phys.* 1973, 11, 493–506.

[4] Shoucri, M.; Gagne, R. R. J. *J. Comput. Phys.* 1976, 22, 238–242.

[5] Klimas, A. J. *J. Comput. Phys.* 1983, 50, 270–306.

[6] Cheng, C. Z.; Knorr, G. *J. Comput. Phys.* 1976, 22, 330–351.

[7] Gagne, R. R. J.; Shoucri, M.M. *J. Comput. Phys.* 1977, 24, 445–449.

[8] Yanenko, N. N.; Holt, M. *The method of fractional steps: The solution of problems of mathematical physics in several variables*, Springer-Verlag: New York, 1971.

[9] Cheng, C. Z. *J. Comput. Phys.* 1977, 24, 348–360.

[10] Ghizzo, A.; Huot, F.; Bertrand, P. *J. Comput. Phys.* 2003, 186, 47–69.

[11] Sonnendrucker, E.; Roche, J.; Bertrand, P.; Ghizzo, A. *J. Comput. Phys.* 1999, 149, 201–220.

[12] Pohn, E.; Shoucri, M.; Kamelander, G.; *Comput. Phys. Commun.* 2005, 166, 81–93.

[13] Utsumi, T.; Kunugi, T.; Koga, J. *Comput. Phys. Commun.* 1998, 108, 159–179.

[14] Nakamura, T.; Yabe, T. *Comput. Phys. Commun.* 1999, 120, 122–154.

[15] Umeda, T., Y. Omura, P. H. Yoon, R. Gaelzer, and H. Matsumoto, *Phys. Plasmas* 2003, 10, 382–391.

[16] Filbet, F.; Sonnendrucker, E.; Bertrand, P. *J. Comput. Phys.* 2001, 172, 166–187.

[17] Arber, T. D.; Vann, R. G. L. *J. Comput. Phys.* 2002, 180, 339–357.

[18] Elkina, N. V.; Buchner, J. *J. Comput. Phys.* 2005, 213, 862–875.

[19] Umeda, T.; Ashour-Abdalla, M.; Schriver, D. *J. Plasma Phys.* 2006, 72, 1057–1060.

[20] Idomura, Y.; Ida, M.; Tokuda, S.; Villard, L. *J. Comput. Phys.* 2007, 226, 244–262.

[21] Umeda, T. *Earth Planets Space* 2008, 60, 773–779.

[22] Crouseilles, N.; Mehrenberger, M.; Sonnendrucker, N. *J. Comput. Phys.* 2010, 229, 1927–1953.

[23] Filbet, F.; Sonnendrucker, E. *Comput. Phys. Commun.* 2003, 150, 247–266.

[24] Begue, M. L.; Ghizzo, A.; Bertrand, P.; *J. Comput. Phys.* 1999, 151, 458–478.

[25] Grandgirard, V.; Brunetti, M.; Bertrand, P.; Besse, N.; Garbet, X.; Ghendrih, P.; Manfredi, G.; Sarazin, Y.; Sauter, O.; Sonnendrucker, E.; Vaclavik, J.; Villard, L. *J. Comput. Phys.* 2006, 217, 395–423.

[26] Idomura, Y.; Ida, M.; Kano, T.; Aiba, N.; Tokuda, S. *Comput. Phys. Commun.* 2008, 179, 391–403.

[27] Valentini, F.; Veltri, P.; Califano, F.; Mangeney, A. *Phys. Rev. Lett.* 2008, 101, 025006.

[28] Valentini, F.; Veltri, P.; Mangeney, A. *J. Comput. Phys.* 2005, 210, 730–751.

[29] Eliasson B. *J. Comput. Phys.* 2007, 225, 1508–1532.

[30] Schmitz, H.; Grauer, R. *Phys. Plasmas* 2006, 13, 092309.

[31] Umeda, T.; Togano, K.; Ogino, T. *Phys. Plasmas* 2010, 17, 052103.

[32] Umeda, T.; Miwa, J.; Matsumoto, Y.; Nakamura, T. K. M.; Togano, K.; Fukazawa, K.; Shinohara, I. *Phys. Plasmas* 2010, 17, 052311.

[33] Umeda, T.; Kimura, T.; Togano, K.; Fukazawa, K.; Matsumoto, Y.; Miyoshi, T.; Terada, N.; Nakamura, T. K. M.; Ogino, T. *Phys. Plasmas* 2011, 18, 012908.

[34] Umeda, T.; Togano, K.; Ogino, T. *Comput. Phys. Commun.* 2009, 180, 365–374.

[35] Schmitz, H.; Grauer, R. *Comput. Phys. Commun.* 2006, 175, 86–92.

[36] Yee, K. S. *IEEE Trans. Antenn. Propagat.* 1966, AP-14, 302–307.

[37] Umeda, T.; Nariyuki, Y.; Kariya, D. A non-oscillatory and conservative semi-Lagrangian scheme with fourth-degree polynomial, submitted.

[38] Birn, J.; Drake, J. F.; Shay, M. A.; Rogers, B. N.; Denton, R. E.; Hesse, M.; Kuznetsova, M.; Ma, Z. W.; Bhattacharjee, A.; Otto, A.; Pritchett, P. L. *J. Geophys. Res.* 2001, 106, 3715–3719.

[39] Karimabadi, H.; Daughton, W.; Scudder, J. *Geophys. Res. Lett.* 2007, 34, L13104.

[40] Shay, M. A.; Drake, J. F.; Swisdak, M. *Phys. Rev. Lett.* 2007, 99, 155002.

[41] Matsumoto, Y.; Hoshino, M. *J. Geophys. Res.* 2006, 111, A05213.

[42] Umeda, T. Effect of ion cyclotron motion on the structure of wakes: A Vlasov simulation, *Earth Planets Space* 2011, in press.

INDEX